FEYNMAN SIMPLIFIED
4A: MATH FOR PHYSICISTS
AND
4B: FEYNMAN'S BEST

EVERYONE'S GUIDE TO THE
FEYNMAN LECTURES ON PHYSICS

BY

ROBERT L. PICCIONI, PH.D.

Real Science Publishing

3949 Freshwind Circle, Westlake Village, CA 91361 USA

Edited by Joan Piccioni

Third Edition

Visit our web site:

www.guidetothecosmos.com

Feynman Simplified Part 4

Everyone's Guide to the Feynman Lectures on Physics

Feynman Simplified gives mere mortals access to the fabled ***Feynman Lectures on Physics***.

Feynman Simplified: 4A explores all the math needed to understand the ***Feynman Lectures*** and much more.

Feynman Simplified 4B is an unprecedented catalog and explanation of every key principle and important equation in all of the ***Feynman Lectures***. This is an encyclopedia of great physics, from the lectures of one of history's most brilliant scientists. In addition, this book explores the major discoveries of physics in the half-century since those lectures.

The topics we explore include:

Feynman Simplified: 4A

Trigonometric Functions & Identitics
Rectilinear, Polar, Cylindrical & Spherical Coordinates
Real & Complex Numbers; Scientific Notation
Polynomial Equations & Solutions
Dimensional Analysis & Approximation Methods
Finite & Infinite Series & Zenos's Paradox
Exponentials, Logarithms & Hyperbolic Functions
Permutations, Combinations & Binomial Coefficients
Discrete & Continuous Probabilities
Poisson, Gaussian & Chi-Squared Distributions
Rotation & Velocity Transformations
Vector Algebra, Identities & Theorems
Calculus: Differential, Integral & Variational
Differential Equations, Tensors & Matices
Numerical Integration & Data Fitting
Fourier Series & Transforms
Monte Carlo & Advanced Data Analysis

Feynman Simplified: 4B

Newtonian Mechanics
Newtonian Gravitation
Special Relativity
General Reltivity
Gravity Waves
Quantum Mechanics
Wave Phenomena
Electromagnetism
Physics of Light
Particle Physics
Harmonic Oscillators
Statistical Mechanics
Thermodynamics
Conservation Laws
Symmetries of Natural Laws
Physics of Solid Matter
Fuild Flow

To learn more about the ***Feynman Simplified*** series, to receive updates, and send us your comments, visit:
www.guidetothecosmos.com

Looking for a specific topic? Visit our website for a free downloadable index to the entire ***Feynman Simplified*** series.

If you enjoy this book please do me the great favor of rating it on Amazon.com.

4A Table of Contents

Chapter 1
Review of Basic Math

For those who have previously studied geometry, trigonometry, and basic algebra, the first few chapters of this book provide a quick review of those topics and definitions of key terms.

4A§1.1 Primary Symbols & Functions

Equality and Inequality Signs:

A = B, means A equals B.
A < B, means A is less than B
A > B, means A is greater than B
A << B, means A is much less than B
A >> B, means A is much greater than B.
A <= B, means A is less than or equal to B.
A >= B, means A is greater than or equal to B.

~ is the **Proportionality** symbol. Variables X and Y are proportional to one another, written X ~ Y, if their ratio (X / Y) is constant.

√ is the **Square Root**. If $y^2 = x$, then $y = \sqrt{x}$. Note that +y and −y are equally valid square roots of y^2.

| **x** | denotes the Absolute Value of x, it is the unsigned magnitude of x. See Section 4A§4.1 and 4A§11.1 for vectors.

Σ is the **Summation** sign denoting the sum of a series of quantities; see Section 4A§6.2.

∫ is the **Integration** symbol denoting a continuous summation; see Section 4A§13.1.

df/dx symbolizes a single entity, the **Derivative**. See Section 4A§12.3.

4A§1.2 Geometry & Trigonometry

Let's begin by reviewing some basic shapes of Euclidean geometry and their key properties. Figure 1-1 shows five two-dimensional (2-D) shapes.

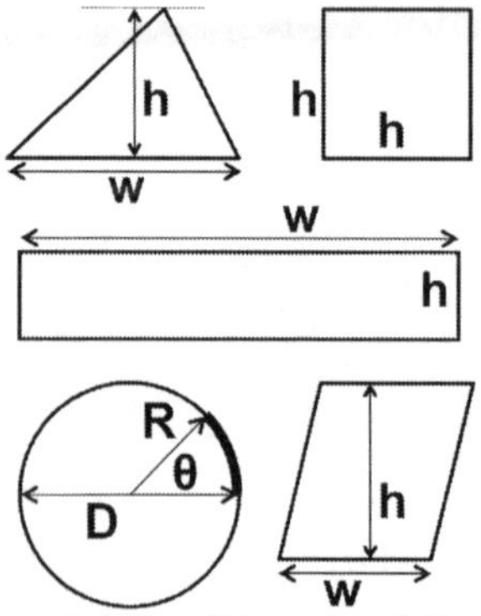

Figure 1-1 Two-Dimensional Shapes

In the upper left is a *triangle*, a figure bounded by three straight line segments; its three internal angles sum to π radians (180 degrees), and its area equals h w / 2, where h is its height, and w is its width; h and w are mutually perpendicular.

In the upper right is a *square*, a figure bounded by four line segments of equal length, with four internal angles that are each 90 degrees ($\pi/2$ radians); its area equals h^2 (h = w).

In the middle is a *rectangle*, a figure bounded by four line segments with opposite sides of equal length, and four internal angles that are each 90 degrees; its area equals hw. (All squares are rectangles, but not all rectangles are squares.)

In the lower left is a *circle* of *radius* R and *diameter* D = 2R, which is the locus (collection) of all points that are a distance R from the circle's center. The circle's circumference (length of its perimeter) equals π D, and its area equals πR^2. The length of the bolded arc that subtends angle θ equals θ R, when θ is measured in radians. This makes sense: for $\theta = 2\pi$, the arc becomes the circle's circumference whose length is 2π R. (This is why God invented radians.) The area enclosed by a circle is called a *disk*.

Lastly, in the lower right is a *parallelogram*, a figure bounded by four line segments with opposite sides of equal length, and opposite angles equal; its area equals hw, the product of its height and width. (All rectangles are parallelograms, but not all parallelograms are rectangles.)

Moving to three dimensions (3-D), Figure 1-2 shows three common shapes.

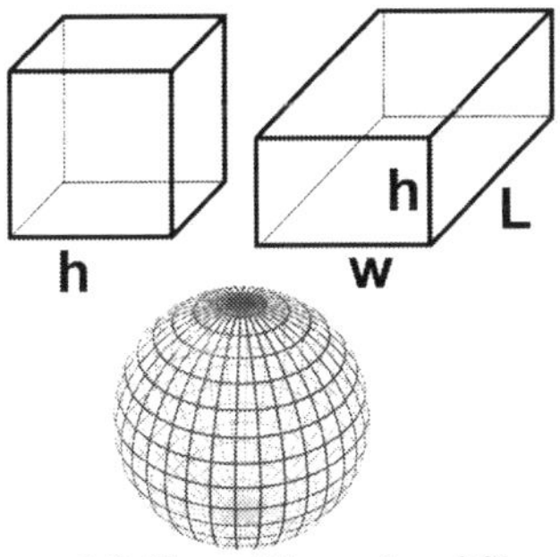

Figure 1-2 Three-Dimensional Shapes

In the upper right is a *cuboid* bounded by six rectangles, with all internal angles being equal. Its height h, width w, and length L may be different. Its enclosed volume equals h w L, and its surface area equals 2 (hw + hL + wL). In the upper left is a *cube*, a cuboid in which h = w = L; its enclosed volume equals h^3 and its surface area equals $6h^2$, where h is the length of any side. (All cubes are cuboids, but not all cuboids are cubes.)

Lastly, in the lower image, is a *sphere*, the locus of all points that are a distance R from the sphere's center. The sphere's area equals $4\pi R^2$, and its enclosed volume equals $4\pi R^3 / 3$. Proper mathematical terminology defines a sphere as the 2-D surface that encloses a 3-D volume called a *ball*.

We wish to examine some quantitative relationships established by trigonometry. But first, we must discuss functions.

4A§1.3 Functions & Fields

In mathematics, functions define relationships among variables. Since physics is all about relationships, functions are the bread and butter of mathematical physics. Variables are quantities whose values change; they can change with location, change over time, or change for some other reason. For example, temperature is a variable that changes with both location and time. We can describe how temperature T varies with location x and time t by using the function f:

$$T = f(x, t)$$

Here, x and t are called *independent variables*, and T is a *dependent variable*. Functions may have one or more independent variables, but they must have exactly one dependent variable. In this case, T is a function of both x and t. As the terms suggest, we are free to choose the values of x and t, and those values uniquely determine the value of T. Some prefer to think of functions as being "black boxes": when x and t are input into f, f outputs T. A more elegant mathematical description is: f *maps* (x, t) to T. A function's independent variables are also called its *arguments*.

The essential characteristic of functions is that for each combination of independent variables there is one and only one value of the dependent variable.

In general, there may be more than one combination of independent variables that produce the same value of the dependent variable. For example, the temperature in Fairbanks, Alaska in mid-August might be the same as the temperature in Miami, Florida in mid-February. We can describe this mathematically by saying: there is a *one-to-one mapping* from (x, t) to T, but the *inverse mapping* from T to (x, t) is not a one-to-one.

Physicists often use the terms *scalar field* and *vector field* when describing functions whose independent variables are the dimensions of space, or space and time. A scalar field is a function whose value is always a simple number. A vector field is a function whose value is always a vector. In Earth's atmosphere, at each location and moment in time, the temperature is a single number, a scalar, but the wind has a certain velocity, which is a vector with magnitude and direction.

4A§1.4 Graphing Functions

Graphs are visual representations that can be extremely helpful in understanding the key properties of functions. Graphs typically plot a function's dependent variable vertically, and the function's independent variable horizontally.

We will discuss *sine* function shortly and the *exponential* function in Chapter 7, but for now suffice it to say that both are very important functions in physics.

Here, we will discuss graphs of these two functions. The upper graph in Figure 1-3 plots the value of Y that corresponds to each value of X, as defined by the exponential function:

$$Y = A + B\,e^X$$

Here, A and B are constants, X is the independent variable and Y is the dependent variable.

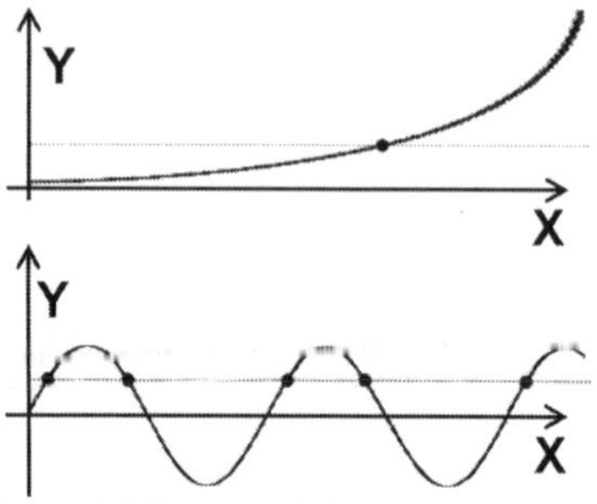

Figure 1-3 Exponential & Sine Functions

The lower graph plots the value of Y that corresponds to each value of X, as defined by the sine function:

$$Y = A \sin(X)$$

In the lower graph, the 5 black dots along the dotted horizontal line indicate 5 values of X for which sin(X) has the same value of Y. Like the prior example of the temperature in Fairbanks and Miami, Y = A sin(X) provides a one-to-one mapping from X to Y, but not a one-to-one mapping from Y to X.

Conversely, in the upper graph, there is only one black dot along the dotted line. In fact, for any Y value there is *only one* value of X for which $Y = A + B\,e^X$. This means exponentials provide one-to-one mappings from X to Y *and* from Y to X. Any function f with this special property has an *inverse function* g, such that:

$$\text{if } y = f(x)$$

$$\text{then } g(y) = g(\,f(x)\,) = x$$

Again, the key property of such functions is that the mapping and the inverse mapping are both one-to-one.

4A§1.5 Trig Functions & Inverses

Trigonometry quantifies the geometric relationships among angles and distances, and is most often employed in analyzing triangles. Let's see how "trig" functions are used. Figure 1-4 shows a triangle, whose longest side has length r, whose vertical side has length y, and whose horizontal side has length x. Because the vertical and horizontal sides are *orthogonal* (perpendicular to one another), this is a *right triangle* and the longest side is the *hypotenuse*.

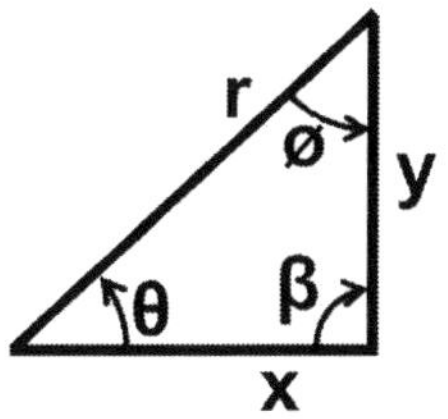

Figure 1-4 Angles & Sides of a Right Triangle

The angle β is a *right angle*, equal to 90 degrees ($\pi/2$ radians). Angles θ and ϕ can have any values that sum to 90 degrees. The three primary trig functions are listed below, with their English names, mathematical notations, and defining equations.

$$\text{sine: } \sin(\theta) = y / r$$

$$\text{cosine: } \cos(\theta) = x / r$$

$$\text{tangent: } \tan(\theta) = y / x$$

As we learned above, the sine function does not have a well-defined inverse function throughout the entire range of all possible angles. Indeed, this applies to all trig functions, because all are *periodic*, meaning that they all repeat exactly at regular intervals. More precisely, for any integer n:

$$\sin(2n\,\pi + \theta) = \sin(2n\,\pi + \pi - \theta) = \sin(\theta)$$

$$\cos(2n\,\pi + \theta) = \cos(2n\,\pi - \theta) = \cos(\theta)$$

$$\tan(n\,\pi + \theta) = \tan(\theta)$$

Well-defined inverse trig functions do exist if we restrict the range of θ. The conventional allowed ranges, English names, mathematical notations, and defining equations of the inverse trig functions are:

$$-\pi/2 < \theta \leq +\pi/2\text{: arc sine: } \arcsin(y / r) = \theta$$

$$+0 \leq \theta < +\pi\text{: arc cosine: } \arccos(x / r) = \theta$$

$$-\pi/2 < \theta \leq +\pi/2\text{: arc tangent: } \arctan(y / x) = \theta$$

These inverse functions are sometimes written:

$$\text{arcsine: } \sin^{-1}(y / r) = \theta$$

$$\text{arccosine: } \cos^{-1}(x / r) = \theta$$

$$\text{arctangent: } \tan^{-1}(y / x) = \theta$$

However, this notation can be confusing: is $\sin^{-1}$ the arcsin or the reciprocal 1/sin? Context often resolves this ambiguity: the argument of arcsin is a ratio of lengths, while the argument of sin is an angle. But since both arguments are dimensionless numbers, it is best to avoid this ambiguity entirely. I will use $\sin^{-1}$ only to reduce clutter in very messy equations, and then only (I hope) to represent 1/sin.

The following reciprocal functions are less commonly used:

$$\text{cotangent: } \cot(\theta) = 1 / \tan(\theta) = x / y$$

$$\text{secant: } \sec(\theta) = 1 / \cos(\theta) = r / x$$

$$\text{cosecant: } \csc(\theta) = 1 / \sin(\theta) = r / y$$

In all equations, angles must be in units of *radians*, with 2π radians equal to 360 degrees.

For the triangle in Figure 1-4, the **Pythagorean theorem** states:

$$r^2 = x^2 + y^2$$

With the above definitions, we replace x with r cos(θ) and y with r sin(θ), yielding the very important equation:

$$1 = \cos^2\theta + \sin^2\theta$$

4A§1.6 Laws of Sines & Cosines

Figure 1-5 shows a triangle whose sides have lengths A, B, and C, and whose opposite angles are a, b, and c, respectively.

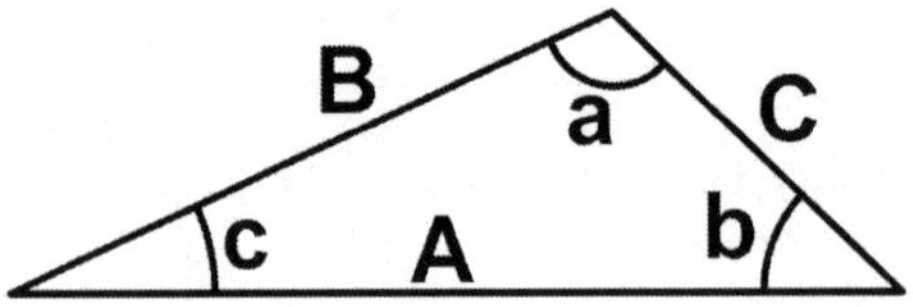

Figure 1-5 Triangle with Sides A, B, C

For any triangle within a plane surface:

$$a + b + c = \pi \text{ radians} = 180 \text{ degrees}$$

Law of Sines: $A / \sin(a) = B / \sin(b) = C / \sin(c)$

Law of Cosines: $A^2 = B^2 + C^2 - 2\,B\,C\cos(a)$

Throughout the rest of this book, and in nearly all physics books, when the argument of a trig function is a single symbol, we typically delete the ( )'s around it. We write: cosx instead to cos(x).

Chapter 2
Coordinate Systems

Coordinates systems are human conventions that facilitate quantifying angles and distances. Nature has no inherent coordinate system. Hence we are free to choose whatever coordinates seem most convenient. Often the best choice in any particular situation is one that matches a natural symmetry.

4A§2.1 Coordinates in 1-D & 2-D

The simplest coordinate system has only one dimension. For example, an object falling straight down in Earth's gravity can be described with only one dimension: let's call it height h. Figure 2-1 shows the *h-axis* pointing straight upward, with *tic-marks* indicating various values of h. A black ball is shown at h = 3.

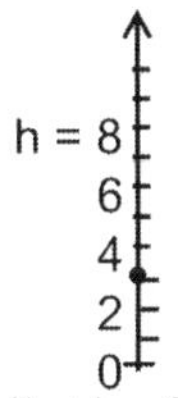

Figure 2-1 Ball at h = 3 along h-axis

With this coordinate system, we can follow the ball's motion as h changes over time.

Moving on to 2-D, let's imagine a basketball player throwing a ball, hoping it goes through the hoop. Here, the motion is two-dimensional. As shown in Figure 2-2, Y is the vertical axis and X is the horizontal axis. This is called a *rectilinear* coordinate system because the axes are orthogonal to one another.

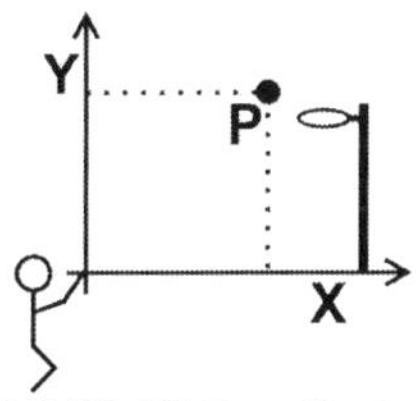

Figure 2-2 (X, Y) Coordinates in 2-D

Any point P in this 2-D space is uniquely specified by how far up P is in the Y-direction, and how far to the right P is in the X-direction. We can choose the *origin* of our coordinate system, the point with coordinates (X=0, Y=0), to be anywhere we wish. Here, we choose the origin to be where the ball was released. At any particular instant in time, the ball has height Y, horizontal distance X, and coordinates (X, Y).

4A§2.2 Polar Coordinates

A quite different situation is the motion of a lone planet around a star. This occurs in three dimensions of course, but due to spherical symmetry, the planet orbits entirely within a single plane. This allows us to analyze its motion in two dimensions. The most convenient approach employs *polar coordinates*, as illustrated in Figure 2-3.

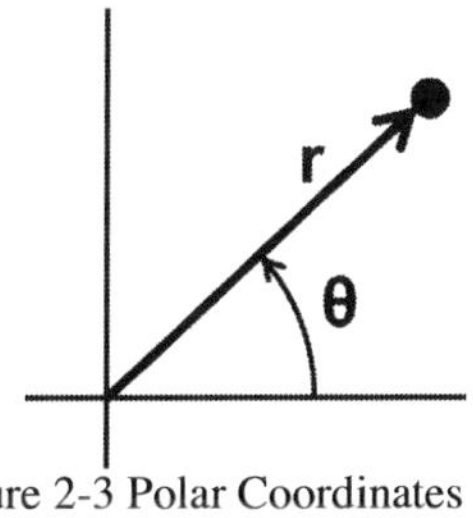

Figure 2-3 Polar Coordinates (r, θ)

Here, r is the planet's distance from the origin, the length of the *radial line* from the origin (the center of the star) to the planet, and θ is the angle between the radial line and a chosen reference direction. In equations, θ ranges from 0 to 2π radians, although in conversation we might say 0 to 360 degrees. The most common reference direction for θ = 0 is the horizontal axis, and the most common convention measures θ in the counterclockwise direction, as shown in Figure 2-3. Those choices are arbitrary, as is the choice of the origin.

If we know a planet's location (r, θ) in 2-D polar coordinates, we can calculate its location (x, y) in 2-D rectilinear coordinates, or we can do the reverse. The conversion equations are:

$$x = r\cos\theta$$

$$y = r\sin\theta$$

$$r = \sqrt{(x^2 + y^2)}$$

$$\tan\theta = y / x$$

4A§2.3 Becoming Three-Dimensional

Computer screens and book pages are two-dimensional, with points defined by their horizontal and vertical positions. A three-dimensional coordinate system adds depth — it has three independent directions. In physics, we most commonly employ a 3-D rectilinear system with three mutually orthogonal coordinate directions labeled x, y, and z, as illustrated in Figure 2-4. Here x is the horizontal right-left direction; y the vertical up-down direction, and z is the in-out direction perpendicular to the page. This is often called a *Cartesian* coordinate system.

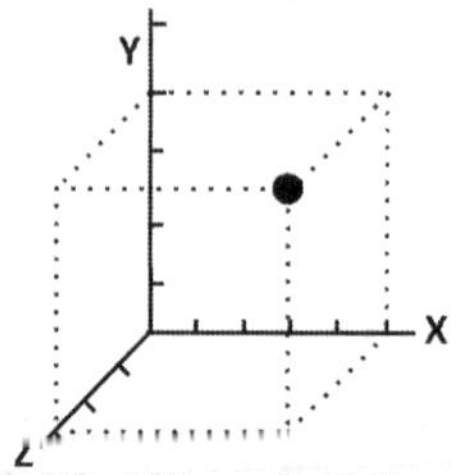

Figure 2-4 Rectilinear 3-D Coordinates

The coordinates of the black ball are its distance from three plane surfaces: x is the distance to the right of the x = 0 plane; y is the distance above the y = 0 plane; and z is the distance out from the z = 0 plane (the page). In this case, counting tic-marks along each axis, we have: (x, y, z) = (5, 4, 3).

4A§2.4 Spherical & Cylindrical Coordinates

Figure 2-5 shows a 3-D *cylindrical* coordinate system that might be useful in analyzing the tip of a corkscrew as it cuts into a cork. The three coordinates are: height z; radial distance from centerline r; and *azimuthal angle* ø.

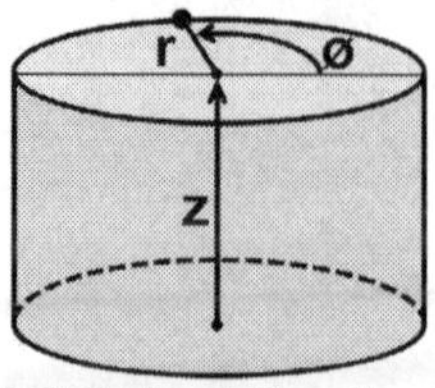

Figure 2-5 Cylindrical Coordinates in 3-D

The tip of the corkscrew has a constant r, an increasing ø, and a decreasing z. The ø and z change at rates whose ratio is determined by the pitch of the corkscrew. Again, ø ranges from 0 to 2π radians. The coordinate transformations between 3-D rectilinear and 3-D cylindrical are:

$$x = r\cos\phi$$

$$y = r\sin\phi$$

$$z = z$$

$$r = \sqrt{(x^2 + y^2)}$$

$$\tan\phi = y / x$$

Another commonly used 3-D coordinate system is *spherical coordinates*, illustrated in Figure 2-6. The three coordinates are: the radial distance from the origin r; the *polar angle* θ; and the *azimuthal angle* ø.

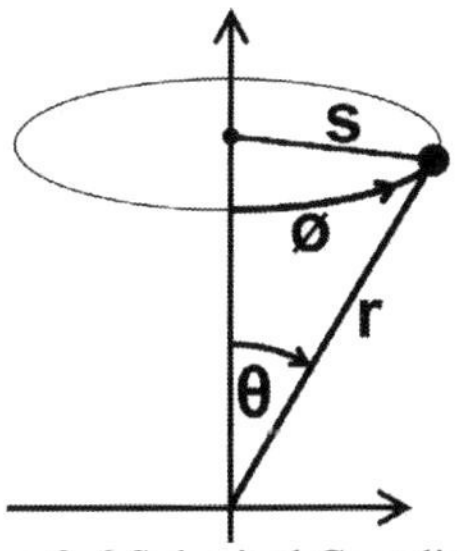

Figure 2-6 Spherical Coordinates

Here, ø ranges from 0 to 2π radians, as before, while θ ranges from 0 to π radians (0 to 180 degrees). Also shown is s, the radius of the indicated circle; s = r sinθ. The coordinate transformations between 3-D rectilinear and 3-D spherical are:

$$x = r \sin\theta \cos\phi$$

$$y = r \sin\theta \sin\phi$$

$$z = r \cos\theta$$

$$r = \sqrt{(x^2 + y^2 + z^2)}$$

$$\cos\theta = z / r$$

$$\tan\phi = y / x$$

A modified version of spherical coordinates is routinely used to specify the position of objects near Earth's surface. In this case: θ is the angle between the equator and the object of interest, and corresponds to latitude; ø is measured relative to the Prime Meridian and corresponds to longitude; and (r minus Earth's radius) corresponds to elevation above sea level.

4A§2.5 4-D Spacetime

Einstein proved the true geometry of our world is four dimensional: the customary three spatial dimensions plus the dimension of time.

Many authors consider time to be the zeroth coordinate and write the location of a *spacetime event* as (ct, x, y, z). Most adopt units in which c, the speed of light, equals 1, and write (t, x, y, z). It was once fashionable to define the time axis as ict, where $i = \sqrt{(-1)}$. The latter was motivated by the fact that the "distance" d between two events in 4-D spacetime is given by:

$$\text{event 1: } (ct_1, x_1, y_1, z_1)$$

$$\text{event 2: } (ct_2, x_2, y_2, z_2)$$

$$d^2 = -c^2(t_2-t_1)^2 + (x_2-x_1)^2 + (y_2-y_1)^2 + (z_2-z_1)^2$$

The minus sign on the time differences is required to ensure the speed of light is the same in all reference frames. Multiplying time values by ic automatically provides the $-c^2$ factor in the distance equation. The modern approach assigns such factors to a metric, as we explore in Chapter 16.

Chapter 3
Numbers

Science strives to understand the fundamental principles that govern nature. Essential to that effort are quantitative observations of natural phenomena, for which we must use and understand numbers.

4A§3.1 Natural to Real

The simplest numbers, *natural numbers*, are the positive integers: 1, 2, 3, … . Mankind employed natural numbers long before the beginning of recorded history; indeed it would be hard to record history without numbers.

In perhaps the fifth century B.C., some societies fully developed the concept of *zero*. That sufficed for most human needs until the advent of credit card debt, which requires *negative numbers*. Debt makes even zero look good.

Commerce spurred an understanding of *fractions*, ultimately opening our eyes to *rational numbers*, those expressible as the ratio of two integers, such as 4/5ths. This was followed by the introduction of *irrational numbers*, such as e and π, which are not equal to the ratio of any pair of integers and whose decimal expansions go on to infinity. Together the rational and irrational numbers comprise the set of *real numbers*.

Real numbers seemed sufficient: every arithmetic operation — addition, subtraction, multiplication, and division — performed on any pair of real numbers resulted in another real number.

That is until someone tried to calculate $\sqrt{(-1)}$, the square root of –1; imagine that.

4A§3.2 Complex Numbers

Imaginary numbers mirror real numbers: there is a one-to-one correspondence between each real number x and its imaginary counterpart $x\sqrt{(-1)}$ that we write ix.

Combining imaginary numbers with real numbers yields the set of *complex numbers*, each of which can be expressed as: x+iy, where x and y are real numbers.

What's next? Nothing, it seems. All known arithmetic operations performed on any combination of complex numbers yield another complex number. Arithmetically, the complex numbers form a *closed set*.

The most convenient way to represent and manipulate complex numbers employs 2-D polar coordinates and exponentials. Figure 3-1 shows the polar and rectilinear representations of the complex number z.

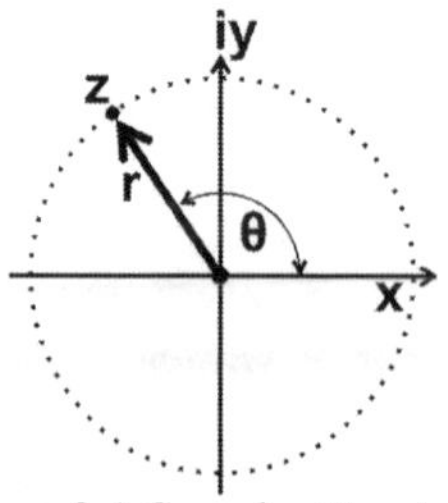

Figure 3-1 Complex Number z

For z = x + iy, z's rectilinear coordinates are (x, y), while z's polar coordinates are (r, θ), where:

$$r = \sqrt{(x^2 + y^2)}$$

$$\theta = \arctan(y / x)$$

$$x = r \cos\theta$$

$$y = r \sin\theta$$

The *complex conjugate* of z, written z*, equals x–iy. To get the complex conjugate of any complex expression, simply replace every "i" with "–i".

Figure 3-2 shows the evolution of the complex number z = (r, θ) as θ goes from 0 to 2π radians. The upper line of images is the polar plot of z for θ = 0, π/2, π, and 3π/2 radians.

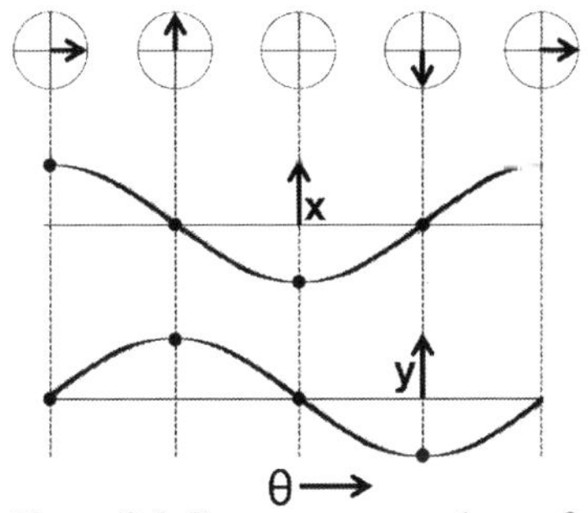

Figure 3-2 Components x and y vs. θ

The middle portion of Figure 3-2 graphs the x-coordinate of z, while the lower portion graphs its y-coordinate.

As we will discuss is Chapter 7, complex numbers can be represented by exponentials with imaginary exponents.

$$z = r\, e^{i\theta} = r \cos\theta + ir \sin\theta$$

One great advantage of $re^{i\theta}$ is that it is very easy to differentiate and integrate, as we will learn in the chapters on calculus. Simplifying the math of physics is always a good thing.

4A§3.3 Scientific Notation

The following are three different ways of writing the number one hundred and twenty three:

$$123 = 1.23\times10^{2} = 1.23\text{E}{+}2$$

The first is the common English format. The second is called **scientific notation**, and the third is a standard computer notation. In the second and third forms, any number of digits may follow the decimal point, but only one digit should precede it. The exponent of 10, 2 in this case, may have any number of digits, and may be positive or negative. An example of a negative exponent is:

$$0.000{,}0123 = 1.23\times10^{-5} = 1.23\text{E}{-}5$$

Scientific and computer notations are convenient for expressing very large and very small quantities.

Ideally, any number quoted for scientific purposes should be accompanied by an explicit uncertainty, such as:

$$123 \pm 4$$

Unless otherwise specified, this means the true value is:

68.3% likely to be between 119 and 127

95.5% likely to be between 115 and 131

99.7% likely to be between 111 and 135...

This assumes a Gaussian distribution with mean 123 and standard deviation 4, as we explore comprehensively in Section 4A§9.5. Often, numbers are quoted without explicit uncertainties. In such cases, we presume a standard scientific convention that numbers are rounded off to their least significant digits. Thus a quantity quoted as being 1200 is presumed to lie between 1150 and 1249; this presumes the two zeros serve only to specify the quantity's magnitude. Conversely, a quantity quoted as being 1203 is presumed to lie between 1202.50 and 1203.49. Thus for integers, we presume the uncertainty equals one-half of a 1 in the least significant nonzero digit. Hence:

1200 means 1200 ± 50

1203 means 1203 ± 0.5

If you wish to specify a quantity that you are confident lies between 1199.50 and 1200.49, you should write: 1200±0.5. Otherwise people will assume the uncertainty is 100 times larger.

When a decimal fraction is specified, we presume the uncertainty equals one-half of a 1 in the least significant digit, whether it is zero or nonzero. Thus:

$$1.20 \text{ means } 1.20 \pm 0.005$$

$$1.2 \text{ means } 1.2 \pm 0.05$$

Chapter 4
Advanced Algebra

4A§4.1 Absolute Value / Magnitude

Often the sign of a quantity is less important than its magnitude. In such cases, we may be interested in its *absolute value*, its unsigned magnitude, denoted by the symbols | |. The absolute value of any real (not complex or imaginary) quantity x is:

$$|x| = +x, \text{ if } x >= 0$$

$$|x| = -x, \text{ if } x < 0$$

The absolute values of any complex quantity z, equal to x + iy, is:

$$|z| = +\sqrt{(x^2 + y^2)}$$

For any Q, |Q| is never negative. We also use the symbols | | to denote the magnitude of a vector, as we discuss in Chapter 11.

4A§4.2 Factorials

The symbol n!, read "n-factorial", equals the product of the first n nonzero integers.

$$n! = 1 \times 2 \times 3 \times 4 \times \ldots \times n$$

For convenience, we define 0! = 1. For large n, we can approximate n! by:

$$n! \sim (n / e)^n \sqrt{(2\pi n)}$$

This approximation is off by only 0.7% for n = 12, and becomes more accurate as n increases.

4A§4.3 Polynomials

Polynomials are sums of *terms*, with each term comprised of a constant multiplied by an independent variable raised to an integral power. The general form is:

$$a_0 + a_1 x + a_2 x^2 + \ldots + a_n x^n$$

Polynomials are ubiquitous in physics equations, and are easily differentiated and integrated, as we will discuss when we get to calculus. Polynomial expressions and equations may not contain square roots, logarithms, trig functions, or other complex mathematical entities. A polynomial is said to be *nth-order* if n is the highest power of the independent variable in the sum. In the prior equation, the constants a_0 through a_{n-1} can have any values, but a_n cannot be zero if this is an nth-order polynomial.

Every nth-order polynomial has n *roots*, n values of the independent variable at which the polynomial equals zero. This means we can *factor* any nth-order polynomial, restating it as the product of n terms, each of which has the form: (x – root). Consider an example:

$$0 = x^4 + x^3 - 6x^2$$

The roots are: x = 0, 0, 2, and –3. The factored equation is:

$$0 = (x)(x)(x - 2)(x + 3)$$

As we see above, some of the roots may have the same value.

Unfortunately, there is no general procedure for finding the roots or factors of an arbitrary polynomial.

One must rely on experience, informed guesswork, and brute force arithmetic.

4A§4.4 Quadratic Equations

A *quadratic equation* is a second-order polynomial equation. The general form is:

$$0 = a\,x^2 + b\,x + c$$

Here, b and c can have any values, but a cannot be zero. Being a second order polynomial, every quadratic equation has two roots that we define as:

$$x = \lambda + \beta$$

$$x = \lambda - \beta$$

With this definition, the quadratic equation becomes:

$$0 = [x - (\lambda + \beta)\,]\,[x - (\lambda - \beta)\,]$$

$$0 = x^2 - 2\lambda\,x + \lambda^2 - \beta^2$$

Matching the last equation with the general form yields:

$$b / a = -2\lambda$$

$$\lambda = -\,b / 2a$$

$$c / a = \lambda^2 - \beta^2$$

$$c / a = (\,-\,b / 2a)^2 - \beta^2$$

$$\beta^2 = b^2 / 4a^2 - c / a$$

Hence the two solutions to the quadratic equation are:

$$x = \lambda \pm \beta$$

$$x = \{-b \pm \sqrt{(b^2 - 4a\,c)}\} / 2a$$

4A§4.5 Circular Orbit Condition

We wish to find the acceleration required to maintain a small satellite in a circular orbit around a much more massive body. Our result will apply equally to any body orbiting a central isotropic force.

Figure 4-1 shows an *arc*, a portion of a circle of radius R that subtends angle θ at the circle's center. The three lines R, R–d, and b form a right triangle, as do the three lines b, d, and s. Figure 4-2 contains an enlarged image of the upper portion of Figure 4-1.

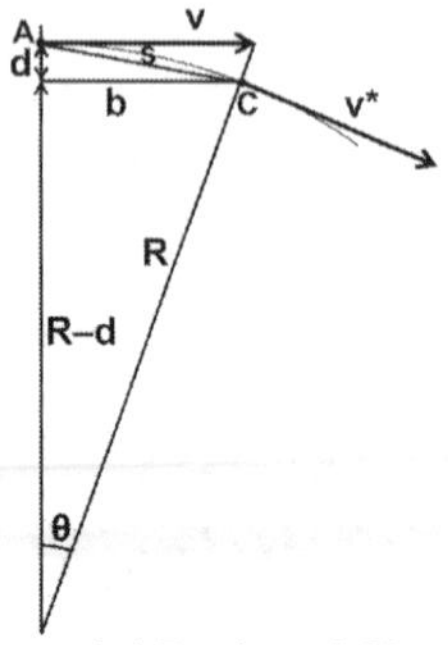

Figure 4-1 Portion of Circular Orbit

In a time interval Δt, the satellite moves from point A to point C, and its velocity changes from ***v*** to ***v****. In a circular orbit, the satellite's velocity has constant magnitude, but its direction of motion is continuously changing. Applying the Pythagorean theorem to the above right triangle R-b-(R–d) yields:

$$R^2 = b^2 + (R - d)^2$$

$$b^2 = R^2 - (R - d)^2$$

$$b^2 = 2R\,d - d^2$$

$$b^2 = d\,(2R - d)$$

This is a useful relationship. You might want to memorize it. However, I find it is easier to remember how such important equations are derived. By re-deriving them each time, I reduce the chance of forgetting a factor of 2, etc.

In the lower image of Figure 4-2, we have moved ***v*** and ***v**** to a common vertex without changing the orientation of either vector. The angle between ***v*** and ***v**** is θ; it must be the same as the angle through which the satellite turns in Figure 4-1. Here, ***Δv*** equals the change in ***v*** during time Δt.

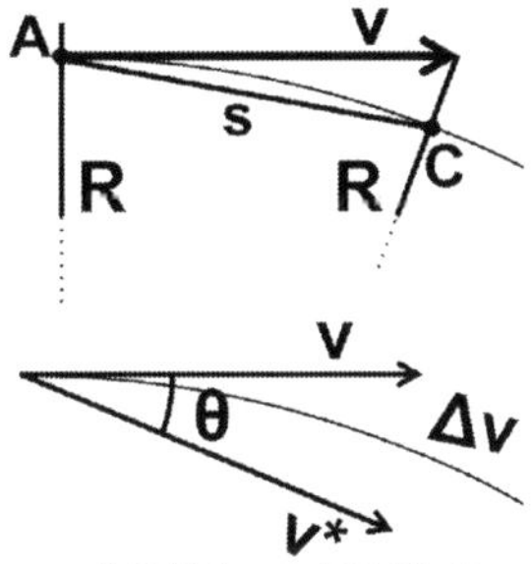

Figure 4-2 Enlarged Orbit Details

In the upper portion of this figure, s is the line that joins points A and C. Comparing Figures 4-1 and 4-2, we can identify two similar isosceles triangles: R-s-R and v-Δv-v*. In both isosceles triangles, the angle between the two equal sides is θ. This means:

$$\Delta v / v = s / R$$

In the upper image of Figure 4-2, the satellite follows the circular arc that is slightly longer than the straight line s. In fact, the arc length is R θ, and the length of s equals 2R sin(θ/2). But for extremely small Δt, these lengths are nearly equal. In that limit:

$$s = v \Delta t$$

$$\Delta v / v = v \Delta t / R$$

$$a = \Delta v / \Delta t = v^2 / R$$

Here, a is the acceleration required to keep the satellite in a circular orbit of radius R.

Chapter 5 Dimensional Analysis

4A§5.1 Units of Measure

This is a good time to talk about equations and units of measure. Quantities of interest in physics commonly are expressed as some number of units appropriate to that type of quantity. For example, distance can be stated as being X microns, Y meters, or Z light-years; each of those units is a standard distance, although their sizes are very different. Clearly, each is appropriate in quite different circumstances.

Time is typically measured in seconds or years. Electric charge is typically measured in coulombs or multiples of the charge of a single proton. Mass is typically measured in grams or kilograms; pound is a poor choice, because that word is used for both mass and force.

Speed, the distance moved divided by the time to move, could be stated as X microns per nanosecond, Y meters per second, or Z light-years per year. In the latter case, we normally expect Z to be less than or equal to 1.

No choice of units is wrong per se. You could even use miles per month, or fathoms per fortnight. But if you follow standard conventions, you will probably minimize mistakes and facilitate communication with others in your field.

The set of units that scientists most commonly use has long been called the *mks* system, which stands for meter-kilogram-second. This system of units has evolved into the *Système International d'unités* metric system, which is abbreviated SI. Most fields of science define additional units appropriate to their unique needs.

4A§5.2 SI System of Units (mks)

Seven *Base Units* are considered the most fundamental. These are:

m: meter, unit of length

kg: kilogram, unit of mass

s: second, unit of time

A: ampere, unit of electric current

K: Kelvin, unit of temperature on absolute scale

mole: number of atoms in 12 grams of carbon-12

cd: candela, unit of luminosity

Additionally, the following are *Derived Units*:

Hz: Hertz, unit of frequency, 1 cycle per second

N: newton, unit of force, $1 \text{ N} = 1 \text{ kg m / s}^2$

Pa: pascal, unit of pressure, $1 \text{ Pa} = 1 \text{ N / m}^2$

J: joule, unit of energy, 1 J = 1 N m

W: watt, unit of power, 1 W = 1 J / s

C: coulomb, unit of charge, 1 C = 1 A s

V: volt, unit of electric field strength, 1 V = 1 W / A

F: farad, unit of capacitance, 1 F = 1 C / V

Ω: ohm, unit of electrical resistance, 1 Ω = 1 V / A

Wb: weber, unit of magnetic flux, 1 Wb = 1 V s

T: tesla, unit of magnetic field strength, $1 \text{ T} = 1 \text{ Wb / m}^2$

H: henry, unit of electrical inductance, 1 H = 1 Wb / A

4A§5.3 Names of Powers of Ten

The key advantage of the metric system is that all units scale by factors of ten — none of this 12 inches to 1 foot, 3 feet to 1 yard, 1760 yards to 1 mile, and all that nonsense. While one can always write 10^3 m or 10^{-9} m, there are standard prefixes for various powers of ten. The same prefixes apply to the names of all units, and the same prefix symbols apply to the symbols for all units.

The more commonly used prefixes are:

P: 10^{+15}: peta, as in Pm, petameter

T: 10^{+12}: tera, as in Ts, terasecond

G: 10^{+9}: giga, as in Gcd, gigacandela

M: 10^{+6}: mega, as in MK, megakelvin

k: 10^{+3}: kilo, as in kg, kilogram

c: 10^{-2}: centi, as in cm, centimeter

m: 10^{-3}: milli, as in mm, millimeter

μ: 10^{-6}: micro, as in μm, micrometer or micron

n: 10^{-9}: nano, as in ns, nanosecond

p: 10^{-12}: pico, as in pA, picoamperes

f: 10^{-15}: femto, as in fm, fentometer

a: 10^{-18}: atto, as in as, attosecond

4A§5.4 Natural Units

Generally, we believe the choices of units and coordinate systems are entirely arbitrary: we are free to choose whatever we please, because nature has no intrinsic units or coordinates. However, there may be some exceptions, some instances in which nature specifies the "right" choice.

One example is the speed of light. Einstein's theory of special relativity, the most precisely and extensively validated concept of all human thought, insists that the speed of light is invariant, that it has the same value c in every circumstance, in every location, and at every moment in time. The speed c is a fundamental property of our universe that establishes a natural unit of velocity.

There are a few other fundamental constants of nature, and Max Planck also found certain combinations that have dimensions of length, time, and mass. These are:

Planck length: $\sqrt{(G \hbar / c^3)} = 1.6162\times10^{-35}$ m

Planck time: $\sqrt{(G \hbar / c^5)} = 5.3911\times10^{-44}$ s

Planck mass: $\sqrt{(\hbar c / G)} = 2.1765\times10^{-8}$ kg

Here, G sets the scale of gravitational phenomena, ħ sets the scale of quantum phenomena, and c sets the scale of relativistic phenomena. These so-called *natural units* are completely independent of people. They have nothing to do with the length of our feet, or our planet's size, day, or year. Indeed, most of these units are fantastically small.

While philosophically intriguing, no one has yet proved a significant physical role for the Planck units. Some cosmologists propose that space and time are quantized on the Planck scale, but that research is still in its infancy.

4A§5.5 Matching Units & Dimensional Analysis

Checking an equation's units is a great way to gain insight and minimize mistakes. This is generally called *dimensional analysis*. In every valid equation, both sides of the equal sign **must** have **exactly** the same units. Consider a simple example:

$$s = g\, t^2 / 2$$

The left side of this equation is a snap; we know that s is a distance, so measuring it in meters makes sense. On the right side, the units are $(m/sec^2) \times (time)^2$. If we measure time in seconds, the units on the right side will be meters, just like the left side. If the right side had t instead of t^2, the units on the right would be $(m/sec^2) \times (sec) = m/sec$, which does not match the left side and is therefore wrong.

There is no partial credit for units that are only slightly wrong; only a perfect match is acceptable.

Note that we could choose to measure time in minutes. But, that would give the wrong answer:

$$1g = 9.8\ m/sec^2$$

$$1g = 35{,}280\ m/min^2$$

These are very different numbers. The units must match *exactly* to get the right answer.

Units also tell us something about the physics: m/sec^2 means meters are changing with the square of time — that is fast.

4A§5.6 Particle Physics Units

For brevity, Feynman sometimes adopts a convention normally employed by particle physicists: distance and time are measured in units that make the speed of light c equal to 1. One can do this by measuring time in seconds and distance in light-seconds (the distance light travels in one second). Astronomers prefer to measure time in years and distances in light-years. Some experimental high energy physicists measure time in nanoseconds and distance in feet. Either way, c = 1.

You can restore missing c's in his electromagnetic equations by:

replacing each "t" with "c t"

replacing each "v" with "v / c"

replacing each "E" with "E / c"

replacing each "ø" with "ø / c"

replacing each "ϱ" with "c ϱ"

After that, check the units on both sides of each equation and add c's as necessary to dimensionally balance the equation.

Let me demonstrate this with an example. Consider an equation relating the magnetic field to changes in the electric field (assuming no electric currents). For now, don't worry about what this equation means; just think about the units.

$$(\Delta B_z / \Delta y) - (\Delta B_y / \Delta z) = (\Delta E_x / \Delta t), \text{ assuming } c = 1$$

Here the notation $\Delta Q / \Delta P$ means the small change in Q due to a small change in P. Using the replacement list above, the equation becomes:

$$(\Delta B_z / \Delta y) - (\Delta B_y / \Delta z) = (\,[\Delta E_x / c] / [c\, \Delta t]\,)$$

$$c^2\,[\,(\Delta B_z / \Delta y) - (\Delta B_y / \Delta z)\,] = (\Delta E_x / \Delta t)$$

The replacements correctly restored the missing c's.

Alternatively, if you forget the replacement list, you can go directly to dimensional analysis. Recall the Lorentz force equation:

$$F_x = q (E_x + v_y B_z - v_z B_y)$$

From this, we know that the units of E must be the same as the units of v B; let's write that [E] = [v B].

In our original equation, the units are:

$$(\Delta B_z / \Delta y) - (\Delta B_y / \Delta z) = (\Delta E_x / \Delta t)$$

$$[B] / [x] = [E] / [t]$$

The units on the left are magnetic field B divided by distance. The units on the right are electric field E divided by time t. Substituting [E] = [v B] yields:

$$[B] = [v\ B]\ [x] / [t] = [v]^2\ [B]$$

The units of distance [x] divided by time [t] are the units of velocity [v]. To balance this equation, we must multiply the left side by a velocity squared. Since we left out the c's, this means the restored equation is:

$$c^2 [(\Delta B_z / \Delta y) - (\Delta B_y / \Delta z)] = (\Delta E_x / \Delta t)$$

Try this on a few equations that you already know. With a little practice, you will find that dimensional analysis is easy and highly effective.

If you want to be doubly sure, keep all the c's, **and** do dimensional analysis on your results.

Chapter 6
Infinite Series

4A§6.1 Zeno's Paradox & Infinite Series

The most famous and perhaps oldest example of an infinite series is *Zeno's paradox*. In the fifth century B.C., the Greek philosopher Zeno proposed a seemingly unresolvable contradiction regarding a hypothetical race between the great Achilles and a tortoise that I will call Yertle.

Zeno said: if Yertle had a head start when the race began, Achilles would *never* catch up. His logic was:

(1) By the time Achilles reaches Yertle's starting point, Yertle has crept ahead by some distance.

(2) By the time Achilles reaches Yertle's new position, Yertle has crept slightly further yet.

(3) This sequence of Achilles reaching where Yertle was, while Yertle advances further, repeats forever.

(4) No matter how far Achilles runs, Yertle will always be slightly ahead.

(5) Hence, Achilles never catches up.

Zeno implicitly assumed that an infinite number of repetitions of step (2) would require an infinite amount of time, so Achilles could never catch up in a finite amount of time. But we all know that if Achilles runs faster than Yertle, he will eventually catch the turtle. Therefore, Zeno's logic must be wrong, which was obvious to everyone, including Zeno.

The ancient Greeks, despite their great wisdom in many fields, could not solve this paradox.

But we can. First, let's show that an infinite series can have a finite sum. Consider the series:

$$1 + 1/10 + 1/100 + 1/1000 = 1.111$$

No matter how many terms are added on the left side, and how many 1's are added to the decimal expansion on the right side, both sides will always be less than 1.12. In fact, as the number of terms increases without limit, the sum comes closer and closer to 10 / 9.

Getting back to Zeno, let's define Yertle's initial head start to be a distance of 1 unit. Let Achilles' speed be v_{ach}, Yertle's speed be v_{yer}, and $x = v_{yer} / v_{ach}$. We assume Achilles is faster than Yertle, so $x < 1$. Whenever Achilles runs a distance D, Yertle creeps a distance x D.

In "stage 1", Achilles runs a distance 1, while Yertle creeps a distance x. In "stage 2", Achilles runs that distance x, while Yertle creeps distance x^2. This cycle continues indefinitely. The total distance Achilles runs during four stages is:

$$D = 1 + x + x^2 + x^3$$

In four stages, Yertle creeps x D. Including its head start, Yertle's distance from Achilles' starting point is:

$$x\,D + 1 = 1 + x + x^2 + x^3 + x^4$$

Hence after four stages, Yertle is ahead by a distance x^4. During each subsequent stage, an additional term with one higher power of x is added to both sums, and Yertle's lead becomes smaller. After an infinite number of stages, Achilles and Yertle are neck and neck. Define S to be the sum after an infinite number of stages.

$$S = 1 + x + x^2 + x^3 + \ldots$$

The "…" means "continue to infinity". Clearly, if x equals 1 or more, the series will have an infinite sum. But $x >= 1$ means Yertle is faster than Achilles, which is wrong. For $0 < x < 1$, we calculate S as follows:

$$S = 1 + x + x^2 + x^3 + \ldots$$

$$x\,S = x + x^2 + x^3 + x^4 + \ldots$$

$$S - x\,S = 1$$

$$S = 1 / (1 - x) = v_{ach} / (v_{ach} - v_{yer})$$

Just like that, we have added an infinite number of terms, and proven:

$$1 + x + x^2 + x^3 + \ldots = 1/(1 - x)$$

Now let's plug in some numbers for $x = v_{yer}/v_{ach}$.

$$\text{for } x = 0.9,\ S = 1 / 0.1 = 10$$

$$\text{for } x = 0.5,\ S = 1 / 0.5 = 2$$

$$\text{for } x = 0.2,\ S = 1 / 0.8 = 1.25$$

$$\text{for } x = 0.1,\ S = 1 / 0.9 = 1.11\ldots$$

For any $x < 1$, S is finite. The total time t required for Achilles to reach S is simply S divided by Achilles' speed:

$$t = S / v_{ach}$$

Since S is finite, so is t. At time t, Achilles has run a total distance S, Yertle has crept a distance $x\,S = S - 1$. So, with Yertle's head start, Achilles catches up at time t.

We solved the problem of an infinite sum in the manner that Zeno presented his paradox.

Now, let's examine a simpler solution, one using an approach that we often employ in relativity problems. The initial separation is 1, and the *closing velocity*, the rate at which that separation decreases, equals the velocity difference: $v_{ach} - v_{yer}$. Hence the separation becomes zero at time t, where:

$$t = 1 / (v_{ach} - v_{yer})$$

In time t, Achilles runs a distance S (just as we found above), where:

$$S = t\, v_{ach} = v_{ach} / (v_{ach} - v_{yer})$$

4A§6.2 Finite Series

We can also calculate the sum of a finite series.

$$\text{For } |x|<1, \text{ let } Q = 1 + x + x^2 + x^3 + \ldots + x^{n-1}$$

We can think of Q being the difference between two infinite sums of powers of x. In the first sum, the lowest power of x is zero, while in the second sum, the lowest power of x is n. We write this:

$$Q = 1 + x + x^2 + x^3 + \ldots$$

$$- (x^n + x^{n+1} + x^{n+2} + \ldots)$$

$$Q = 1 + x + x^2 + x^3 + \ldots$$

$$- x^n (1 + x + x^2 + x^3 + \ldots)$$

$$Q = [\, 1 / (1 - x) \,] - [\, x^n / (1 - x) \,]$$

$$Q = (1 - x^n) / (1 - x)$$

We can do this more elegantly, using $\Sigma_{k=a}{}^{b}\,\{Y\}$ to denote the sum of {Y} from k = a to k = b.

$$Q = \Sigma_{k=0}{}^{n-1}\,\{x^k\}$$

$$Q = \Sigma_{k=0}{}^{\infty}\,\{x^k\} - \Sigma_{k=n}{}^{\infty}\,\{x^k\}$$

$$Q = \Sigma_{k=0}{}^{\infty}\,\{x^k\} - x^n\, \Sigma_{k=0}{}^{\infty}\,\{x^k\}$$

$$Q = (1 - x^n) / (1 - x)$$

A very important finite series is the sum of the integers 1 through n:

$$S_n = 1 + 2 + 3 + \ldots. + n$$

We can calculate S_n geometrically. Imagine placing 1 block in row 1, 2 blocks in row 2, ..., and finally n blocks in row n. The total number of blocks equals S_n. We can calculate S_n from the area covered by blocks whose dimensions are 1 × 1. Now imagine a diagonal line slicing midway through the last block on each row. The diagonal is the hypotenuse of a triangle that is covered by blocks. The triangle's other two sides are each of length n, and its area equals $n^2 / 2$. To that we add the total area of the half-blocks cut by the diagonal that lie outside the triangle, which equals (1/2) times the number of rows. The total area covered by blocks, and therefore S_n equals:

$$S_n = n^2 / 2 + n / 2 = n\,(n + 1) / 2$$

4A§6.3 Monthly Loan Payments

Installment payment loans, such as car loans, student loans, and home mortgages, are features of modern life. You borrow principal amount P at annual interest rate i, and make n monthly payments of M. To be clear, if the annual nominal (uncompounded) interest rate is 6%, i = 0.06 and the monthly interest rate is i/12 = 0.005.

At the end of month #1, the amount you owe equals principal P, plus interest iP/12, minus your first payment M, which equals:

$$\text{Balance at end of month \#1: } P\,(1 + i/12) - M = P\,x - M$$

Here, we define $x = 1 + i/12$ to reduce clutter. At the end of month #7, the amount you owe equals (amount owed at the end of month #6) multiplied by x and then reduced by M. Each month, we multiply the prior balance by x and then subtract M.

The month ending balances are:

$$\text{Month 1: } Px - M$$

$$\text{Month 2: } Px^2 - Mx - M$$

$$\text{Month 3: } Px^3 - Mx^2 - Mx - M$$

$$\text{Month 4: } Px^4 - Mx^3 - Mx^2 - Mx - M$$

$$\text{Month k: } Px^k - M(x^{k-1} + x^{k-2} + \ldots + x + 1)$$

The balance at the end of month #n, when the loan is paid off, is zero. So:

$$0 = P\,x^n - M\,(x^{n-1} + x^{n-2} + \ldots + x + 1)$$

We do not know how to calculate this partial sum directly, because $x > 1$. So, we use a little trick. Define $y = 1 / x$, and multiply everything by y^n, which is the same as dividing everything by x^n.

$$P = M\,(y + y^2 + y^3 + \ldots + y^n)$$

$$P = M\,y\,(1 + y + y^2 + \ldots + y^{n-1})$$

Now, since $y < 1$, we can use the sum calculated above for this partial series:

$$P = M\,y\,(1 - y^n) / (1 - y)$$

Now multiply the right side numerator and denominator by x^{n+1}, and then solve for M:

$$P = M\,(x^n - 1) / x^n\,(x - 1)$$

$$M = P\,(i/12)\,x^n / (x^n - 1)$$

4A§6.4 Sums of Integers Squared

Let's try some more interesting sums. Consider the sum of the squares of all integers up to n:

$$S_n = 1 + 2^2 + 3^2 + \ldots + n^2 = \Sigma_{j=1}^{\ n} \{ j^2 \}$$

For a few values of n, S_n is:

$$S_1 = 1$$

$$S_2 = 1 + 4 = 5$$

$$S_3 = 1 + 4 + 9 = 14$$

$$S_4 = 1 + 4 + 9 + 16 = 30$$

We seek an equation for S_n that is valid for all n. A clue comes from looking for prime factors. We find primes in S_2 (5) and S_3 (7), both of which equal $(2n + 1)$. If we factor $(2n + 1)$ out of each S_n, we can rewrite all these sums in a consistent form.

$$S_1 = 1 = (2n + 1)\,[1] / 3$$

$$S_2 = 5 = (2n + 1)\,[3] / 3$$

$$S_3 = 14 = (2n + 1)\,[6] / 3$$

$$S_4 = 30 = (2n + 1)\,[10] / 3$$

For each n, the numbers in [ ]'s are the sums of the integers from 1 to n, which equals $n\,(n + 1) / 2$. We therefore have:

$S_n = (2n + 1)\ n\ (n + 1) / 6$

That works for n up to 4. To prove it for all n, we use the principle of *mathematical induction*. To use mathematical induction to prove that a statement Ω is true for all integers, we must prove two things:

(1) Ω is true for some integer n; and

(2) Ω being true for any n proves Ω is true for n + 1.

Let's see how this works. We showed above that S_n is the true sum for n = 4. Let's evaluate S for n + 1:

$S_n = n\ (n + 1)\ (2n + 1) / 6$

$S_{n+1} = (n + 1)\ (n + 2)\ (2n + 3) / 6$

$S_{n+1} = (n + 1)\ \{\ 2n^2 + 7n + 6\ \} / 6$

$S_{n+1} = (n + 1)\ \{\ (2n^2 + n) + 6\ (n + 1)\ \}/ 6$

$S_{n+1} = (n + 1)\ \{\ n\ (2n + 1)\ \} / 6 + (n + 1)\ (n + 1)$

$S_{n+1} = S_n + (n + 1)^2$

Hence, since S_n is correct for n = 4, it is correct for n = 5. Since S_n is correct for n = 5, it is correct for n = 6 … and so on ad infinitum.

QED (Quod Erat Demonstrandum — Latin for "High-5")

Next consider the sum of the squares of even integers:

$E_n = 2^2 + 4^2 + 6^2 + \ldots + n^2$

$E_n = 4\{\ 1^2 + 2^2 + 3^2 + \ldots + (n/2)^2\ \}$

$E_n = 4\ S_{n/2}$

$E_n = 4\ (n/2)\ (1+n/2)\ (n+1) / 6$

$E_n = n\ (n+1)\ (n+2) / 6$

Next consider the sum of the squares of odd integers:

$O_n = 1^2 + 3^2 + 5^2 + \ldots + n^2$

This is the sum of all squares minus the sum of even squares.

$O_n = \{1^2 + 2^2 + 3^2 + 4^2 + \ldots + n^2\}$
$\quad - \{2^2 + 4^2 + 6^2 + \ldots + (n-1)^2\}$

$O_n = S_n - E_{n-1}$

$6O_n = n\ (n+1)\ (2n+1) - (n-1)\ (n)\ (n+1)$

$6O_n = n\ (n+1)\ \{\ (2n+1) - (n-1)\ \}$

$O_n = n\ (n+1)\ (n+2) / 6$

Perhaps surprisingly, $O_n = E_n$ — the sum of odd squares and the sum of even squares have the same equation, but of course with different values of n.

4A§6.5 Bessel Functions

Bessel functions often arise in problems involving cylindrical symmetry. *Bessel functions of the first kind* are denoted $J_n(x)$, and are also called *cylindrical harmonics*. For integer, non-negative n, J_n can be written as a power series:

$J_n(2x) = \Sigma_{k=0}^{\infty}\ \{\ (-1)^k\ x^{2k+n} / k!\ (k+n)!\ \}$

$J_{-n}(y) = (-1)^n\ J_n(y)$

For $n = +1/2$, $J_n(y) = \sin(y)\ \sqrt{(2/\pi y)}$

For $n = -1/2$, $J_n(y) = \cos(y)\ \sqrt{(2/\pi y)}$

For $n = 0$, $J_0(2x) = 1 - x^2 / (1!)^2 + x^4 / (2!)^2 - x^6 / (3!)^2 + \ldots$

Chapter 7
Exponentials

Most physics equations define how things *change*. For example, the laws of nature do not specify an object's velocity, but rather how rapidly that velocity changes. Each second, the velocity of a ball falling from the *Leaning Tower of Pisa increases* by a fixed amount: 9.8 meters per second.

4A§7.1 Proportional Change

Other changes are *proportional*. Each second, the population of bacteria in a Petri dish might increase by a fixed *multiple* of the population, such as 1%. As the population grows, it changes by an ever-increasing amount — this is an *exponential* phenomenon.

Proportionality is the essential characteristic of an exponential: each exponential function f(t) always changes by the same percentage for a given change in t. Proportional change describes many diverse natural phenomena. A bacterial population might double in one hour, exhibiting exponential growth. Conversely, the number of surviving radioactive particles might decrease by 50% in one nanosecond, exhibiting exponential decay.

Exponentials are normally written:

$$b^u$$

Here, the *base* b is raised to the u *power*, with the exponent u being any constant or any function. You are probably familiar with expressions like:

$$10^6 = 1{,}000{,}000$$

Exponentials have several wonderful advantages.

Firstly, they are easy to integrate and differentiate, as we discuss in the chapters on calculus.

Secondly, exponentials with the same base are easy to multiply and divide, as shown by:

$$b^u \times b^w = b^{u+w}$$

$$b^u / b^w = b^{u-w}$$

Thirdly, exponentials with complex exponents greatly simplify the analysis of wave and harmonic phenomena, as we discuss in the chapter on differential equations.

In physics, the most common base is e, the base of the natural logarithm that we will get to shortly. This is normally written:

$$e^u$$

But popular modern communication technologies are limited to one level each of superscript or subscript. That is a problem because exponents in physics equations can sometimes be quite elaborate. Therefore, I typically write exponentials with base e as:

$$\exp\{u\}$$

4A§7.2 Definition of e

The definition of e is:

$$e = \text{limit as } n\text{–>}\infty \text{ of } \{ (1 + 1/n)^n \}$$

e is an irrational number whose numerical value is 2.718281828459045…

When I was learning how to use computers, I calculated the first 40,000 digits of e, so let me know if you need more digits.

4A§7.3 Exponential & Trig Series

Let's now consider e raised to the power x.

$$e^x = \exp(x) = \text{limit as } n \to \infty\ (1 + 1/n)^{xn}$$

Let's evaluate $(1 + 1/n)^{xn}$ using the binomial expansion. The terms with the smallest powers of the infinitesimal quantity 1/n are:

$$\exp(x) = 1$$
$$+ (xn)\,(1/n)$$
$$+ (xn)\,(xn - 1) / (2!\, n^2)$$
$$+ (xn)(xn - 1)\,(xn - 2) / (3!\, n^3)$$
$$+ \ldots$$

As xn –> ∞, this becomes:

$$\exp(x) = 1 + x + x^2 / 2 + x^3 / 3! + \ldots$$
$$e^x = \Sigma_n\, x^n / n!$$

This is the polynomial series expansion for e^x; it is a very useful equation.

Let's substitute x with iθ, recalling that $i = \sqrt{(-1)}$.

$$\exp\{i\theta\} = \Sigma_n\, (i\theta)^n / n!$$
$$\exp\{i\theta\} = 1 + i\theta - \theta^2 / 2! - i\theta^3 / 3! + \theta^4 / 4! - \ldots$$

Now separate the series into real terms and imaginary terms.

$$\exp\{i\theta\} = \{1 - \theta^2 / 2! + \theta^4 / 4! - \ldots\}$$
$$+ i\,\{\theta - \theta^3 / 3! + \theta^5 / 5! - \ldots\}$$

Let's look again at the plot in Figure 3-1 of a complex number z in the x-iy plane.

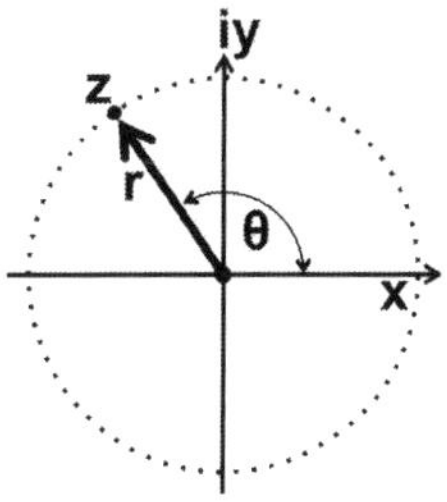

Recall that for z = x + iy, z's rectilinear coordinates (x,y) are:

$$r = \sqrt{(x^2 + y^2)}$$
$$\theta = \arctan(y / x)$$
$$x = r \cos\theta$$
$$y = r \sin\theta$$

For r = 1, compare z with exp{iθ}:

$$z = \cos\theta + i \sin\theta$$
$$\exp\{i\theta\} = \{1 - \theta^2 / 2! + \theta^4 / 4! - \ldots\}$$
$$+ i\,\{\theta - \theta^3 / 3! + \theta^5 / 5! - \ldots\}$$

We can equate z with exp{iθ} if:

$$\sin\theta = \theta - \theta^3 / 3! + \theta^5 / 5! - \ldots$$
$$\cos\theta = 1 - \theta^2 / 2! + \theta^4 / 4! - \ldots$$
$$\tan\theta = \theta + \theta^3 / 3! + 2\theta^5 / 15 - \ldots$$

These are three more very useful equations, with the last equation being derived from the prior two.

The *reciprocal* of an exponential is obtained by simply inverting the sign of the exponent.

$$\exp(x)\ \exp\{-x\} = \exp\{x - x\} = 1$$

$$\exp\{-x\} = 1 / \exp\{x\}$$

4A§7.4 Natural & Based 10 Logarithms

Any nice function has an inverse, a way to get back to your starting point. (You wouldn't want to fly from Honolulu to Fairbanks in January without a return ticket, would you?)

The inverse function of e^x is the **natural logarithm** ln(x).

$$\exp\{\ \ln(x)\ \} = x = \ln(\ \exp\{x\}\ )$$

The natural logarithm and the ordinary base-10 logarithm are related by:

$$\ln\{x\} = \log_{10}\{x\} * 2.3026\ldots$$

$$\log_{10}\{x\} = \ln\{x\} * 0.43429\ldots$$

For $-1 < x <= +1$, the polynomial series for the natural logarithm is:

$$\ln\{1+x\} = x - x^2/2 + x^3/3 - x^4/4 + x^5/5 - \ldots$$

Some useful logarithmic identities are:

$$\ln\{x\ y\} = \ln\{x\} + \ln\{y\}$$

$$\ln\{x / y\} = \ln\{x\} - \ln\{y\}$$

$$\ln\{x^a\} = a \ln\{x\}$$

Before everyone had calculators and computers, logarithms facilitated the arithmetic of many physics calculations by replacing multiplication and division with addition and subtraction, as in:

$$A\ B / C = \exp\{\ \ln(A) + \ln(B) - \ln(C)\ \}$$

Feynman once told me that he memorized the logarithms, rather than the actual values, of key constants like ħ. He said this helped him do his calculations much faster. To make this work, he also memorized logarithm tables and interpolated between the listed values. I bought a calculator.

4A§7.5 Exponential Decay

Consider an example of radioactive decay. Assume we start at time t = 0 with 512 atoms that have a *half-life* of 2 seconds. While radioactive decay is probabilistic and therefore subject to statistical fluctuations, let's simplify our example by assuming that *exactly* 50% of all existing atoms decay during each half-life. This means, the number of atoms at various times will be:

t = 0 sec, N = 512

t = 1 sec, N = 362

t = 2 sec, N = 256

t = 4 sec, N = 128

t = 10 sec, N = 16

t = 18 sec, N = 1

This behavior is described by the equation:

$$N(t) = N(0) \exp\{\ -t / T\}$$

Here, T is *mean lifetime*, which is related to the half-life by:

$$\text{mean lifetime} = \text{half-life} \times \ln\{2\}$$

In many contexts, including particle physics, physicists generally quote mean lifetimes. In other contexts, including radioactive nuclear decay, we generally quote the half-life.

During any small time interval Δt, some small number of radioactive atoms will decay. For a small enough time interval, the number of decays is proportional to Δt. We can show this by letting ΔN be the change in the number of atoms during Δt. The equations are:

$$\Delta N = N(0) \exp\{-(t+\Delta t) / T\} - N(0) [\exp\{-t /T\}$$

$$\Delta N = N(0) \exp\{-t / T\} [\exp\{-\Delta t / T\} - 1]$$

$$\Delta N = N(t) [(1 - \Delta t / T + \dots) \quad 1]$$

$$\Delta N = - N(t) (\Delta t / T)$$

Above, "…" represents terms proportional to higher powers of $\Delta t / T$, which can be neglected for small enough Δt. This equation shows the proportional behavior; the rate of change of N is at all times a constant factor times N.

4A§7.6 Hyperbolic Trig Functions

Related to the exponential are:

$$\text{hyperbolic sine: } \sinh(x) = (\exp\{+x\} - \exp\{-x\}) / 2$$

$$\text{hyperbolic cosine: } \cosh(x) = (\exp\{+x\} + \exp\{-x\}) / 2$$

$$\text{hyperbolic tangent: } \tanh(x) = \sinh(x) / \cosh(x)$$

Some useful hyperbolic identities are:

$$\cosh(ix) = \cos x$$

$$\sinh(ix) = i \sin x$$

$$\tanh(ix) = i \tan x$$

$$\cosh^2 x - \sinh^2 x = 1$$

$$\sinh x = x + x^3 / 3! + x^5 / 5! - \dots$$

$$\cosh x = 1 + x^2 / 2! + x^4 / 4! - \dots$$

$$\tanh x = x - x^3 / 3 + 2x^5 / 15 - \dots$$

$$\sinh(x \pm y) = \sinh x \cosh y \pm \sinh y \cosh x$$

$$\cosh(x \pm y) = \cosh x \cosh y \pm \sinh x \sinh y$$

$$2 \sinh x \sinh y = \cosh(x + y) - \cosh(x - y)$$

$$2 \cosh x \cosh y = \cosh(x + y) + \cosh(x - y)$$

$$2 \sinh x \cosh y = \sinh(x + y) + \sinh(x - y)$$

$$\sinh x + \sinh y = 2 \sinh[(x + y) / 2] \cosh[(x - y) / 2]$$

$$\sinh x - \sinh y = 2 \sinh[(x - y) / 2] \cosh[(x + y) / 2]$$

$$\cosh x + \cosh y = 2 \cosh[(x + y) / 2] \cosh[(x - y) / 2]$$

$$\cosh x - \cosh y = 2 \sinh[(x + y) / 2] \sinh[(x - y) / 2]$$

Chapter 8
Approximation Techniques

Virtually every mathematical analysis of natural phenomena employs numerous approximations. Successful physicists must know when and how to make the right approximations in the right way.

Even something as simple as studying Earth's gravity by dropping balls from the *Leaning Tower of Pisa* cannot be accomplished without approximations. General relativity says time proceeds more quickly at the Tower's top than at its base. The Sun's gravity retards the balls' fall at midday, but accelerates them at midnight. These effects are extremely small — too small to impact any practical measurement — but they are not zero.

Physics courses are intended to teach students essential principles of science, and how these principles can be employed to understand a myriad of natural phenomena. If we attempt to include all effects that impact every situation, an ocean of minutia would obscure the key principles and make any mathematical analysis impossible. No one can solve the equations for everything all at once.

Approximations allow us to focus on just a few important issues at a time.

The key to making good approximations is knowing which terms and which effects are small enough to be neglected in any particular situation. In this chapter, we will explore some common approximation techniques.

4A§8.1 Taylor Series

Taylor series are a common and easy approach to successful approximations. They provide a convenient means of simplifying complicated, potentially unsolvable functions. Any physically realistic function f(x) can be represented by a Taylor series, a polynomial of the form:

$$f(x) = a_0 + a_1 x + a_2 x^2 + a_3 x^3 + \dots$$

Here, the a's are some set, possibly an infinite set, of constants.

Polynomials have a great advantage over some other functions: all polynomials are solvable, and all are easily differentiated and integrated, as we discover in the chapters on calculus. While there may be an infinite number of terms in the Taylor series, often it is possible to approximate f(x), perhaps within some limited range of x, by just a few terms. In such cases, we might write:

$$f(x) = a_0 + a_1 x + a_2 x^2 + O(x^3)$$

Here, $O(x^3)$ denotes any collection of terms, each of which is proportional to x to the third or higher power. If $|x| > 1$, $O(x^3)$ could be enormous, perhaps even infinite. But if $|x| << 1$, terms proportional to higher powers of x will be much less than terms proportional to lower powers. If $|x|$ is small enough, we can make the approximation that those higher power terms are zero. We might even be able to neglect the x^2 term as well, leaving us with a simpler function:

$$f(x) = a_0 + a_1 x$$

In such cases, f(x) becomes linear, greatly simplifying the math.

Let's consider two examples. First, we examine e^x and its Taylor series.

$$f(x) = \exp\{x\} = 1 + x + x^2 / 2! + x^3 / 3! + \dots$$

For $|x| << 1$, we can often approximate this as:

$$f(x) = 1 + x$$

Now consider:

$$g(x) = \exp\{x\} - x = 1 + x^2 / 2! + x^3 / 3! + \dots$$

If, for $|x| << 1$, we drop all terms proportional to x^2 or higher powers, we would be left with $g(x) = 1$. An approximation that crude might eliminate all the interesting physics this problem may entail.

In general, we should avoid approximations that completely eliminate key variables. Let's instead make the approximation that $O(x^3)$ is zero, but $O(x^2)$ is not.

We then have:

$$g(x) = \exp\{x\} - x = 1 + x^2 / 2$$

This shows that g(x) has a quadratic dependence on x near x = 0 that could illuminate some interesting effects that a constant g(x) would not.

Let's next examine some common cases that lend themselves to Taylor series approximations. Many of these series are discussed in Chapter 6. In each series, we assume $|x| << 1$, and I provide here one more term than you will generally need.

$$1/(1-x) = 1 + x + x^2 / 2 + O(x^3)$$

$$1/(1+x) = 1 - x + x^2 / 2 - O(x^3)$$

$$\sqrt{(1+x)} = 1 + x / 2 - x^2 / 8 + O(x^3)$$

$$1/\sqrt{(1+x)} = 1 - x / 2 + 3x^2 / 8 - O(x^3)$$

$$\exp\{x\} = 1 + x + x^2 / 2 + O(x^3)$$

$$\ln\{1+x\} = x - x^2 / 2 + x^3 / 3 - O(x^4)$$

The following approximations are often used in relativistic calculations:

For modest velocities, where $v / c = \beta << 1$:

$$\gamma = 1 / \sqrt{(1 - \beta^2)}$$

$$\gamma = 1 + \beta^2 / 2 + 3\beta^4 / 8 + O(\beta^6)$$

For v very close to c, and $(1 - \beta) << 1$:

$$\gamma^2 = 1 / [(1 + \beta) (1 - \beta)]$$

$$\gamma^2 = 1 / [(2) (1 - \beta)]$$

$$2\gamma^2 = 1 / (1 - \beta)$$

$$1 / 2\gamma^2 = 1 - \beta$$

$$\beta = 1 - 1 / 2\gamma^2$$

4A§8.2 Interpolation & Extrapolation

Imagine that we want to measure the orbit of Mercury. Since Mercury is so close to our Sun, we can only briefly observe it shortly after sunset or shortly before sunrise. Of all the Sun's planets, Mercury has the fastest orbital velocity (averaging 170,000 km/hr or 107,000 mph), and the shortest orbital period (88 Earth days). Thus, Mercury's position changes more between measurement opportunities than any other major body in our Solar System.

Figure 8-1 illustrates a set of three consecutive hypothetical measurements of Mercury's position (black dots), taken near *aphelion*, its farthest distance from the Sun. Define the three measurement times to be t = –1, 0, and +1.

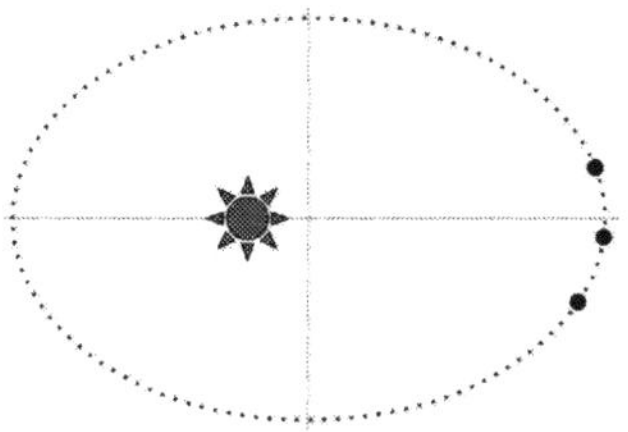

Figure 8-1 Mercury at 3 Times

To determine Mercury's aphelion, we employ *interpolation*, estimating the value of a function *between* measured values. The first step in interpolation is fitting a function to the measured data. For the sake of discussion, let's forget that Kepler taught us that planets orbit in ellipses, and fit Mercury's motion to a polynomial.

With three points, we choose a function of the form:

$$r(t) = A + B\,t + C\,t^2$$

We calculate A, B, and C, from the measured points as follows:

$$r(0) = A$$

$$r(-1) = A - B + C$$

$$r(+1) = A + B + C$$

$$A = r(0)$$

$$B = [r(+1) - r(-1)] / 2$$

$$C = [r(+1) + r(-1)] / 2 - r(0)$$

The next step is finding t_{aph}, the value of t at which r(t) reaches its maximum value. When r(t) is at a maximum value, r does not increase when t changes by a small amount, either a positive or negative amount. If r(t) did increase for any small change in t, r(t) would not be the maximum. Consider r(t) and $r(t + \Delta t)$:

$$r(t) = A + B\,t + C\,t^2$$

$$r(t+\Delta t) = A + B\,(t+\Delta t) + C\,(t+\Delta t)^2$$

We now require that r(t) at t_{aph} is unchanged for small Δt:

$$0 = r(t+\Delta t) - r(t)$$

$$0 = B\,\Delta t + 2C\,t_{aph}\,\Delta t + C\,\Delta t^2$$

For very small Δt, we can approximate Δt^2 as being zero. This yields:

$$0 = B + 2\,C\,t_{aph}$$

$$t_{aph} = -B / 2C$$

Interpolation provides the best possible *estimate* based on the available data. But, remember that we have not directly measured Mercury's actual aphelion; we relied on the assumption that its orbit is perfectly described by r(t). The actual orbit R(t) may deviate from r(t) by some unknown function Q. We can then write:

$$R(t) = r(t) + Q$$

$$R(t) = A + B\,t + C\,t^2 + Q$$

What can we say abut Q? Assuming the validity of our measurements, Q must be zero at t = –1, 0, and +1. We can therefore replace Q as follows:

$$R(t) = A + B\,t + C\,t^2 + \{\,(t+1)\,t\,(t-1)\}\;g(t)$$

The expression in { }s, which equals $\{t^3 - t\}$, ensures that Q is zero at the measured times. Also, g(t) is another unknown function that we expand below in a Taylor series.

$$R(t) = A + B\,t + C\,t^2 + \{t^3 - t\}\;(a_0 + a_1 t + a_2 t^2 + \ldots)$$

If we interpolate in the region near the center of the measured points, so that $|t| < 1/2$, the maximum deviation from our fitted function r(t) is:

$$|Q| < \{0.188\}\;(a_0 + a_1/2 + a_2/4 + \ldots)$$

While some functions do oscillate wildly, most physical functions are relatively smooth. This means a_0, a_1, a_2, ... are not likely to be enormous numbers. Furthermore, the terms in g(t) that are higher order in t are depressed by $|t| < 0.5$. Finally, the entire deviation is depressed by the {0.188} fitting factor. We therefore have a reasonable expectation that our interpolated estimate is not far off the true value.

By contrast, if we need the value of r(t) at t = 2, we must *extrapolate*, estimate the value of r(t) *outside the range* of its measured values. We then have:

$$Q = \{6\}\;(a_0 + 2a_1 + 4a_2 + \ldots)$$

Here, the potential deviation is much greater. The { }-factor is 32 times larger, and the higher order terms of g(t) have much larger coefficients. Depending on the coefficients, extrapolating to t = 2 could result in an error 100 times larger than interpolating to t = 0.5. The farther we extrapolate beyond measured data, the greater the probability of very bad estimate. The figure of merit (or demerit) is t divided by (the distance between the farthest data points).

Let's consider a real example: precisely calculating the position of Pluto when the New Horizons spacecraft flew by in July 2015.

This $700 million mission was literally a shot in the dark that was saved by old-school astronomy. When New Horizons launched in 2006, Pluto's orbit was known with a precision of *only* 0.002%. That seems fine, but is actually not nearly good enough. Since Clyde Tombaugh discovered it in 1930, Pluto has completed only 1/3 of a single 248-year orbit. It has not been where it is now since 1769.

Modern high-precision observations of Pluto began fairly recently. Extrapolating this data to the flyby date left Pluto's location uncertain by 62,000 miles, 44 times Pluto's diameter. To save weight, New Horizons was not equipped with any image analysis or search capabilities. It relied on mission control personnel on Earth to tell it where to aim and when to collect images. Flying at 32,000 mph (51,000 km/hr), New Horizons moves one Pluto diameter in under 3 minutes. In a 9.5-year mission, this is a time window of 0.6 parts per million.

Six years after launch, a desperate Dr. Marc Buie visited the Lowell Observatory where Tombaugh discovered Pluto. Tombaugh's images were too blurry, but Buie happened to find archived photographs taken by another astronomer named Carl Lampland. For reasons that had nothing to do with Pluto, Lampland took 1000 pictures of the same piece of sky over a 20-year period ending in the 1950's. By sheer luck, Pluto just happened to be in Lampland's photographs.

Using the best current technology, Buie meticulously digitized 1000 very old photographs, and — just in time — calculated Pluto's orbit with unprecedented precision. On July 14, 2015, New Horizons took this image of Pluto that awed the world.

Buie still had to extrapolate, but thanks to Lampland, he had good data over a much longer orbital segment.

4A§8.3 Functions near Extrema

Next, consider the behavior of an arbitrary function f(x) near an extremum, at a value of x at which f has a local minimum or maximum. Let's shift the x-axis so that this extremum occurs at x = 0, and expand f in a Taylor series.

$$f(x) = a_0 + a_1 x + a_2 x^2 + a_3 x^3 + \ldots$$

Let's now prove that $a_1 = 0$, because f has an extremum at x = 0. Examine what happens at infinitesimal values of x, either positive or negative. For $|x| \ll 1$, we can make the approximation that all terms proportional to x^2 and higher powers are negligible. In that limit:

$$\text{for } |x| \ll 1,\ f(x) = a_0 + a_1 x$$

If a_1 were > 0, f(x) would be > f(0) for positive x; hence f(0) would not be a maximum. If a_1 were < 0, f(x) would be > f(0) for negative x, and f(0) would not be a maximum. You can satisfy yourself that the same logic prevents f(0) from being a minimum for any nonzero a_1. Only if $a_1 = 0$, can f have an extremum at x = 0. This reduces our original Taylor series to:

$$f(x) = a_0 + a_2 x^2 + a_3 x^3 + \ldots$$

Let's now consider somewhat larger values of x, but still require that |x| is small. For small enough |x|, each term proportional to x^2 and higher powers is very small. Hence, f changes very slowly near its extrema.

The mathematically elegant conclusion is:

Near any function's extrema,

its changes are only *second order*.

"Second order" means the change in f(x) is proportional to x^2.

This principle has great importance in several areas of physics.

In ***Feynman Simplified 2D***, Chapter 36, we explore the *Principle of Least Action*, which governs: the motion of objects subject to forces; the distribution of electromagnetic fields; and the curvature of spacetime in general relativity. In each case, the solution to complex problems can be reduced to finding where the action reaches an extremum and varies most slowly.

The same mathematics arises in wave interference. The paths along which total phase change reaches an extremum are the paths with constructive wave interference. Those are the paths traversed by light and by elementary particles.

Chapter 9
Probability & Statistics

Successful scientists need to master probability, statistics, permutations, and combinations, because these are everywhere in nature and in our analyses of observations. All quantum theory is built on probabilities. All experimental data contain statistical fluctuations that must be understood and properly interpreted.

4A§9.1 Permutations

Permutations are re-orderings of a set of objects. Consider a *set* S comprised of a finite number n of *elements*. For example, let S be the first five letters of the alphabet:

$$S = \{A, B, C, D, E\}$$

Each of the following is a *permutation* of S.

$$P1 = B\ A\ C\ D\ E$$

$$P2 = B\ C\ A\ D\ E$$

$$P5 = B\ C\ E\ D\ A$$

P1 is obtained from S with one exchange of adjacent pairs (A B goes to B A). P2 is obtained from S with two exchanges (A B C goes to B A C goes to B C A). P5 is obtained from S with five exchanges. P1 and P5 are *odd* permutations and P2 is an *even* permutation, because their numbers of exchanges of adjacent pairs are odd and even respectively.

If all elements of a set are different (no duplicates), the number of possible permutations P(n) of n elements is:

$$\text{for n distinct elements, } P(n) = n!$$

Recall that n! is n-factorial, the product of the integers from 1 to n. This is easily proven. The first element of a permutation can be chosen in n different ways, because there are n distinct elements to choose from. The second element can be chosen in n–1 ways, because that is the number of remaining unchosen elements. This logic continues until the nth element of the permutation, which can only be chosen in 1 way. Since each choice is independent, the number of possible sequences of choices is:

$$n \times (n-1) \times \ldots \times 2 \times 1 = n!$$

4A§9.2 Combinations

Combinatorics describes the number of ways in which k elements can be chosen from a set S of n elements. Note that we are not concerned here about the order of the elements: ABC and ACB are the same combination, although they are different permutations. For example, consider again set S = {A, B, C, D, E}. In how many ways can we choose 2 of the 5 elements? The ten distinct *combinations* are:

AB, BC, AC, AD, AE

BD, BE, CD, CE, DE

We define C(n,k) to be the number of ways of choosing k elements from a set of n elements. Let's calculate C(n,k) using the example of five elements taken 2 at a time. First, we show the 5! permutations of S in Figure 9-1.

S = {A, B, C, D, E}

1	2	3	←Permutations→	5!
A	B	B		E
B	A	C		D
C	C	A	· · · · · · · · · ·	C
D	D	D		B
E	E	E		A

Figure 9-1 The 5! Permutations of 5 Elements

In general a set S of n distinct elements has n! permutations. Next, we divide each permutation into two parts: (1) the first two elements, and (2) the remaining three elements. We then sort all n! permutations into M groups, with each group containing all permutations that have the same first two elements in any order. As yet, we do not know the value of M.

The first **key point** is that each permutation belongs to *one and only one* group. The sorted groups are illustrated in Figure 9-2.

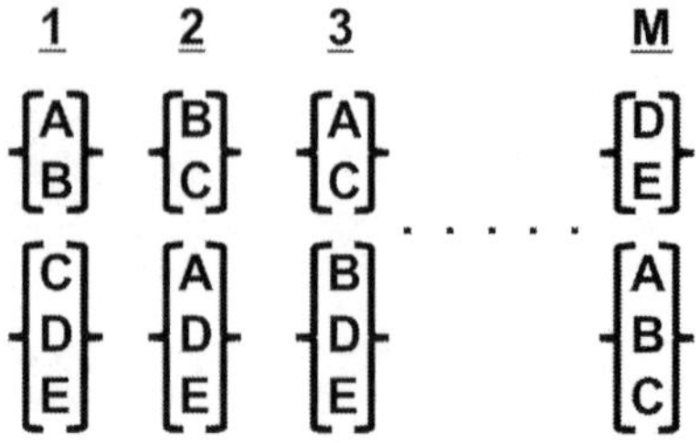

Figure 9-2 Groups Sorted By First Two Elements

The second **key point** is that *each group contains the same number of permutations*. Since the elements are all equivalent, we can relabel them in any way we please without changing M and without changing the number of permutations in any group. Therefore the groups must be the same except for the identity of the their elements.

Let's next determine the number of permutations per group. Figure 9-3 shows all the permutations in group #3. There are two permutations of the first two elements (shown horizontally), and 6 permutations of the remaining three elements (shown vertically).

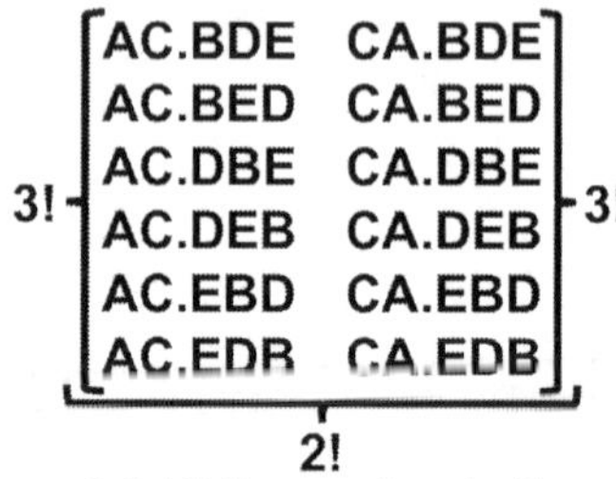

Figure 9-3 All Permutations in Group #3

The total number of permutations in group #3 is:

$$2! \times 3! = 2 \times 6 = 12$$

In general, in selecting k elements from n elements, there are k! permutations of the selected elements, and (n – k)! permutations of the remaining elements. This means there are k!(n – k)! permutations in each of the M groups. Since each of the n! permutations of set S is a member of one and only one group, we have:

$$n! = M\, k!\, (n - k)!$$

$$M = C(n, k) = n! / [\, k!\, (n - k)!\,]$$

Here, we note that M = C(n, k) = the number of ways of taking k elements from a set of n elements. For 2 elements chosen from a set of 5, k = 2 and n = 5, hence:

$$C(5, 2) = 5! / [2!\, 3!] = 120 / [2 \times 6] = 10$$

Note that C(n, k) = C(n, n–k). The equation for C(n, k) demonstrates this, but logic also demands it. There is no mathematical distinction between: (a) choosing 2 elements from a set of 5, and leaving 3 unchosen; or (b) choosing 3 elements from a set of 5, and leaving 2 unchosen. Both (a) and (b) divide the original set of 5 elements into subsets of 2 and 3 elements. The distinction between elements that are "chosen" and those that are "unchosen" is only in our minds, not in the math.

4A§9.3 Binomial Coefficients

An important result of combinatorics is the set of *binomial coefficients*. Consider the following *binomial equation* with n being an integer.

$$Z = (p + q)^n$$

Let's expand $(p + q)^n$ for several values of n.

$$\text{For } n = 0\text{: } Z = (p + q)^n = 1$$

$$\text{For } n = 1\text{: } Z = (p + q)^1 = p + q$$

$$\text{For } n = 2\text{: } Z = (p + q)^2 = (p + q)\,(p + q)$$

$$= p^2 + p\,q + q\,p + q^2$$

$$= p^2 + 2\,p\,q + q^2$$

The last Z is the sum of 4 *terms*. Recall that a term is any group of quantities that are multiplied or divided, but not added or subtracted. The sum of the exponents in each term is 2. Let's do one more.

$$\text{For } n = 3\text{: } Z = (p + q)^3 = (p + q)\,(p + q)^2$$

$$= p^3 + p^2\,q + q\,p^2 + p\,q^2 + q\,p^2 + p\,q^2 + q^2\,p + q^3$$

$$= p^3 + 3\,p^2\,q + 3\,p\,q^2 + q^3$$

Here, Z is the sum of 8 terms, several of which are equal, and the sum of the exponents in each term is 3.

In general, the expansion of $(p + q)^n$ is the sum of 2^n terms, many of which are equal. Also, the sum of the exponents in each term must be n; this is because there are n ( )'s, and each ( ) contributes one and only one factor, either p or q, to each term. Because there are 2 ways to pick one factor from each ( )'s, there are 2^n possible ways to pick the n factors that comprise each term. The 2^n terms contain powers of p that range from 0 to n, and the same is true for q. But in all terms, the (exponent of q) equals (n minus the exponent of p). There are therefore only n + 1 distinctly different terms; the rest are duplicates. Thus Z must have this form:

$$Z = \Sigma_{k=0}{}^{n}\, A_k\, p^k\, q^{n-k}$$

The set of A_k are the *binomial coefficients*. To calculate A_k, we employ combinatorics. In the original equation for Z:

$$Z = (p+q)\,(p+q)\,(p+q)\,\ldots\,(p+q)$$

There are n factors of p. The number of distinct ways to pick k p's from a set of n p's is C(n, k). Hence:

$$A_k = C(n, k) = n!\,/\,[\,k!\,(n - k)!\,]$$

$$Z = \Sigma_{k=0}{}^{n}\, p^k\, q^{n-k}\, n!\,/\,[\,k!\,(n - k)!\,]$$

Below is a table of the first few binomial coefficients, with each row having the same n and the C(n, k) listed in order from k = 0 to k = n. Note the pattern that each entry is the sum of two entries in the preceding line: the entry slightly to the left and the entry slightly to the right. This pattern is replicated for all values of n.

n = 0:	1
n = 1:	1 1
n = 2:	1 2 1
n = 3:	1 3 3 1
n = 4:	1 4 6 4 1

4A§9.4 Discrete Probabilities

The simplest probabilistic situation has only two possible outcomes. Let's consider flipping a coin that has probability p of coming up heads and probability q = 1 – p of coming up tails. A fair coin has p = q = 0.5. However, the real world isn't always fair. Using the binomial expansion, P(n, k) the probability of k heads occurring in n flips is:

$$P(n, k) = p^k\, q^{n-k}\, n!\,/\,[\,k!\,(n - k)!\,]$$

Note that we have the correct normalization, the probability of any number of heads is:

$$\Sigma_{k=0}{}^{n}\, P(n, k) = \Sigma_{k=0}{}^{n}\, p^k\, q^{n-k}\, n!\,/\,[\,k!\,(n - k)!\,]$$

$$\Sigma_{k=0}{}^{n}\, P(n, k) = (p + q)^n = 1^n = 1$$

For 4 flips of a fair coin, the P(4, k) are:

$$k = 0\text{: } 1 / 16 = 6.25\%$$

$$k = 1\text{: } 4 / 16 = 25\%$$

$$k = 2\text{: } 6 / 16 = 37.5\%$$

$$k = 3\text{: } 4 / 16 = 25\%$$

$$k = 4\text{: } 1 / 16 = 6.25\%$$

The sum of the probabilities of the five possible outcomes is 100%. As expected, the most probable result is 2 heads out of 4, and the probability of k heads is the same as the probability of k tails, for any k.

By contrast, if the probability of a head is p = 0.25, the P(4, k) are:

$$k = 0\text{: } p^0\, q^4 \times 1 = 31.64\%$$

$$k = 1\text{: } p^1\, q^3 \times 4 = 42.19\%$$

$$k = 2\text{: } p^2\, q^2 \times 6 = 21.09\%$$

$$k = 3\text{: } p^3\, q^1 \times 4 = 4.69\%$$

$$k = 4\text{: } p^4\, q^0 \times 1 = 3.91\%$$

Again, the sum of probabilities is 100%. Here, the most probable result is 1 head.

It is no coincidence that:

$$1 = p \times n = 0.25 \times 4$$

This brings us to the *expectation value* EV, also called the *mean* μ. The expectation value is the average result expected over a very large number of repetitions. For example, let a single trial consist of flipping a coin n times. After an enormous number of trials, the expectation value EV is:

$$EV = \mu = \Sigma\, k\, P(n, k)$$

Here, the sum is over all possible outcomes: k = 0 to k = n. For a binomial distribution this is:

$$\mu = \Sigma_{k=0}{}^{n}\, k\, p^k\, q^{n-k}\, n! / [\, k!\, (n - k)!]$$

Note that the k = 0 term does not contribute to the sum.

Deleting the k = 0 term, the sum now ranges over n values of k. Let's make these substitutions:

$$m = n - 1$$

$$j = k - 1$$

Note that:

$$n - k = m - j$$

$$n! = n\, m!$$

$$k! = k\, j!$$

We then have:

$$\mu = \Sigma_{j=0}{}^{m}\, k\, p^{j+1}\, q^{m-j}\, n\, m! / [\, k\, j!\, (m - j)!\,]$$

$$\mu = n\, p\, \Sigma_{j=0}{}^{m}\, p^j\, q^{m-j}\, m! / [\, j!\, (m - j)!\,]$$

$$\mu = n\, p\, \Sigma_{j=0}{}^{m}\, P(m, j)$$

$$\mu = n\, p\, [1]$$

For a binomial distribution, the mean number (expectation value) of A outcomes in n trials equals n p, where p is the probability of outcome A in one trial.

The other major characteristic of a probability distribution is its *variance* σ^2 defined by:

$$\sigma^2 = \Sigma_{k=0}^{n} (k - \mu)^2 P(n, k)$$

For a population of independent *trials* or measurements, the *standard deviation* σ is defined to be the square root of the variance σ^2. For a binomial distribution, σ^2 is:

$$\sigma^2 = \Sigma_{k=0}^{n} (k - n\,p)^2 P(n, k)$$

$$\sigma^2 = \Sigma_{k=0}^{n} (k^2 - 2k\,n\,p + n^2 p^2) P(n, k)$$

The $n^2 p^2$ factor is a constant that we can move outside the summation; it is multiplied by the sum of P(n, k) that equals 1. Also, we previously derived $n\,p = \Sigma_k\{k\,P(n, k)\}$, so the equation becomes:

$$\sigma^2 = -2\,(n\,p)\,n\,p + n^2 p^2 + \Sigma_{k=0}^{n} k^2 P(n, k)$$

Now make a strange substitution: replace k^2 by $k\,(k-1) + k$.

$$\sigma^2 = -n^2 p^2 + \Sigma_{k=0}^{n} \{k\,(k-1) + k\} P(n, k)$$

$$\sigma^2 = -n^2 p^2 + n\,p + \Sigma_{k=0}^{n} k\,(k-1) P(n, k)$$

Let's deal with the last sum. Note that the $k = 0$ and $k = 1$ terms are both zero. As above, we replace n with $m + 2$, and replace k with $j + 2$, yielding:

$$\Sigma_{k=2}^{n} k\,(k-1) P(n, k) = \Sigma_{j=0}^{m} Q\, p^{j+2} q^{m-j}$$

$$\text{with } Q = k\,(k-1)\,n! / [k!\,(n-k)!]$$

$$Q = k\,(k-1)\,n\,(n-1)\,m! / [k\,(k-1)\,j!\,(m-j)!]$$

$$Q = n\,(n-1)\,m! / [\,j!\,(m-j)!]$$

$$\sigma^2 = -n^2 p^2 + n\,p + p^2 n\,(n-1)\,\Sigma_{j=0}^{m} P(m,j)$$

$$\sigma^2 = -n^2 p^2 + n\,p + (n^2 p^2 - n\,p^2)\,(1)$$

$$\sigma^2 = +n\,p - n\,p^2$$

$$\sigma^2 = n\,p\,(1 - p)$$

For a binomial distribution, the variance of the number of A outcomes in n trials equals n p q, where p is the probability of outcome A in one trial, and $q = 1 - p$. That was fun.

Next, let's consider a rare event with $p \ll 1$. We have previously set $q = 1 - p$. Here, it will be more mathematically convenient to set $q = 1$. The binomial equation is still valid, but the sum has changed.

$$\text{for } q{=}1{:}\ Z = \Sigma_{k=0}^{n} p^k\, n! / [\,k!\,(n-k)!\,] = (1 + p)^n$$

Since the mean $\mu = n\,p$, we have in the limit of very large n:

$$Z = \{1 + \mu / n\}^n \to \exp\{\mu\} \text{ as } n \to \infty$$

Also, for $k \ll n$, in the limit that $n \to \infty$:

$$n! / (n-k)! = n \times (n-1) \times \ldots \times (n-k+1) \to n^k$$

This means the probability of k occurrences of a rare outcome A becomes:

$$P(k) = p^k n^k \exp\{-\mu\} / k!$$

$$P(k) = \mu^k \exp\{-\mu\} / k!$$

This is the *Poisson* probability distribution, illustrated in Figure 9-4, what the universally correct binomial distribution becomes in the limit of $p \ll 1$ and very large n. Note the asymmetry and the long tail extending to large k values.

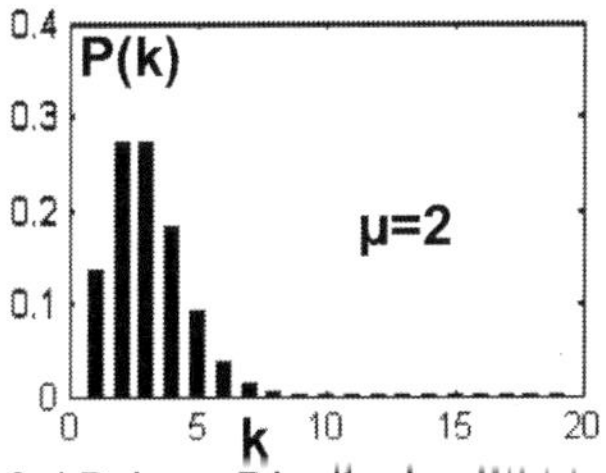

Figure 9-4 Poisson Distribution P(k) for $\mu = 2$

Some early applications of Poisson's probability distribution were:

- estimating the number of wrongful criminal convictions (1837); and
- investigating the number of soldiers accidentally killed by horse kicks (1898).

The mean and variance of the Poisson distribution are:

$$\mu = n\,p$$

$$\sigma^2 = n\,p$$

This is a **very important** result for physicists studying rare events. Whether counting radioactive decays, photons from remote quasars, or groups of particles with a combined effective mass near 127 GeV (Higgs boson), each event is extremely rare. Hence, $p \ll 1$, and the variance equals the number of detected events. In most stochastic processes, such as those mentioned here, one normally quotes σ as the *one standard deviation statistical uncertainty* (more on this later). Hence, if the number of detected events is N, one normally quotes the result with its statistical uncertainty as:

$$N \pm \sqrt{N}$$

This provides a standard measure of the level of confidence in the result.

For example, compare two cases: one with N = 100 and the other with N = 10,000.

100 ± 10 has 10% precision

10,000 ± 100 has 1% precision

To make the precision 10 times better, one must collect 100 times more data. No one said this was easy.

We have so far discussed stochastic processes with two possible outcomes. Now let's consider a situation with multiple possible outcomes. We turn to dice, and roll two dice, each a cube with faces numbered 1 through 6.

The table below shows the sum of the numbers on the top faces of die X (vertically) and die Y (horizontally) for each of the 36 possible outcomes of one roll.

X↓	1	2	3	4	5	6 ←Y
1:	2	3	4	5	6	7
2:	3	4	5	6	7	8
3:	4	5	6	7	8	9
4:	5	6	7	8	9	10
5:	6	7	8	9	10	11
6:	7	8	9	10	11	12

As you see, there are 11 possible sums: the integers from 2 to 12. The probability of a specific sum equals the number of combinations that yield that sum divided by 36. The probabilities are:

Sum = 2: 1/36 = 2.78%

Sum = 3: 2/36 = 5.55%

Sum = 4: 3/36 = 8.33%

Sum = 5: 4/36 = 11.11%

Sum = 6: 5/36 = 13.89%

Sum = 7: 6/36 = 16.67%, most probable

Sum = 8: 5/36 = 13.89%

Sum = 9: 4/36 = 11.11%

Sum=10: 3/36 = 8.33%

Sum=11: 2/36 = 5.55%

Sum=12: 1/36 = 2.78%

4A§9.5 Continuous Probabilities

Coins and dice produce a small number of discrete outcomes.

By contrast, most scientific measurements — the energy of cosmic rays, the luminosity of quasars, the mass of the Higgs boson, or the mass of a black hole — result in continuous outcomes.

These continuous outcomes can be distributed in many different ways that are determined by the natural phenomena themselves and by how we measure them.

Let's take a "simple" example: measuring the mass of a black hole. While the measurement may be extremely challenging in practice, it is simple conceptually because we seek one quantity that has a specific value that changes imperceptibly over the time scale of our measurement. Our measurement strategy is to identify objects that orbit the black hole, measure their orbital radii and periods, and use Kepler's "123" law:

$$M^1 T^2 = R^3$$

Here, M is the black hole's mass divided by the mass of our Sun, T is the orbital period in years, and R is the orbit's semi-major axis in AU (1 Astronomical Unit = Earth's mean orbital radius). The measurement challenges include precisely imaging very remote objects in obscured locations, and repeating those observations on a consistent basis over the many-year span of celestial orbits.

In nearly all scientific measurements, our instruments have limited precision. In this case, our measurements of T and R are inevitably uncertain to some degree. Experience, and the *central limit theorem* of probability theory, show that if the uncertainties are *stochastic* — random, independent fluctuations — a large number of measurements have a *Gaussian distribution*, also called a *normal distribution*, as illustrated in Figure 9-5.

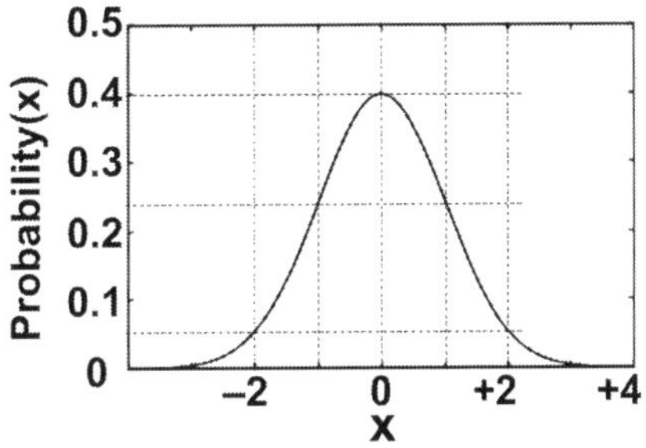

Figure 9-5 Gaussian Distribution

Here, Prob(x) is the probability that a quantity whose true value is zero will be measured to have value x if the measurement variance is 1. The equation for the curve in Figure 9-5 is:

$$\text{Prob}(x) = \exp\{-x^2 / 2\} / \sqrt{(2\pi)}$$

The constants in this equation are chosen so that: the total probability of all x values equals 1; and the variance of x also equals 1. You may enjoy proofs of these statements that are presented in Section 4A§13.4, Integration by Parts. Note that the distribution is symmetric about its most probable value, x = 0. Also, the probability of x decreases monotonically and substantially as |x| increases. At x = 0, the *probability density* is nearly 40%. For an infinitesimal dx, this means:

$$\text{Prob}(-dx/2 < x < +dx/2) = \text{Prob}(x = 0)\, dx$$

$$\text{Prob}(-dx/2 < x < +dx/2) = 0.3989\, dx$$

The good news is: the most probable measured value is the true value. The bad news is: actual measurements are spread over a significant range. As the deviation from the true value increases, the probability density decreases. Some examples are:

$$\text{Prob}(x = \pm 1) = 24.20\%$$

$$\text{Prob}(x = \pm 2) = 5.40\%$$

$$\text{Prob}(x = \pm 3) = 0.44\%$$

$$\text{Prob}(x = \pm 4) = 0.013\%$$

Note that at large deviations, the probability drops precipitously as x increases. From x = 0 to x = 1, the probability density drops by a factor of 1.6; from x = 1 to x = 2, it drops by a factor of 4.5; from x = 2 to x = 3 it drops by a factor of 12; and from x = 3 to x = 4, it drops by a factor of 34.

What is generally most important to physicists is the total probability that a measurement deviates, in either direction, from the true value (zero in this case) by more than some quantity Z. That probability is:

$$\text{Prob}(\,|x| > Z\,) = \text{Prob}(\,x > +Z\,) + \text{Prob}(\,x < -Z\,)$$

$$\text{Prob}(\,|x| > Z\,) = 1 - \text{Prob}(\,-Z < x < +Z\,)$$

This is often expressed in terms of erf, the *error function*, as:

$$\text{Prob}(\,|x| > Z\,) = 1 - \text{erf}(\,Z / \sqrt{2}\,)$$

The prior equation says erf(Y) equals the probability that any measured x value deviates from the mean of a Gaussian distribution by less than $Y / \sqrt{2}$ standard deviations. We show here the definition of the error function, which involves an integral. We will learn all about integrals in Chapter 13.

$$\text{erf}(Y) = (2 / \sqrt{\pi}) \int_0^Y \exp\{-u^2\}\, du$$

The equation for erf(Y) cannot be reduced to anything simple. One must obtain its values from tables.

In almost all cases, we are interested in probabilities rather than error functions. I provide below two short tables that are also given in Appendix 3. More comprehensive tables are readily available online.

Some values for Z, Prob(|x| > Z), and the odds of |x| > Z are:

Z	P(|x|>Z)	Odds of |x|>Z
0.5	0.61708	1 in 1.6204
0.7	0.48392	1 in 2.0665
1	0.31731	1 in 3.1515
2	0.04550	1 in 21.978
3	0.00270	1 in 370.40
4	6.33E–5	1 in 15787.
5	5.73E–7	1 in 1.74E+6
6	1.97E–9	1 in 5.07E+8

In the first line, Z = 0.5 and P(|x| > Z) is the probability of a measurement x deviating, either positively or negatively, from the mean of a Gaussian distribution by *more* than 0.5 standard deviations; P(|x| > 0.5) = 61.708%, and the odds of x deviating by *more* than 0.5 are 1 in 1.6204 = 1 / 0.61708. The last line says |x| exceeds 6 in 1.97 of one billion data sets, and the odds that |x| exceeds 6 are 1 in 507 million.

Often physicists assume that for all practical purposes |x| never exceeds 6: you would have to make half a billion measurements before expecting to have one |x| > 6. For example, the discovery of the Higgs boson was almost universally accepted when the probability of it *not existing* corresponded to x > 6. This condition is sometimes described as a "six sigma" confidence level. I say "almost", because in a democracy, there is always at least one dissenting opinion.

The following table lists Z values corresponding to various values of Prob(|x| < Z). For example, the top entry on the left side states that 80% of all measured x values are in the interval: –1.28156 < x < +1.28156.

P(|x|<Z)	Z	P(|x|<Z)	Z
0.80	1.28156	3 9's	3.29053
0.90	1.64485	4 9's	3.89059
0.95	1.95996	5 9's	4.41717
0.98	2.32635	6 9's	4.89164
0.99	2.57583	7 9's	5.32672
0.995	2.80703	8 9's	5.73073
0.998	3.09023	9 9's	6.10941

On the right side, "5 9's" means 0.99999 — five nines following the decimal point. The bottom entry on the right side means that 99.9999999% of all measured x values are in the interval: –6.10941 < x < +6.10941.

We have so far discussed a Gaussian distribution with zero mean and unit variance. In general, a Gaussian distribution has mean μ and variance σ^2, with this equation:

$$\text{Prob}(y) = \exp\{\, -(y - \mu)^2 / 2\sigma^2\,\} / \sqrt{(2\pi\,\sigma^2)}$$

Figure 9-6 shows four Gaussian distributions, and lists the mean μ and variance σ^2 of each.

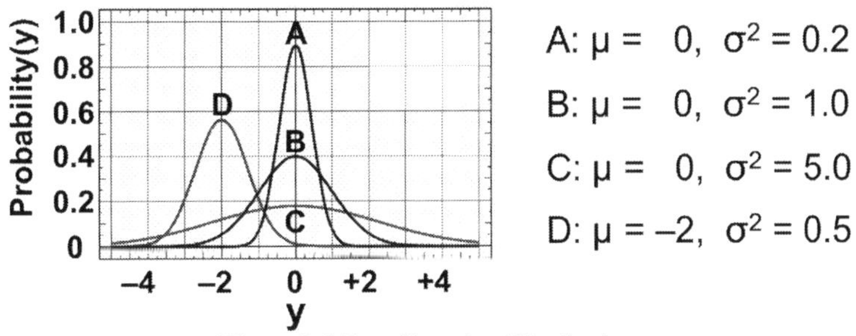

Figure 9-6 Four Gaussian Distributions

The effect of μ is simply to shift the curve left for $\mu > 0$, or right for $\mu < 0$. The effect of σ is to stretch the curve horizontally for $\sigma > 1$, or compress it for $\sigma < 1$.

To maintain proper normalization — to keep the probabilities of all y values summing to 100% — the peak rises for $\sigma < 1$ and drops for $\sigma > 1$.

Everything that we said above about Prob(x) is equally valid for Prob($[y - \mu] / \sigma$).

Let's review the **key points** of Gaussian distributions that are most important to physicists. Firstly, averages of independent measurements approach a Gaussian distribution as the number of measurements n becomes large. Secondly, as n increases, the average measured value μ approaches the true value of the quantity being measured. Thirdly, the uncertainty in μ equals $\sigma / \sqrt{n}$, where the standard deviation σ is the square root of the variance σ^2. The standard format for quoting the result of a group of measurements is:

$$\text{Result} = \mu \pm \varepsilon, \text{ with } \varepsilon = \sigma / \sqrt{n}$$

For a measurement whose uncertainties are entirely stochastic, the true value is within 1ε of μ 68% of the time, within 2ε 95% of the time, within 3ε 99.7% of the time, and within 4ε 99.994% of the time.

Mathematics cannot address *systematic* measurement errors or uncertainties. Math cannot fix faulty equipment or wrong interpretations.

4A§9.6 Combining Uncertainties

Virtually every scientific measurement process entails some degree of uncertainty. Quantum mechanics insists that uncertainty is a fundamental characteristic of nature. All instruments have limited precision, and the many measurements involved in the of counting rare events are all subject to statistical fluctuations.

Often these uncertainties are improperly called "errors" — scientists speak of the error function, error bars, and standard errors — but none of these are due to human mistakes. Rather, all of these are truly random statistical variations that are intrinsically unavoidable.

We have discussed the uncertainties that arise from statistical fluctuations in the measurement of specific quantities. Now let's examine the impact of such uncertainties in more complex analyses. What happens when we combine two quantities, each with their own uncertainties?

Let's assume two quantities m and n are the results of two different measurement processes. Assume each process has a finite number of discrete outcomes, with probabilities p(m) and q(n). Let p and q be normalized so that:

$$\Sigma_m \, p(m) = \Sigma_n \, q(n) = 1$$

We repeatedly measured one m and one n, and then sum those values. We wish to know the statistical properties of that result. The arithmetic operation is:

$$Q = m + n$$

We seek the mean and variance of Q. The variance of the left side of this equation must equal the variance of the right side, which we can calculate.

Recall the general equations for mean and variance:

$$\text{mean}(m): < m > = \Sigma_m\, m\, p(m)$$

$$\text{var}(m) = \Sigma_m\, (m - < m >)^2\, p(m)$$

$$\text{mean}(n): < n > = \Sigma_n\, n\, q(n)$$

$$\text{var}(n) = \Sigma_n\, (n - < n >)^2\, q(n)$$

The mean of $Q = m + n$ is:

$$\text{mean}(Q) = \Sigma_{mn}\, (m + n)\, p(m)\, q(n)$$

$$\text{mean}(Q) = \Sigma_{mn}\, m\, p(m)\, q(n) + \Sigma_{mn}\, n\, p(m)\, q(n)$$

$$\text{mean}(Q) = \Sigma_m\, m\, p(m) + \Sigma_n\, n\, q(n)$$

$$\text{mean}(Q) = < m > + < n >$$

The variance of Q is:

$$\text{var}(Q) = \Sigma_{mn}\, [(m + n) - (< m > + < n >)]^2\, p(m)\, q(n)$$

$$\text{var}(Q) = \Sigma_{mn}\, [(m - < m >) + (n - < n >)]^2\, p(m)\, q(n)$$

Expanding the prior equation yields:

$$\text{var}(Q) = \Sigma_{mn}\, (m - < m >)^2\, p(m)\, q(n) + \Sigma_{mn}\, (n - < n >)^2\, p(m)\, q(n)$$
$$+ 2\, \Sigma_{mn}\, (m - < m >)\, (n - < n >)\, p(m)\, q(n)$$

$$\text{var}(Q) = \text{var}(m) + \text{var}(n) + 2\, \{\Sigma_m\, (m - < m >)\, p(m)\}\, \{\Sigma_n\, (n - < n >)\, q(n)\}$$

$$\text{var}(Q) = \text{var}(m) + \text{var}(n) + 2\, (< m > - < m >)\, (< n > - < n >)$$

$$\text{var}(Q) = \text{var}(m) + \text{var}(n)$$

Thus we have the key results: (1) the mean of the sum equals the sum of the means; and (2) the variance of the sum equals the sum of the variances.

$$\text{mean}(Q) = \text{mean}(m) + \text{mean}(n)$$

$$\text{var}(Q) = \text{var}(m) + \text{var}(n)$$

$$\sigma_Q^2 = \sigma_m^2 + \sigma_n^2$$

If m and n are the counts of rare events, their variances are m and n, and we can write the summation equation as:

$$(m + n) \pm \sigma, \text{ with } \sigma = \sqrt{(m + n)}$$

Here, σ is the one standard deviation uncertainty in Q, assuming all uncertainties are independent and purely random fluctuations.

Subtraction is equivalent to adding m and $(-n)$. Inverting the polarity of n inverts the polarity of its mean but leaves its variance unchanged. The result for subtraction is: (1) the mean of the difference equals the difference of the means; and (2) the variance of the difference equals the *sum* of the variances.

We have covered addition and subtraction; let's turn next to multiplication and division of quantities with uncertainties. Consider the quantities A and B that have standard deviations a and b:

$$Q = (A \pm a)\, (B \pm b)$$

$$Q = A\, B\, (1 \pm a/A)\, (1 \pm b/B)$$

$$Q = A\, B\, (1 \pm a/A \pm b/B \pm ab/AB)$$

If we assume $a \ll A$ and $b \ll B$, we can drop the last term, reducing this to:

$$Q = A\, B\, (1 \pm a/A \pm b/B)$$

From the rule for adding quantities with uncertainties, the rule for multiplication becomes:

$$Q = AB\, \{1 \pm \sqrt{[\, (a^2/A^2) + (b^2/B^2)\,]}\, \}$$

The rule for division is:

$$Q = (A \pm a) / (B \pm b)$$

$$Q = (A / B) \{1 \pm \sqrt{[(a^2/A^2) + (b^2/B^2)]} \}$$

Next, let's examine powers.

$$Q = (A \pm a)^n$$

This time the uncertainties in the product terms are not independent; they are in fact all identical. We can evaluate Q by expanding the right side using binomial coefficients.

$$Q = A^n \pm n A^{n-1} a + n (n-1) A^{n-2} a^2 / 2 + \dots$$

Dropping terms of order $(a/A)^2$ and higher yields:

$$Q = A^n (1 \pm n a / A)$$

Now let's try some more complex functions: sine, cosine, and tangent. For these examples, we assume $|A|<<1$ and $|a|<<1$, and we will keep only the most significant terms in the small quantities.

$$\sin(A \pm a) = A \pm a$$

$$\cos(A \pm a) = 1 - (A \pm a)^2 / 2$$

$$\cos(A \pm a) = 1 - A^2 / 2 \pm A a - a^2 / 2$$

$$\cos(A \pm a) = \cos(A) \pm A a - a^2 / 2$$

$$\tan(A \pm a + \pi/2) = ?$$

The last equation shows that, for small ε, $\tan(\pm\varepsilon + \pi/2)$ can be anything from $-\infty$ to $+\infty$. Even a tiny uncertainty in an angle near $\pi/2$ leads to a completely uncertain value of the tangent. I am sure you see that uncertainties can be subtle.

In very complex situations, the propagation of uncertainties from the raw measurements to the final results are often evaluated using *Monte Carlo* methods, as we explore in Chapter 20.

Recall that when we subtracted two quantities with uncertainties, their variances added. This has important consequences, particularly when subtracting quantities of similar magnitude. For example, consider the difference in intensity of two gamma ray bursts that have 100 photon counts and 95 photon counts.

$$Q = (100 \pm 10.0) - (95 \pm 9.7)$$

$$Q = +5 \pm 14$$

The initial two numbers, known with 10% precision, have a difference that has an uncertainty of 280%. We do not even know which gamma ray burst was brighter.

This highlights a critical issue in experimental science. Often it is important to compare two quantities A and B that are very similar, but that have a crucial distinction. We want to know if they are exactly equal, or if one is slightly larger. As we have just shown, it is very difficult to measure A, measure B, subtract the measurements, and obtain an extremely precise result. It is far better, if possible, to directly measure A – B.

Let's consider two real world examples. The famous 1887 Michelson-Morley experiment sent two light beams along two identical but orthogonal paths, and determined the difference in their travel times. If Michelson had measured each travel time and subtracted those measurements, measurement uncertainties would have dominated his results, rendering them meaningless. Michelson's genius was devising a method of directly measuring the *difference* in travel time by using wave interference. His high-precision measurements prove that Earth's velocity through the luminiferous ether, if it exists, is no more than 2.5% of Earth's orbital velocity. Michelson's measurement thus definitively refutes the model of ether as the medium of light, although it took most scientists decades to accept that conclusion.

Conversely, my thesis experiment confirmed the violation of matter-antimatter symmetry by showing that neutral kaons are 0.28% more likely to decay to $\pi^- \mu^+ \nu$ than to $\pi^+ \mu^- \underline{\nu}$. Unable to measure the 0.28% difference directly, we were forced to collect 10 million decays of each type in order to measure each decay rate with 0.035% precision. That determined their difference to a precision of only 18%. Forty years ago, that was the best measurement of this quantity, and it is still the second best measurement. After collecting all that data, we were able to prove that the decay rates are different, but it was very painful to see 0.035% precision degrade to 18% when two nearly equal numbers were subtracted.

4A§9.7 Chi-Square Analysis

Chi-square (χ^2) analysis is the standard method of comparing data sets to one another, or comparing a data set to a theoretical hypothesis.

Imagine taking N measurements of a quantity X, with each measured value x within a fixed range: xmin < x < xmax. Let's now sort our measured x values into K *bins*, with the bin number j of each measurement given by:

$$j = 1 + K\,(x - xmin) / (xmax - xmin)$$

We define n(j) to be the number of measurements in bin #j. This means p(j), the probability of a measured value falling into bin #j is given by:

$$p(j) = n(j) / N$$

$$q(j) = 1 - p(j) = [N - n(j)] / N$$

As discussed above, for large N, the one standard deviation uncertainty in n(j) equals:

$$\sigma \text{ of } n(j) = \sqrt{\{N\, p(j)\, q(j)\}}$$

$$\sigma \text{ of } n(j) = \sqrt{\{ n(j)\, [N - n(j)] / N\}}$$

$$\text{if } n(j) << N, \sigma \text{ of } n(j) = \sqrt{\{ n(j) \}}$$

Let's now compare the set of numbers n(j), with a theoretical hypothesis $\hat{H}$, assuming for simplicity that all n(j) << N. Typically, $\hat{H}$ predicts a probability distribution from which we can calculate p(j), the probability that a single measured x falls into bin #j. For N total measurements, $\hat{H}$ predicts that bin #j will contain N p(j) measurements. We then compute:

$$\chi^2 = \Sigma_j \{ N\, p(j) - n(j) \}^2 / N\, p(j)$$

Here, we sum from j = 1 to j = K, excluding any j for which p(j) = 0. Define k to be the number of bins for which p(j) > 0.

We now define #df to be the *number of degrees of freedom*: #df equals the number of independent comparisons in the χ^2 summation minus the number of adjustable parameters.

In the χ^2 summation, there are k terms — k comparisons between data and theory. But, each theoretical prediction p(j) is multiplied by N, and N is determined by the data: [$N = \Sigma_j n(j)$]. Hence there is one adjustable parameter, making #df = k – 1.

To clarify this point, consider the special case of k = 1, where the χ^2 summation reduces to:

$$\text{for } k = 1: \chi^2 = \{ N - N \}^2 / N = 0$$

Here, there are zero degrees of freedom, and no meaningful comparison between data and theory.

The figure of merit in a chi-squared analysis is:

$$\chi^2 / \#df$$

If $\hat{H}$ were a perfect description of nature, if the data were free of systematic errors, and if k >> 1, the summation terms would on average equal 1, and χ^2 would on average equal #df.

If χ^2/#df is less than or equal to 1, theory $\hat{H}$ is considered to be consistent with the measured data. Consistency is not sufficient to claim that $\hat{H}$ is proven, but it is a significant step in that direction. Consistency with multiple sets of data by multiple observers testing multiple predictions of $\hat{H}$ leads to wider and wider acceptance of $\hat{H}$ as an effective model.

Conversely, if χ^2/#df is much more than 1, the theory and the data are inconsistent, which typically means one or both are wrong. A theory is considered *falsified* if it is strongly inconsistent with several independent sets of data from different observers, or is strongly inconsistent with one indisputable set of data.

Clearly, there is a substantial gray area between acceptance and falsification. Physicists seek to resolve such quandaries with more data, taken by more observers, testing more predictions. "We need more data" is a common refrain.

Figure 9-7, created by Mikael Haggstrom, plots χ^2 horizontally vs. *p-value* vertically for five indicated values of #df. The horizontal dotted line is at a p-value of 0.5.

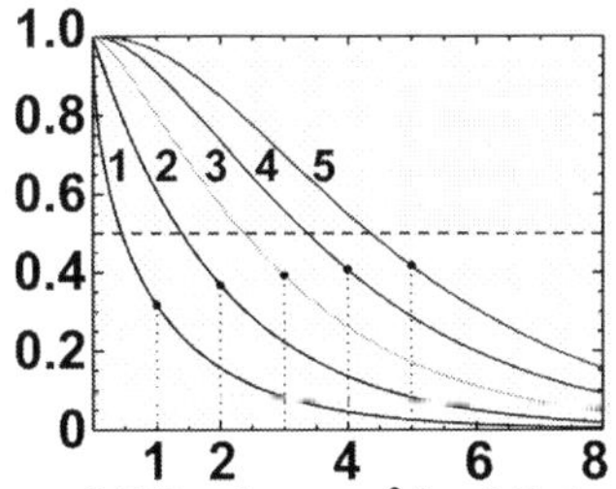

Figure 9-7 P-value vs. χ^2 for #df=1 to 5

Appendix 4 contains a more comprehensive table of χ^2 / #df.

P-value(χ^2) is the fraction of independent data sets that have a larger χ^2. The fraction of data sets with χ^2 > #df is:

#df = 1, 31% have $\chi^2 > 1$

#df = 2, 36% have $\chi^2 > 2$

#df = 3, 39% have $\chi^2 > 3$

#df = 4, 40% have $\chi^2 > 4$

#df = 5, 41% have $\chi^2 > 5$

for #df >> 1, 50% of data sets have χ^2 > #df

We can also use χ^2 to compare two data sets and test the hypothesis that they consistently measure the same natural phenomenon. In comparing data with theory, we assumed above that the theoretical predictions had zero statistical uncertainties. This is often, but not always, true. If we measure muon lifetimes versus velocity, and compare our data to the predictions of special relativity, the predicted values have zero uncertainty. But if we measure Alnico V's magnetization versus an external field, the theoretical predictions will necessarily incorporate numerous measured material properties whose uncertainties make the theoretical predictions uncertain, sometimes in very complex ways.

Whether the uncertainties are instrumental or theoretical, we can compare data set n(j) with data set m(j) by computing:

$$\chi^2 = \Sigma_j \{ m(j) - n(j) \}^2 / [\sigma_m(j)^2 + \sigma_n(j)^2]$$

If both m(j) and n(j) are numbers of rare events with purely statistical uncertainties, this equation reduces to:

$$\chi^2 = \Sigma_j \{ m(j) - n(j) \}^2 / [m(j) + n(j)]$$

Again, the figure of merit is given by χ^2 / #df. If χ^2 / #df >> 1, the data sets may be measuring different phenomena, or they may have different systematic errors.

Chapter 10
Rotation & Velocity Transformations

This chapter explores transforming coordinate systems. Two types of coordinate transformations are common in physics: (1) rotating the coordinate axes; and (2) changing the velocity of a coordinate system. Each such transformation has an inverse: a second transformation that goes back to the original coordinate system.

Imagine that we wish to analyze a rocket's trajectory during a one-hour flight from one point on Earth's surface to another. We could analyze this motion using Earth-bound coordinates: latitude, longitude, and elevation. That coordinate system continuously rotates as Earth spins on its axis. Earth's meridians, lines of constant longitude, turn 15 degrees per hour. Alternatively, we could analyze the rocket's motion in a coordinate system whose axes remain stationary as Earth turns.

One transformation converts the rocket's coordinates in the Earth-bound system into coordinates in the stationary system. An inverse transformation converts stationary coordinates into Earth-bound coordinates.

Since the laws of nature are independent of our choice of coordinate systems, we can employ any coordinate system that seems convenient. Proper transformations ensure that measurements in any coordinate system are consistent with those laws.

We can view transformations in two equivalent ways. We can rotate the coordinate axes by angle +θ while keeping the rocket stationary, or we can rotate the rocket by angle –θ while keeping the axes stationary. In the case of boosts, we can boost our reference frame by velocity +***v*** while keeping the rocket's velocity fixed, or we can boost the rocket by velocity –***v*** while keeping our reference frame fixed. Throughout this chapter, and almost all of the ***Feynman Lectures***, transformations change the coordinate axes and reference frames while leaving fixed the physical entities that we are describing.

4A§10.1 Simple Rotations in 2-D

The simplest coordinate rotation is shown in Figure 10-1. Here, point P remains stationary while the xy-axes are rotated by angle θ to become the x*y*-axes.

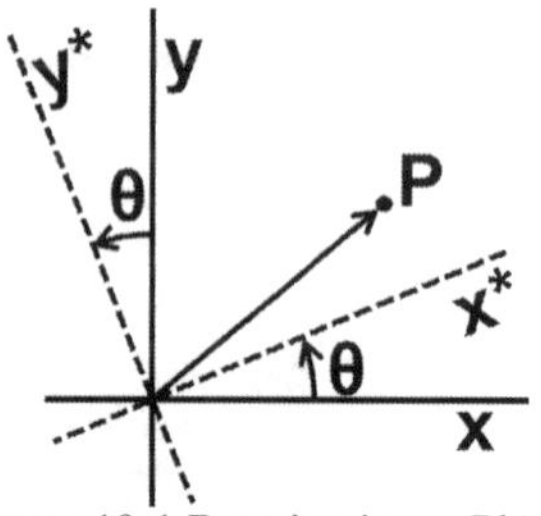

Figure 10-1 Rotation in xy-Plane

We can also describe this in 3-D, with the z-axis perpendicular to the page, and pointing outward in accordance with the right hand rule (see 4A§11.2). We say the coordinate system is rotated about the z-axis by the positive angle θ. Again employing the right hand rule, we point our right thumb in the +z-direction, and note that our right fingers curl counterclockwise, which defines the direction of positive rotation angle.

Note that both x and y axes rotate in the same direction.

For a rotation of the xy-axes by **angle θ about the z-axis**, the transformation equations that relate the coordinates of point P in the xyz system to P's coordinates in the x*y*z* system are:

$$x^* = + x \cos\theta + y \sin\theta$$

$$y^* = - x \sin\theta + y \cos\theta$$

$$z^* = z$$

Note that the transformation equations contain one minus sign.

As we should expect, the **inverse transformation** is the same, but with θ replaced by –θ:

$$x = + x^* \cos\theta - y^* \sin\theta$$
$$y = + x^* \sin\theta + y^* \cos\theta$$
$$z = z^*$$

For a rotation of the yz-axes by **angle θ about the x-axis**, the transformation equations are:

$$x^* = x$$
$$y^* = + y \cos\theta + z \sin\theta$$
$$z^* = - y \sin\theta + z \cos\theta$$

For a rotation of the xz-axes by **angle θ about the y-axis**, the transformation equations are:

$$x^* = + x \cos\theta - z \sin\theta$$
$$y^* = y$$
$$z^* = + x \sin\theta + z \cos\theta$$

4A§10.2 Rotation by Euler Angles in 3-D

Any arbitrary rotation — any angles about any axes — can be achieved with three simple rotations of the form described above. Such rotations are defined using the *Euler angles* shown in Figure 10-2.

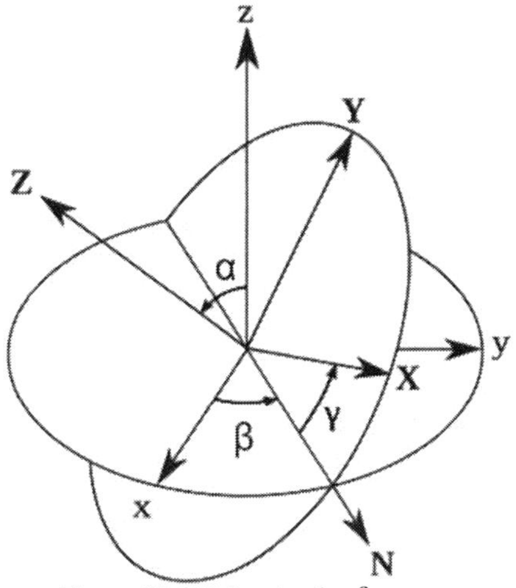

Figure 10-2 Euler Angles β, α, γ

An Euler rotation consists of three steps in this specific sequence:

(1) rotate about the z-axis by angle β

(2) rotate about N the new x-axis by angle α

(3) rotate about Z the new z-axis by angle γ

The transformation equations relating the coordinates of point P in the (x, y, z) system to P's coordinates in the (x*, y*, z*) system are:

$$x^* = x\,(\cos\gamma \cos\beta - \cos\alpha \sin\gamma \sin\beta) + y\,(\cos\gamma \sin\beta + \cos\alpha \cos\beta \sin\gamma) + z\,(\sin\alpha \sin\gamma)$$
$$y^* = x\,(-\cos\beta \sin\gamma - \cos\alpha \cos\gamma \sin\beta) + y\,(\cos\gamma \cos\alpha \cos\beta - \sin\gamma \sin\beta) + z\,(\cos\gamma \sin\alpha)$$
$$z^* = x\,(\sin\beta \sin\alpha) + y\,(-\cos\beta \sin\alpha) + z\,(\cos\alpha)$$

4A§10.3 Relativistic Boosts — Lorentz Transform

A *boost* is a change in velocity — reference frame S is transformed to reference frame S* that is moving at velocity $+v$ relative to S.

Figure 10-3 shows a boost in which v is in the +x-direction. Here, point P remains stationary, while the x-ct axes are boosted to become the x*-ct* axes.

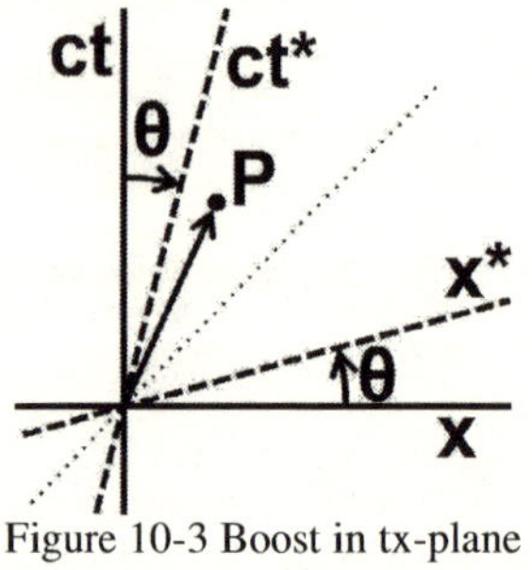

Figure 10-3 Boost in tx-plane

Note that the x* and ct* axes tilt in opposite directions: both tilt toward the 45-degree line by the same angle θ.

The transformation equations that relate the coordinates of point P in the x-ct frame to P's coordinates in the x*-ct* frame are called the *Lorentz transform*. These are:

$$x^* = +x\,\gamma - ct\,\gamma\,\beta$$

$$ct^* = -x\,\gamma\,\beta + ct\,\gamma$$

Here, $\beta = v / c$ and $\gamma = 1 / \sqrt{(1 - \beta^2)}$ are the usual relativistic factors.

As we should expect, the inverse transformation is the same, but with β replaced by –β:

$$x = +x^*\,\gamma + ct^*\,\gamma\,\beta$$

$$ct = +x^*\,\gamma\,\beta + ct^*\,\gamma$$

Note that the transformation equations contain an even number of minus signs. The angle θ is given by:

$$\tanh(\theta) = \beta$$

In most advanced physics books and papers, authors adopt units in which c = 1, so the corresponding graphs and equations do not include c.

4A§10.4 Rotations of Quantum Spin States

We tabulate here the equations for several rotational transformations between a set S of quantum basis states and a rotated set T of basis states. Basis states in quantum mechanics have a similar role to normal coordinate axes.

For **spin 1/2** particles, each basis set has two linearly independent states, denoted + and –.

To rotate about the **z-axis by angle ø**:

$$|+T> = \exp\{+i\text{ø}/2\}\,|+S>$$

$$|-T> = \exp\{-i\text{ø}/2\}\,|-S>$$

To rotate about the **y-axis by angle θ**:

$$|+T> = +\cos(\theta/2)\,|+S> + \sin(\theta/2)\,|-S>$$

$$|-T> = -\sin(\theta/2)\,|+S> + \cos(\theta/2)\,|-S>$$

To rotate about the **x-axis by angle θ**:

$$|+T> = \cos(\theta/2)\,|+S> + i\sin(\theta/2)\,|-S>$$

$$|-T> = i\sin(\theta/2)\,|+S> + \cos(\theta/2)\,|-S>$$

For **Euler angles**: rotate first about the **z-axis by angle β**,
then rotate about the **new x-axis by angle α**,
then rotate about the **new z-axis by angle γ**:

$$<+T \mid +S> = \cos(\alpha/2)\,\exp\{+i(\beta+\gamma)/2\}$$
$$<+T \mid -S> = i\sin(\alpha/2)\,\exp\{-i(\beta-\gamma)/2\}$$
$$<-T \mid +S> = i\sin(\alpha/2)\,\exp\{+i(\beta-\gamma)/2\}$$
$$<-T \mid -S> = \cos(\alpha/2)\,\exp\{-i(\beta+\gamma)/2\}$$

For **spin 1 particles**, each basis set has three linearly independent states, denoted +, 0, –.

To rotate about the **y-axis by angle θ**:

$$<+T \mid +S> = (1+\cos\theta)/2$$
$$<0T \mid +S> = -(\sin\theta)/\sqrt{2}$$
$$<-T \mid +S> = (1-\cos\theta)/2$$
$$<+T \mid 0S> = +(\sin\theta)/\sqrt{2}$$
$$<0T \mid 0S> = +\cos\theta$$
$$<-T \mid 0S> = -(\sin\theta)/\sqrt{2}$$
$$<+T \mid -S> = (1-\cos\theta)/2$$
$$<0T \mid -S> = +(\sin\theta)/\sqrt{2}$$
$$<-T \mid -S> = (1+\cos\theta)/2$$

To rotate about the **z-axis by angle θ**:

$$<+T \mid +S> = \exp\{+i\theta\}$$
$$<0T \mid 0S> = 1$$
$$<-T \mid -S> = \exp\{-i\theta\}$$

The other six amplitudes are zero.

To rotate about the **x-axis by angle θ**:

$$<+T \mid +S> = (\cos\theta+1)/2$$
$$<0T \mid +S> = +i(\sin\theta)/\sqrt{2}$$
$$<-T \mid +S> = (\cos\theta-1)/2$$
$$<+T \mid 0S> = +i(\sin\theta)/\sqrt{2}$$
$$<0T \mid 0S> = +\cos\theta$$
$$<-T \mid 0S> = +i(\sin\theta)/\sqrt{2}$$
$$<+T \mid -S> = (\cos\theta-1)/2$$
$$<0T \mid -S> = +i(\sin\theta)/\sqrt{2}$$
$$<-T \mid -S> = (\cos\theta+1)/2$$

Chapter 11
Vector Algebra

4A§11.1 Vectors in 3-D

In three dimensions, we can analyze generalized motion with three independent sets of equations, one for each coordinate:

Along the X-axis, location is specified by the coordinate value x, velocity by v_x, and acceleration by a_x. Velocity v_x is the rate at which x increases, and a_x is the rate at which v_x increases. Similarly for y and z.

In a very brief time interval dt, let the tiny coordinate changes be dx, dy, and dz. The 3-D velocity v and 3-D distance traveled ds are:

$$ds = \sqrt{[\, dx^2 + dy^2 + dz^2]}$$

$$ds = dt \sqrt{[(dx/dt)^2 + (dy/dt)^2 + (dz/dt)^2]}$$

$$ds = dt \sqrt{[(v_x)^2 + (v_y)^2 + (v_z)^2]}$$

$$v = \sqrt{[v_x^2 + v_y^2 + v_z^2]}$$

$$ds = v\, dt$$

While we could write any physics equation explicitly, listing all the components in each dimension, mathematicians have devised a more compact notation: *vectors*. Compare Newton's force equations below that are first written in component notation and then in vector notation. The fourth equation contains the same information as do the first three equations combined.

$$F_x = m\, a_x$$

$$F_y = m\, a_y$$

$$F_z = m\, a_z$$

$$\boldsymbol{F} = m\, \boldsymbol{a}$$

In this book, we denote vectors with bold italics.

In three dimensions, a vector is an ordered triplet of quantities, one component for each dimension of space, that *transform properly under rotations*. To understand what "transform properly" means, consider the position vector ***r***:

Position vector: $\boldsymbol{r} = (x, y, z)$

For a point P whose rectilinear coordinates are x, y, and z, ***r*** is the vector from the origin of the coordinate system to point P, as shown in Figure 11-1.

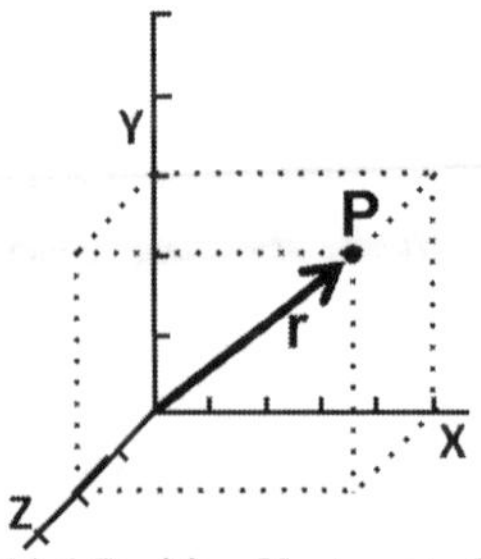

Figure 11-1 Position Vector ***r*** to Point P

The distance between the origin and point P is the length of ***r***, which is given by:

$$r = |\boldsymbol{r}| = \sqrt{(\boldsymbol{r} \bullet \boldsymbol{r})} = \sqrt{(x^2 + y^2 + z^2)}$$

Note that for vectors, we use | | to denote the vector's magnitude (length).

Now imagine what happens when the coordinate axes rotate, as they would if the axes are defined relative to Earth's surface. As Earth turns, the coordinates of point P, which are the components of ***r***, change in a very well defined manner, as we discussed in Chapter 10.

An essential property of vector is that rotations never change the length of any vector, nor do rotations change the relative orientations of one vector to another. If the angle between two vectors is θ before rotation, it will be θ after any rotation. This is what we mean by transforming properly.

Not all quantities can be combined to form vectors. Your height H, weight W, and age A do not transform properly under rotation (they each remain constant), hence (H, W, A) is not a vector.

Other important vectors are:

Velocity vector: $\boldsymbol{v} = (v_x,\ v_y,\ v_z)$

Acceleration vector: $\boldsymbol{a} = (a_x,\ a_y,\ a_z)$

Force vector: $\boldsymbol{F} = m\,\boldsymbol{a} = (F_x,\ F_y,\ F_z)$

The force equation states that each of the three components of ***F*** equals m multiplied by the corresponding component of ***a***. This demonstrates two advantages of using vectors: (1) less writing; and (2) less clutter to distract our focus from the important physics underlying the equations.

Two important vector operations are the *dot product* (***A • B***) and *cross product* (***A × B***):

$$\boldsymbol{A} \bullet \boldsymbol{B} = A_x B_x + A_y B_y + A_z B_z$$
$$\boldsymbol{A} \bullet \boldsymbol{B} = \boldsymbol{B} \bullet \boldsymbol{A}$$

$$\boldsymbol{A} \times \boldsymbol{B} = (A_y B_z - A_z B_y,\quad A_z B_x - A_x B_z,\quad A_x B_y - A_y B_x)$$
$$\boldsymbol{A} \times \boldsymbol{B} = -\boldsymbol{B} \times \boldsymbol{A}$$

The cross product produces a vector that has graphic significance, as shown on the left side of Figure 11-2. If one moves vectors ***A*** and ***B*** to a common origin, without changing their directions, the two vectors define a parallelogram. Vector ***A × B*** is perpendicular to the plane of the parallelogram, and its length equals the parallelogram's area: A B sinθ.

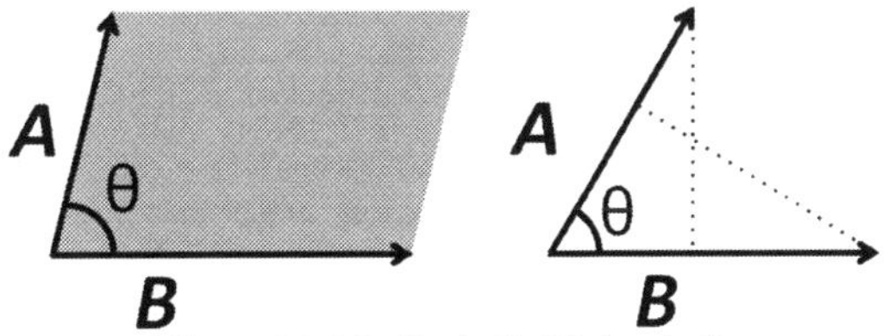

Figure 11-2 Left: ***A×B***. Right: ***A•B***

The dot product ***A • B*** equals A B cosθ, which is a scalar, an ordinary quantity without dimensionality. The geometric significance of the dot product is shown on the right side of Figure 11-2. The dot product is the normal projection of ***A*** onto ***B*** (indicated by the vertical dotted line), or equivalently, the normal projection of ***B*** onto ***A*** (indicated by the diagonal dotted line).

To determine whether or not ***s*** is a proper vector: calculate the dot product ***s • r***, for any known proper vector ***r***. This dot product is an invariant scalar that is unchanged by any rotation if and only if ***s*** is a proper vector. We prove this statement in Chapter 16.

4A§11.2 Right Hand Rule

In many mathematical analyses — including those of vector algebra, trigonometry, and rotations — an ambiguity arises between two equally valid choices: one "left-handed" and the other "right-handed". In Figure 11-2 for example, vector ***A × B*** is perpendicular to both ***A*** and ***B***, which means it is perpendicular to this page. But, does it point toward you or away from you?

While both alternatives are equally valid mathematically, we must choose such alternatives consistently in order to avoid errors. Mathematicians and scientists have universally adopted the *right hand rule*.

You lefties out there may feel umbrage, but you should note one advantage that lefties enjoy from this convention: you can use your right hand to determine answers while simultaneously writing equations with your left hand. Righties have to switch back and forth. More than once have I seen right-handed students groan in anguish midway through an exam during which they had repeatedly twisted their left hands in various directions.

The left image below shows how I employ the right hand rule for the cross product: $\boldsymbol{F} = q\, \boldsymbol{v} \times \boldsymbol{B}$. The right image shows how I employ the right hand rule to determine positive circulation or angle of rotation.

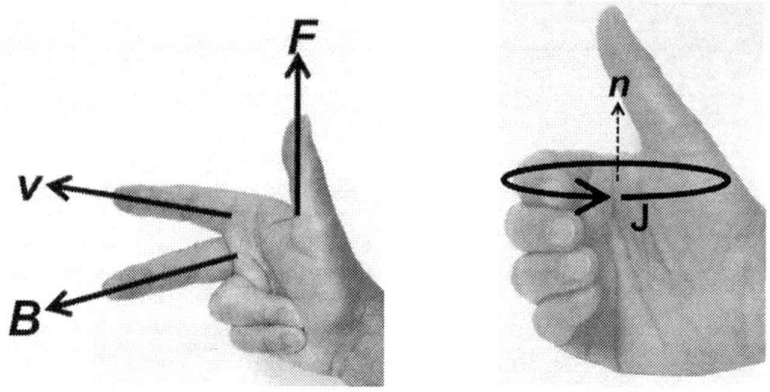

For cross products, I point my index finger along the left vector of $\boldsymbol{v} \times \boldsymbol{B}$, and point my middle finger along the right vector of that product; my thumb now points in the direction of the resultant vector $\boldsymbol{F}$.

For rotational problems, I point my thumb along the rotational axis, and curl my fingers; the tips of my fingers now point in the direction of increasing angle, the direction of positive rotation.

For a circulating current (or orbiting mass), I curl my fingers in the direction of circulation (or orbital motion); my thumb then points in the direction of the magnetic field (or angular momentum).

This works every time, but it can sometimes be a bit awkward. You should practice this until it becomes automatic.

Be sure to use your **right hand**. If you are unable to use your right hand, use your left and then take the direction opposite to your thumb.

Nothing in physics employs a left hand rule.

4A§11.3 Polar & Axial Vectors

The vectors we have so far discussed are all *polar vectors*. There is another type: *axial vectors*. The two types behave identically under rotations, but behave oppositely under coordinate reflections or inversions. Axial vectors, often called *pseudovectors*, generally represent mathematical concepts rather than physical observables, and are typically defined by vector cross products. Torque and angular momentum are examples of axial vectors.

Let's examine a reflection of the x-axis. Let a mirror lie in the yz-plane, perpendicular to the x-axis, and assume real actions occur to the left of the mirror, as illustrated in Figure 11-3. From the left side, we see the action's reflected image, with all the same y- and z- coordinates, but with all x-coordinates having the opposite polarity. In the upper real image, a black dot moves with velocity v in the +y, +x direction; in the mirror image, the black dot moves in the +y, –x direction. Velocity, like all other polar vectors, transforms under x-axis reflection according to:

<u>Polar Vector</u>

Real Action: $\boldsymbol{v} = (v_x, v_y, v_z)$

Mirror Image: $\boldsymbol{v} = (-v_x, v_y, v_z)$

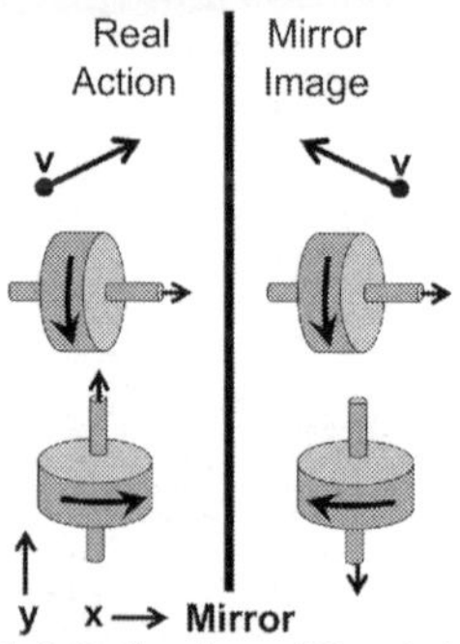

Figure 11-3 Reflection of X-axis in Mirror

Now let's examine the angular momenta of the two flywheels shown in Figure 11-3. Angular momentum $\boldsymbol{L}$ equals (position $\boldsymbol{r}$) × (linear momentum $\boldsymbol{p}$). The flywheel's axis is horizontal in the middle image and vertical in the lower image.

In the middle real image, the flywheel's near side moves downward (long arrow); so $\boldsymbol{L}$ (short arrow) points toward +x, by the right hand rule. In the mirror image, the flywheel's near side still moves downward, because the y-component of its velocity does not change sign; so $\boldsymbol{L}$ still points toward +x.

In the lower real image, the flywheel's near side moves toward +x; so $\boldsymbol{L}$ points toward +y. In the mirror image, the flywheel's near side moves toward –x, because the x-component of its velocity does change sign; so $\boldsymbol{L}$ points toward –y.

Under x-axis reflection:

the x-components of polar vectors flip sign, but the y- and z-components do not

the y- and z-components of axial vectors flip sign, but the x-component does not

Let's see how the math works for a different axial vector: the angular velocity defined according to the right-hand rule as:

$$\boldsymbol{\omega} = \boldsymbol{v} \times \boldsymbol{r}$$

In component notation this is:

$$\omega_x = +v_y r_z - v_z r_y$$

$$\omega_y = +v_z r_x - v_x r_z$$

$$\omega_z = +v_x r_y - v_y r_x$$

Exchanging x with –x changes v_x to $-v_x$, and r_x to $-r_x$. The mirror image ω, call it ω^*, is:

$$\omega^*_x = +v_y r_z - v_z r_y$$

$$\omega^*_y = +v_z (-r_x) - (-v_x) r_z$$

$$\omega^*_z = +(-v_x) r_y - v_y (-r_x)$$

Axial Vector

Real Action: $\boldsymbol{\omega} = (\omega_x,\ \omega_y,\ \omega_z)$

Mirror Image: $\boldsymbol{\omega} = (\omega_x, -\omega_y, -\omega_z)$

For comparison, recall the result for polar vectors:

Polar Vector

Real Action: $\boldsymbol{v} = (\ v_x,\ v_y,\ v_z)$

Mirror Image: $\boldsymbol{v} = (-v_x,\ v_y,\ v_z)$

All that was for mirror reflections that reverse only one axis. A *coordinate inversion* reverses all three axes. For polar and axial vectors, the results of an inversion are:

Polar Vector

Original: $\boldsymbol{v} = (\ v_x,\ v_y,\ v_z)$

Inversion: $\boldsymbol{v} = (-v_x, -v_y, -v_z)$

Axial Vector

Original: $\boldsymbol{\omega} = (\omega_x,\ \omega_y,\ \omega_z)$

Inversion: $\boldsymbol{\omega} = (\omega_x,\ \omega_y,\ \omega_z)$

Chapter 12
Differential Calculus

Welcome to Calculus!

In the next three chapters, we explore the most basic procedures of calculus: derivatives and integrals. This branch of mathematics is both beautiful and essential to scientific literacy.

For physicists, knowing how to differentiate and integrate is a survival skill. Calculus will not only help you pass exams and succeed in your careers, but it will also change your worldview. After learning calculus, you will appreciate the majesty of the universe more profoundly than ever before.

For those discovering calculus for the first time, this may be exciting, but perhaps a bit frightening. Do not expect to fully appreciate calculus after just one reading. Changing your worldview will take some time. For me, and for others I know, calculus was a dense fog that one-day seemed to clear in a single instant. I still vividly recall that one moment.

Enjoy this. Learning calculus is one of the great experiences of a scientific education.

Thank you for sharing this experience with me.

4A§12.1 The Need For Speed

Derivatives describe rate of change. Since everything in the universe changes, derivatives are the foundation upon which science strives to describe all natural phenomena.

Most physical laws describe how quantities change over time. Consider, for example, a ball dropped from the *Leaning Tower of Pisa*. Newton's law of gravity does not tell us the ball's height or speed, but it does tell us how rapidly its speed is changing. To make use of these laws, we must be able to quantify rate of change, and that requires *differential calculus*.

Let's talk about speed: what exactly do physicists mean by "speed"? Speed is distance traveled divided by travel time. The speedometer in your car tells you the car's speed. If it reads 100 kilometers per hour (km/hr), it means you would travel 100 km in the next hour if you maintained a constant speed. How does the speedometer know how far you would travel in one hour if you just started driving? If you happen to be only 2 km from your destination, how can you be going 100 km/hr? Clearly, "100 km/hr" cannot be the whole story.

What the speedometer measures is how far the car's tires turn in some small interval of time, such as one second. The turning rate, combined with the known tire circumference, determines how many meters your car travels in one second. If the speedometer reads 100 km/hr, what it really means is that your car is moving, at this instant, at the rate of about 28 meters per second (m/sec).

Averaging over one second might be good enough for a car, but what about a ball falling from a great height? It turns out, that 9 seconds after being dropped, the ball's speed is 88 m/sec. This increases to 98 m/sec one second later. Since that is a substantial change, we should probably compute its speed over a time interval that is even less than one second.

The ultimate answer is to define speed in terms of the *infinitesimal distance* traveled during an *infinitesimal time* interval. This concept, developed independently by Isaac Newton (1643-1727) and Gottfried Leibniz (1646-1716), is the basis of *differential calculus*, the first "new math."

Calculus is a rich and beautiful branch of mathematics that was essential to the development of physics. It allows us to deal sensibly with infinitesimal quantities and also to properly accumulate them to describe macroscopic results.

4A§12.2 Going to the Limit

By convention, we use the symbol *ds* for an infinitesimal change in distance s and *dt* for an infinitesimal change in time. Calculus provides a precise definition of speed that we denote with the letter v:

$$v = \text{the limit as } dt \rightarrow 0 \text{ of } (ds / dt)$$

A ratio of two infinitesimals, such as (ds / dt), is well-behaved if the ratio comes closer and closer to a final value v as we compute the ratio for smaller and smaller time intervals dt. In which case, v is the *asymptotic limit* of (ds / dt).

Not all ratios are well behaved. The ratio 1/x is not well behaved as x approaches 0, because it grows ever-larger and is infinite at x = 0. But the ratio (sinx) / x is well behaved as x approaches 0, because the numerator and denominator both approach 0 *at the same rate*. For every infinitesimal but nonzero value of x, (sinx) / x equals 1. As we get closer and closer to x = 0, but not at x = 0, the ratio remains 1. This means the *limit* of this ratio equals 1.

Let's consider another example of this concept of limits: what is the speed of a falling ball 8 seconds after its release? From Newton's laws, we know that the equation relating s, the distance dropped, and t, the time lapsed since release, is:

$$s(t) = g\, t^2 / 2$$

Here, g is the acceleration of gravity near Earth's surface, which equals 9.8 m/sec^2. We write s(t) to denote the distance s at time t, emphasizing that s is a *function* of t. To compute speed, we calculate the distance dropped at time t, and also at an infinitesimally later time t + dt.

$$s(t) = g\, t^2 / 2$$

$$s(t + dt) = g\, (t + dt)^2 / 2$$

The infinitesimal distance traveled, ds, is the change in distance during dt.

$$ds = s(t + dt) - s(t)$$

$$ds = g\, [\, (t + dt)^2 - t^2\,] / 2$$

$$ds = g\, [\, t^2 + 2t\, dt + dt^2 - t^2\,] / 2$$

$$ds = g\, [\, 2t\, dt + dt^2\,] / 2$$

$$ds / dt = g\, [2t + dt] / 2$$

Now, we take advantage of dt being extremely small, dt << 1. We can discard the dt term in the [ ]'s in the last equation because dt << 2t in the limit that dt goes to zero. Now we can compute the speed as a function of time, v(t), and evaluate it at t = 8 sec:

$$v(t) = ds/dt\ = g\, [2t] / 2 = g\, t$$

$$v(8\text{ sec}) = [9.8\text{ m/sec}^2]\ [8\text{ sec}] = 78\text{ m/sec}$$

4A§12.3 Differentiation

The procedure performed above on s(t) is called *differentiation* or more specifically *taking the derivative of* s *with respect to* t. Differentiation is important enough to merit an entire symbology: dq is not d times q, but is rather a single symbol denoting a *differential*, a tiny increment in the variable q or a tiny range of values of q. The d-symbology applies to any variable, but coordinate differentials — such as dt, dx, dy, dz — are the most common.

The expression ds/dt is the ratio of two differentials, and is called the *derivative* of s with respect to t. The two d's in ds/dt **do not** cancel one another to leave s/t — ds/dt and s/t are entirely different expressions.

Any normal function or equation in physics can be differentiated. To differentiate X with respect to z, compute:

$$dX/dz = [\, X(z + dz) - X(z)\,] / dz$$

and take the limit as dz goes to zero.

Derivatives have graphic significance. Figure 12-1 shows a plot of y(x) = sinx. The derivative of y(x) at each value of x is the slope of sinx at that x.

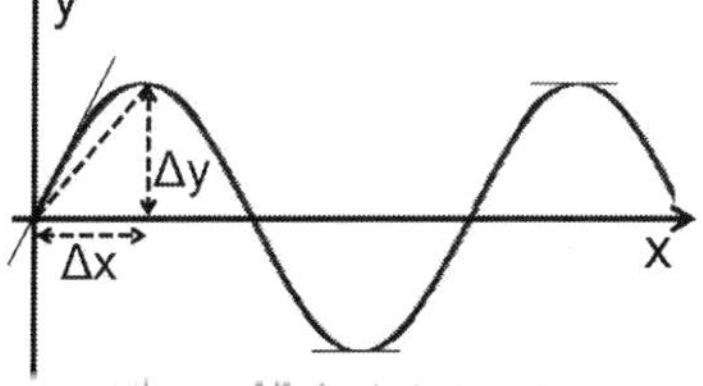

Figure 12-1 y(x) = sin(x)

At two values of x, the slope of y(x) is zero: at the *minimum* value of y(x) about halfway along the x-axis; and also at the *maximum* value of y(x) near the right end of the x-axis. At both points, the tangent lines are horizontal, and the derivative, dy(x)/dx, equals 0. Figure 12-1 also shows two increments from x = y = 0, Δy and Δx that are indicated by dashed lines. We see that for a substantial value of Δx, the ratio (Δy / Δx), the slanted dashed line, differs from the tangent line at x = 0. We show below that (Δy / Δx) comes closer and closer to the tangent line as Δx gets closer and closer to zero.

In Chapter 7, we found that the sine function can be expressed as an infinite series of which the first three terms are:

$$\sin(x) = x - x^3 / 6 + x^5 / 120 - \ldots.$$

The derivative at x = 0 is the limit as Δx goes to zero of:

$$d\sin(0)/dx = \{ \sin(\Delta x) - \sin(0) \} / \Delta x$$

$$d\sin(0)/dx = \sin(\Delta x) / \Delta x$$

$$d\sin(0)/dx = \{\Delta x - \Delta x^3 / 6 + \Delta x^5 / 120 - \ldots.\} / \Delta x$$

$$d\sin(0)/dx = 1 - \Delta x^2 / 6 + \Delta x^4 / 120 - \ldots.$$

We see that for moderate values of Δx, the right side of the last equation is less than 1. But as Δx goes to zero, Δx^2 and higher powers of Δx become negligible, and the expression approaches 1, the true derivative.

The maxima and minima of any function are always at points at which the derivative of the function is zero.

Now, let's step this up a notch, and consider acceleration a, the derivative of speed with respect to time. Continuing from above:

$$v(t) = g\,t$$

$$v(t + dt) = g\,[t + dt]$$

$$dv = g\,[t + dt] - g\,t = g\,dt$$

$$a = dv/dt = g = 9.8\text{m/sec}^2$$

Since acceleration is the derivative of speed, and speed is the derivative of distance, we say acceleration is the *second derivative* of distance. Here, "second" means we differentiate twice. The symbology denoting the second derivative of s with respect to t is:

$$a = dv/dt = d^2s/dt^2$$

First derivatives, like ds/dt, are ubiquitous in physics, and we often omit the word "first." Second derivatives, like d^2s/dt^2, are less common; third and higher order derivatives are rare. The third derivative of distance, d^3s/dt^3, is called *jerk*. If your car speeds up with constant acceleration, you will be pressed back in your seat, but will not be unduly uncomfortable. But if the driver alternately floors and releases the gas pedal, the resulting *changes* in acceleration will toss you back and forth, which is very uncomfortable. That discomfort is due to jerk. Smooth rides are all about minimizing jerk.

4A§12.4 Partial Derivatives

Physicists often deal with functions that vary with two or more variables. Temperature is a simple example: it is a function of the three dimensions of space and also time. Earth's highest recorded air temperature was 134ºF (56.7ºC) in mid-afternoon, on July 10, 1913 at Furnace Creek, in Death Valley, California. But, we might be more interested in knowing *where* was the highest-ever noon temperature, or we might wish to know *when* was the highest-ever temperature at the North Pole. The highest-ever air temperature yields neither. To get what we want, we need to find the maximum temperature at a fixed position or at a fixed time. This is what partial derivatives do.

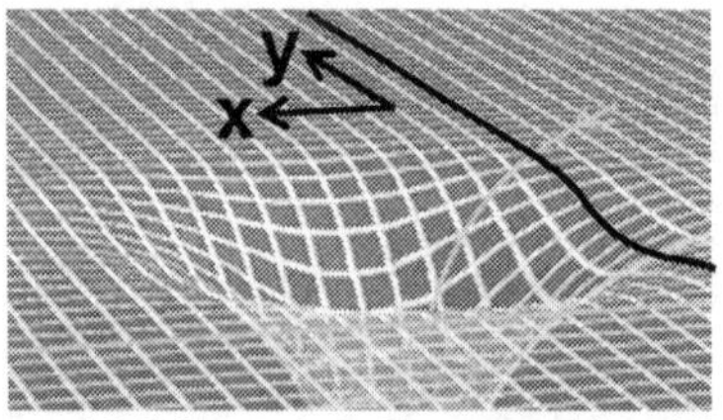

Figure 12-2 Spacetime Curvature Near Mass

Next, consider an example from general relativity shown above in Figure 12-2: the curved space near a massive body. The spatial curvature, shown as a downward deflection, is clearly a function of both x and y.

The maximum curvature is deep down within the chasm. But we might be interested in the maximum curvature along the indicated black line parallel to the y-axis. We can obtain that by taking the partial derivative of the curvature function with respect to y, *while holding x fixed*. Holding x fixed ensures finding the greatest curvature for that value of x only, thus avoiding the even greater curvature deep within the chasm.

Mathematically, taking a partial derivative is identical to taking the derivative of a function of only one variable. Said another way: for functions of only one variable, there is no difference between the normal derivative and the partial derivative. Consider a function f of n+1 variables:

$$f(x, y_1, y_2, \ldots, y_n)$$

Here, x is one independent variable, and $y_1, y_2, \ldots, y_n$ are n other independent variables. The partial derivative of f with respect to x is defined to be:

$$\partial f/\partial x = \partial f(x, y_1, y_2, \ldots, y_n)/\partial x$$

$$\partial f/\partial x = \text{limit of } \{Q/dx\} \text{ as dx goes to 0, with}$$

$$Q = f(x + dx, y_1, y_2, \ldots, y_n) - f(x, y_1, y_2, \ldots, y_n)$$

Note the notation: $\partial/\partial x$ rather than d/dx. Also note that none of the y's change.

4A§12.5 General Rules of Differentiation

Differentiation is a *linear* operation that obeys these rules:

For any constant a and variable q:

$$da/dq = 0$$

For any functions F & G, constants a & b, and variable q:

$$d(a F + b G)/dq = a\, dF/dq + b\, dG/dq$$

$$d(F G)/dq = G\, dF/dq + F\, dG/dq$$

$$d(F G)/dq = d(G F)/dq$$

$$d(F / G)/dq = (1 / G)\, dF/dq - (F / G^2)\, dG/dq$$

$$d(F^b)/dq = b F^{b-1}\, dF/dq$$

4A§12.6 Derivatives of Common Functions

Here, we list the derivatives of the most common functions of physics: polynomials, trig functions, and exponentials. In the following sections, we show the proofs of these equations. Learning requires practice, so if this material is new to you, study the proofs carefully, and then try some similar functions on your own.

$$d\, x^b / dt = b\, x^{b-1}\, dx/dt, \text{ for any constant b}$$

$$d \sin(x) / dt = + \cos(x)\, dx/dt$$

$$d \cos(x) / dt = - \sin(x)\, dx/dt$$

$$d \tan(x) / dt = dx/dt / \cos^2(x)$$

$$d\, e^x / dt = e^x\, dx/dt$$

$$d \ln\{x\} / dt = dx/dt / x$$

$$d \sinh(x) / dt = \cosh(x)\, dx/dt$$

$$d \cosh(x) / dt = \sinh(x)\, dx/dt$$

$$d \tanh(x) / dt = dx/dt / \cosh^2(x)$$

4A§12.7 Vector Differential Operators

We define the vector operator ***Ď***, in rectilinear coordinates, as:

$$\check{\boldsymbol{D}} = (\partial/\partial x,\ \partial/\partial y,\ \partial/\partial z)$$

This operator, called *del*, is most often denoted by an inverted Δ, the Greek capital letter delta. However, that symbol is not supported in all media, so I use ***Ď***.

We consider here the operation of ***Ď*** in 3-D rectilinear coordinates.

Applying ***Ď*** to any scalar function f yields the *gradient* of f:

$$\check{\boldsymbol{D}}\, f = (\partial f/\partial x,\ \partial f/\partial y,\ \partial f/\partial z)$$

The gradient is a vector that points in the direction in which f is increasing most rapidly with distance; the magnitude of that vector equals the derivative of f in that direction.

Taking the dot product of ***Ď*** with any vector field ***A*** yields a scalar, the *divergence* of ***A***:

$$\check{\boldsymbol{D}} \bullet \boldsymbol{A} = \partial A_x/\partial x + \partial A_y/\partial y + \partial A_z/\partial z$$

Taking the cross product of ***Ď*** with any vector field ***A*** yields a vector, the *curl* of ***A***:

$$\check{\boldsymbol{D}} \times \boldsymbol{A} = (\ \partial A_z/\partial y - \partial A_y/\partial z,\ \partial A_x/\partial z - \partial A_z/\partial x,\ \partial A_y/\partial x - \partial A_x/\partial y\ )$$

Lastly, we apply the *Laplacian* ***Ď*** • ***Ď*** = $\check{D}^2$ to any scalar field f, yielding:

$$\check{D}^2 f = \partial^2 f/\partial x^2 + \partial^2 f/\partial y^2 + \partial^2 f/\partial z^2$$

These operators have different forms in other coordinate systems. Appendix 5 provides these operators in the two most common non-rectilinear systems. Due to their greater complexity in other coordinate systems, it is best to use these operators in rectilinear coordinates whenever possible.

4A§12.8 Directional Derivatives

We now know how to take derivatives along any coordinate axis. But sometimes, we also need to take derivatives along other directions, such as the direction specified by a vector ***v***. The derivative of function f(***r***) along ***v*** is defined using the gradient ***Ď*** by:

$$\boldsymbol{v} \bullet \check{\boldsymbol{D}}\, f(\boldsymbol{r}) = \lim ds \rightarrow 0\ \{\ [f(\boldsymbol{r} + \boldsymbol{v}\, ds) - f(\boldsymbol{r})]\ /ds\}$$

This equation is valid even if ***v*** varies with location. Often ***v*** is chosen to be a unit vector, such as the unit normal to a surface.

4A§12.9 Derivative Proofs & Exercises

The remainder of this chapter proves the general rules of derivatives and derives the derivatives of the most common functions used by physicists. A listing of common derivatives is in Appendix 6.

You might review some of these and then try to prove the others yourself, and compare your results with mine.

For brevity, I will write "P→Q" to denote "in the limit that P goes to Q", and use F and G to denote F(q) and G(q).

GENERAL RULES

The definition of the derivative of F with respect to q is:

$$\text{as } dq \rightarrow 0: dF/dq = \{ F(q + dq) - F(q) \} /dq$$

Hence,

$$F(q + dq) = F(q) + dq\, dF/dq$$

$$\mathbf{d(a F + b G)/dq} = \{ a F(q+dq) - a F + b G(q+dq) - b G\} / dq$$

$$d(a F + b G)/dq = \{ a F + dq\, a\, dF/dq - a F + b G + dq\, b\, dG/dq - b G \} / dq$$

$$d(a F + b G)/dq = a\, dF/dq + b\, dG/dq$$

$$\mathbf{d(F G)/dq} = \{ F(q+dq) G(q+dq) - F G\} / dq$$

$$d(F G)/dq = \{ [F + dq\, dF/dq] [G + dq\, dG/dq] - F G\} / dq$$

$$d(F G)/dq = \{ F G + dq\, G\, dF/dq + dq\, dG/dq + dq^2 dF/dq\, dG/dq - F G\} / dq$$

As $dq \rightarrow 0$, we can drop terms of order dq^2, yielding:

$$d(F G)/dq = \{ dq\, G\, dF/dq + dq\, F\, dG/dq \} / dq$$

$$d(F G)/dq = G\, dF/dq + F\, dG/dq$$

$$\mathbf{d(G F)/dq} = \{ G(q+dq) F(q+dq) - G F\}/dq$$

$$d(G F)/dq = \{ F(q+dq) G(q+dq) - F G\}/dq$$

$$d(G F)/dq = d(F G)/dq$$

$$\mathbf{d(F / G)/dq} = \{ F(q+dq) / G(q+dq) - F / G\} / dq$$

$$d(F / G)/dq = \{ [F + dq\, dF/dq] / [G + dq\, dG/dq] - F / G\} / dq$$

$$d(F / G)/dq = \{ [F + dq\, dF/dq] G - F [G + dq\, dG/dq] \} / \{ G [G + dq\, dG/dq] dq\}$$

As $dq \rightarrow 0$, the denominator goes to $G^2 dq$.

$$d(F / G)/dq = \{ F G + dq\, G\, dF/dq - F G - dq\, F\, dG/dq \} / \{ G^2 dq\}$$

$$d(F / G)/dq = (1 / G)\, dF/dq - (F / G^2)\, dG/dq$$

Derivative of x^n

$$dx^n/dx = \text{limit } dx \rightarrow 0 \{ [(x + dx)^n - x^n] / dx\}$$

As we found in Chapter 9, we can rewrite $(x+dx)^n$ as:

$$(x + dx)^n = \Sigma_k \{x^{n-k} dx^k n! / k! (n-k)! \}$$

Here, Σ_k denotes the sum over k from k = 0 to k = n. Dropping all terms of order dx^2 and higher reduces the sum to:

$$(x + dx)^n = x^n + n x^{n-1} dx$$

We now put that into the derivative equation.

$$dx^n/dx = \text{limit } dx \rightarrow 0 \{ [(x + dx)^n - x^n] / dx\}$$

$$= \{ [x^n + n x^{n-1} dx - x^n] / dx\}$$

$$dx^n/dx = n x^{n-1}$$

$$(dx^n/dx) (dx/dt) = n x^{n-1} (dx/dt)$$

$$dx^n/dt = n x^{n-1} dx/dt$$

Derviatives of Trig Functions

Recall the trig relations:

$$\sin(A + B) = \sin(A) \cos(B) + \sin(B) \cos(A)$$

$$\cos(A + B) = \cos(A) \cos(B) – \sin(B) \sin(A)$$

Sine

$$d \sin(x)/dx = \{\sin(x + dx) – \sin(x) \} / dx$$

$$\sin(x + dx) = \sin(x) \cos(dx) + \sin(dx) \cos(x)$$

As $dx \rightarrow 0$, $\cos(dx) \rightarrow 1$ and $\sin(dx) \rightarrow dx$. Hence, as $dx \rightarrow 0$:

$$\sin(x+dx)–\sin(x) \rightarrow \sin(x)+dx \cos(x) – \sin(x)$$

$$\sin(x+dx)–\sin(x) \rightarrow dx \cos(x)$$

$$d \sin(x)/dx = \cos(x)$$

Cosine

$$d \cos(x)/dx = \{\cos(x+dx) – \cos(x) \} / dx$$

$$\cos(x+dx) = \cos(x) \cos(dx) – \sin(dx) \sin(x)$$

As $dx \rightarrow 0$, $\cos(dx) \rightarrow 1$ and $\sin(dx) \rightarrow dx$. Hence:

$$\cos(x+dx) \rightarrow \cos(x) – dx \sin(x)$$

$$\cos(x+dx) – \cos(x) \rightarrow – dx \sin(x)$$

$$d \cos(x)/dx = – \sin(x)$$

Tangent

$$d \tan(x)/dx = d [\sin(x)/\cos(x)] / dx$$

$$= [\cos^{-1}(x)] d\sin(x)/dx + [\sin(x)] d\cos^{-1}(x)/dx$$

$$= [\cos^{-1}(x)] \cos(x) + (–1) [\sin(x)] \cos^{-2}(x) [–\sin(x)]$$

$$= \cos^{-2}(x) \{\cos^2(x) + \sin^2(x)\}$$

$$d \tan(x) / dx = \cos^{-2}(x) = 1 / \cos^2(x)$$

Derivative of $e^x = \exp\{x\}$

The definition of e is: $e = \text{limit } n \rightarrow \infty \{ (1 + 1/n)^n \}$.

Hence:

$$e^x = \exp(x) = \lim n \rightarrow \infty (1 + 1/n)^{xn}$$

We now evaluate $(1 + 1/n)^{xn}$ using the binomial expansion. In that expansion, the terms with the lowest powers of the infinitesimal quantity 1/n are:

$$\exp(x) = 1 + (xn) (1/n) + (xn) (xn – 1) / 2n^2 + (xn) (xn – 1) (xn – 2) / 3!\, n^3 + \ldots$$

As $xn \rightarrow \infty$, this becomes:

$$\exp(x) = 1 + x + x^2 / 2 + x^3 / 3! + x^4 / 4! + \ldots$$

$$d \exp(x) /dx = 0 + 1 + 2x / 2 + 3x^2 / 3! + 4x^3 / 4! + \ldots$$

$$d \exp(x) /dx = 1 + x + x^2 / 2 + x^3 / 3! + \ldots$$

$$d \exp(x) /dx = \exp(x)$$

Derivative of Natural Logarithm

By definition of the natural logarithm ln:

$$x = \exp(\ \ln[x]\ )$$

Taking the derivative of both sides with respect to x yields:

$$dx/dx = \exp(\ln[x])\ d(\ \ln[x]\ )/dx$$

$$1 = x\ d(\ \ln[x]\ )/dx$$

$$d(\ \ln[x]\ )/dx = 1\ /\ x$$

Derivatives of Hyperbolics

Recall the definitions of sinh, cosh, and tanh:

$$\sinh(x) = (\exp\{x\} - \exp\{-x\})\ /\ 2$$

$$\cosh(x) = (\exp\{x\} + \exp\{-x\})\ /\ 2$$

$$\tanh(x) = \sinh(x)\ /\ \cosh(x)$$

Sinh

$$d \sinh(x)\ /dx = (d \exp\{x\}\ /dx - d \exp\{-x\}\ /dx)\ /\ 2$$

$$d \sinh(x)\ /dx = (\exp\{x\} + \exp\{-x\})\ /\ 2$$

$$d \sinh(x)\ /dx = \cosh(x)$$

Cosh

$$d \cosh(x)\ /\ dx = (d \exp\{x\}\ /dx + d \exp\{-x\}\ /dx)\ /\ 2$$

$$d \cosh(x)\ /\ dx = (\exp\{x\} - \exp\{-x\})\ /\ 2$$

$$d \cosh(x)\ /dx = \sinh(x)$$

Tanh

$$d \tanh(x)\ /dx = d\ [\sinh(x)/\cosh(x)]\ /\ dx$$

$$= [1/\cosh(x)]\ d\sinh(x)/dx + [\sinh(x)]\ d\cosh^{-1}(x)/dx$$

$$= [1/\cosh(x)] \cosh(x) + (-1)[\sinh(x)] \cosh^{-2}(x)\ [\sinh(x)]$$

$$= \cosh^{-2}(x)\ \{\cosh^2(x) - \sinh^2(x)\}$$

$$d \tan(x)\ /dx = \cosh^{-2}(x) = 1\ /\ \cosh^2(x)$$

Derivative of x^b

For any x and constant b:

$$d\ (x^b)\ /dx = d\ (\ \exp\{\ b \ln[x]\ \}\ )\ /dx$$

$$d\ (x^b)\ /dx = (\exp\{\ b \ln[x]\ \})\ d(\ b \ln[x]\ )/dx$$

$$d\ (x^b)\ /dx = (x^b)\ b\ /\ x = b\ x^{b-1}$$

Chapter 13
Integral Calculus

4A§13.1 It All Adds Up

Just as addition is the inverse of subtraction, integration is the inverse of differentiation. If the derivative of X equals Y, then the integral of Y equals X — well almost. Since the derivative of any constant is zero, a more precise statement is:

If the derivative of (X + any constant) = Y,

then the integral of Y = (X + any constant)

Generally, the so-called *arbitrary constant of integration* is determined by initial conditions, as we shall soon see.

Integrals solve problems that are the reverse of the problems that derivatives solve.

For example, consider a falling ball. In the prior chapter, we calculated the ball's velocity at time t from the equation for the distance the ball has dropped by time t. Now, let's ask: how far does the ball drop between time t = 4 seconds and t = 8 seconds? Let's first use an equation quoted in the prior chapter for the distance s that a ball drops in time t.

$$s(t) = g\, t^2 /2$$

The distance fallen between t = 4 seconds to t = 8 seconds is:

$$s(8\text{ sec}) - s(4\text{ sec}) = (g / 2)\ [64\text{ sec}^2 - 16\text{ sec}^2]$$

$$s(8\text{ sec}) - s(4\text{ sec}) = 4.9\text{ m/sec}^2\ [48\text{ sec}^2]$$

$$s(8\text{ sec}) - s(4\text{ sec}) = 235.2\text{ m}$$

That is the correct answer, but now let's pretend that we know only the ball's velocity equation and not its distance equation. We know that distance equals velocity multiplied by time, but how can we calculate distance if the velocity is continuously changing? The velocity equation is:

$$v(t) = g\, t, \text{ with } g = 9.8\text{ m/sec}^2$$

Let's imagine that the velocity is constant for a short time, and then changes to some other velocity for another short time, and so on. In that case, we could sum (velocity multiplied by time) for each "short time" and get the total distance traveled. Let's see how this works out, assuming the "short time" is 2 seconds.

Figure 13-1 shows a plot of the speed of a falling ball versus time, with the dotted line representing the true equation: v(t) = g t.

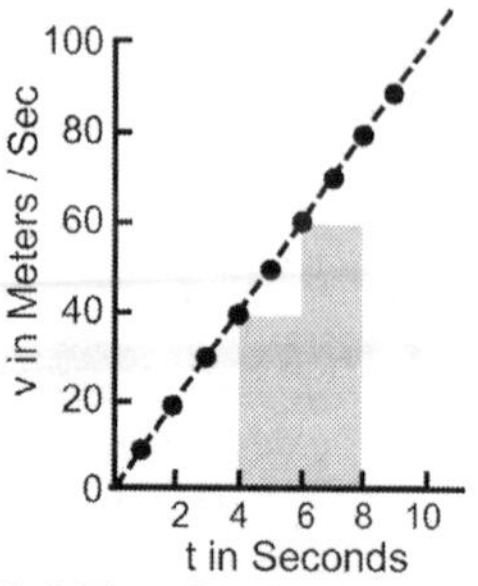

Figure 13-1 Plot of Velocity versus Time

The two shaded rectangles have heights of v(4sec) and v(6sec), and both have widths of 2 seconds. The area of each rectangle has units of meters/sec × sec = meters. Since we seek the distance traveled, we should sum the areas of these rectangles. The total area of both rectangles is:

$$\text{Area} = v(4\text{sec}) \times 2\text{sec} + v(6\text{sec}) \times 2\text{sec}$$

$$\text{Area} = (g \times 4\text{sec}) \times 2\text{sec} + (g \times 6\text{sec}) \times 2\text{sec}$$

$$\text{Area} = 9.8\text{m} \times (8 + 12) = 196\text{m}$$

This total area of 196m is a rough approximation to the distance the ball actually traveled between 4 and 8 seconds, which we calculated above to be 235.2m. The area is only approximate because the two rectangles do not cover the entire region under the line in Figure 13-1. The ball's velocity changes substantially during 2 seconds, leaving gaps above the rectangles.

We can do better using shorter time intervals. With 4 rectangles each 1 second wide, the total area covered is:

$$\text{Area} = [v(4) + v(5) + v(6) + v(7)] \times 1\text{sec}$$

$$\text{Area} = g \times [4 + 5 + 6 + 7] \times 1$$

$$\text{Area} = 9.8 \times 22 = 216\text{m}$$

With 8 rectangles, each 1/2 second wide, the area is 9.8 x 23 = 225m. With 16 rectangles, each 1/4 second wide, the area is 9.8 x23.5 = 230m. We are getting closer.

Clearly the thing to do is to use an infinitesimal time interval dt, as shown in Figure 13-2.

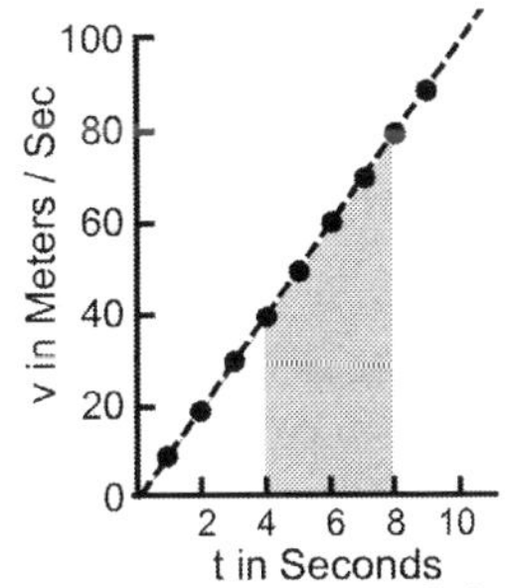

Figure 13-2 Shaded Area = Definite Integral

The shaded area under the line is the sum of an extremely large number of extremely thin rectangles, each of width Δt that we label n = 1, 2, 3, … . We can now write:

$$\text{Area} = \text{sum of } v(t_n) \times \Delta t$$

Here, the sum is over all values of n. If we let n go to infinity, which means taking the limit as Δt goes to zero, the sum of the areas of all the rectangles becomes an *integral* yielding the correct distance.

The operation of *integration* is denoted with the symbol $\int$, an enlarged S derived from the Latin word *summa*. The general form of an integral is:

$$Q = \int f(u)\, du$$

Here, f(u) is the *integrand*, the quantity being summed, u is the independent variable, du is the infinitesimal change in u, and Q is the *integral*, the sum of the areas of an infinite number of infinitesimal rectangles, each of height f(u) and width du.

The equation for finding the distance s traveled by an object with speed v(t), which need not be constant, is:

$$s(t) = \int ds$$

$$s(t) = \int (ds/dt)\, dt = \int v(t)\, dt$$

For the case of a falling ball:

$$s(t) = \int g\, t\, dt = g\, t^2 /2 + C$$

The arbitrary integration constant C represents our arbitrary choice in defining the location of s = 0 and t = 0. The equation says the ball accelerates with the same time dependence from any initial height. In each specific situation, we set that initial height with C. Here, we will choose s = 0 at time t = 0, which makes C = 0.

We know the value of the above integral because we found in Chapter 12 that the derivative of t^2 equals 2t, thus the integral of t equals $t^2 / 2$.

Integrals correspond to areas "under the curve" of a function, whereas derivatives correspond to the slope of that curve.

4A§13.2 Definite & Indefinite Integrals

There are two types of integrals that are closely related. The integral that we just discussed:

$$s(t) = \int v(t)\, dt$$

is called an *indefinite integral*. We integrate the function v(t) with respect to t, and get another function s(t) that like any normal function has a value at each value of t.

The other type of integral is called a *definite integral*. Here we select two values of t, A and B, and the definite integral yields the area under the curve between t = A and t = B. The definite integral is written:

$$\int_A^B v(t)\, dt = s(t{=}B) - s(t{=}A)$$

In the example in Figure 13-2, the definite integral from t = 4 sec to t = 8 sec is the shaded area under the curve that we compute as follows:

$$\int_{t=4}^{t=8} v(t)\, dt = s(8) - s(4)$$

$$= g \{ (8 \text{ sec})^2 - (4 \text{ sec})^2 \} / 2$$

$$= (4.9 \text{ m/sec}^2) \{ 48 \text{ sec}^2 \}$$

$$\int_{t=4}^{t=8} v(t)\, dt = 235.2\text{m}$$

The result of an indefinite integral includes an arbitrary constant. The result of a definite integral has no arbitrary constant; the constant is the same at both limits A and B, and therefore cancels.

4A§13.3 How to Integrate

Summing rectangles, as we did above, is an integration procedure that works, but it must be done numerically, generally using a computer. We discuss the best approaches to numerical integration in Chapter 17. Numerical integration yields the area under the curve, but it cannot yield an analytical function. In our example, numerical integration of the ball's velocity yields a drop distance of 235.2 meters, but does not yield the equation for distance versus time:

$$s(t) = g\, t^2 / 2$$

In the prior chapter, we described an analytical procedure to differentiate any expression. Unfortunately, there is no corresponding general analytical procedure for integration. We learn how to do integrals with a haphazard reverse process: if we know that B is the derivative of A, then we know that A is the integral of B. Mathematicians have differentiated a vast menagerie of functions and tabulated their results. You can search these tables hoping to find the answer to a challenging integral. If it is not listed, you must employ trickery or resort to numerical integration.

Most physicists memorize many common integrals and keep extensive tables handy for others.

Here are two simple but very useful integrals (without the arbitrary constants):

$$\int x^b\, dx = x^{b+1} / (b + 1), \text{ for b not} = -1$$

$$\int \exp\{b\, u\}\, du = \exp\{u\} / b$$

Appendix 7 contains a short table of common integrals. In additional, I will present some useful tricks.

4A§13.4 Integration by Parts

Our first trick is *integration by parts*. This allows us to transfer a derivative from one function in an integrand to another. For any two functions u and v:

$$d(u\, v)/dx = u\, dv/dx + v\, du/dx$$

$$\int [d(u\, v)/dx]\, dx = \int [u\, dv/dx]\, dx + \int [v\, du/dx]\, dx$$

$$\int d(u\, v) = \int u\, dv + \int v\, du$$

$$u\, v - \int u\, dv = \int v\, du$$

An example of integration by parts is the calculation of the variance σ^2 of the Gaussian distribution that we discuss in Section 4A§9.5. Recall the probability distribution:

$$\text{Prob}(x) = \exp\{-x^2 / 2\} / \sqrt{(2\pi)}$$

We first show that the total probability of all x values equals 1.

$$\text{Let } Z = \int_{-\infty}^{+\infty} \exp\{-x^2 / 2\}\, dx / \sqrt{(2\pi)}$$

As it turns out, Z^2 is easier to calculate than Z.

$$Z^2 = \{\int \exp\{-x^2/2\}\, dx\} \{\int \exp\{-y^2/2\}\, dy\} / 2\pi$$

$$2\pi Z^2 = \iint_{-\infty}^{+\infty} \exp\{ -(x^2 + y^2) / 2 \}\, dx\, dy$$

Now switch to polar coordinates with:

$$x = r \cos\theta$$

$$y = r \sin\theta$$

$$r^2 = x^2 + y^2$$

r goes from 0 to $+\infty$; θ goes from 0 to 2π

In Section 4A§14.4 on Volume Elements, we discover that:

dx dy gets replaced by r $d\theta$ dr

The integral in polar coordinates becomes:

$$2\pi Z^2 = \int_0^{+\infty} \int_0^{2\pi} \exp\{-r^2 / 2\}\, r\, d\theta\, dr$$

$$2\pi Z^2 = 2\pi \int_0^{+\infty} r \exp\{-r^2 / 2\}\, dr$$

Now let $u = -r^2 / 2$, with $du = -r\, dr$.

$$Z^2 = -\int_0^{+\infty} \exp\{u\}\, du$$

$$Z^2 = -\exp\{u\} |_0^{+\infty} = 1$$

$$Z = 1$$

This confirms the normalization factor.

Now let's calculate the variance.

$$\sigma^2 = \int x^2\, \text{Prob}(x)\, dx$$

Let's begin by defining W:

$$W = \sigma^2 \sqrt{(2\pi)} = \int_{-\infty}^{+\infty} x^2 \exp\{-x^2 / 2\}\, dx$$

Choose u and v such that:

$$v = x$$

$$dv = dx$$

$$u = -\exp\{-x^2 / 2\}$$

$$du = +x \exp\{-x^2 / 2\}\, dx$$

Now integrate W by parts:

$$W = \int v\, [du] = v\, [u] - \int u\, dv$$

$$W = \int_{-\infty}^{+\infty} x\, [x \exp\{-x^2 / 2\}]\, dx$$

$$W = x\, [-\exp\{-x^2 / 2\}]\, |_{-\infty}^{+\infty} + \int \exp\{-x^2 / 2\}\, dx$$

$$W = 0 + \sqrt{(2\pi)}$$

$$\sigma^2 = W / \sqrt{(2\pi)} = 1$$

QED

4A§13.5 Completing the Square

Another trick applies to integrals like:

$$Q = \int \exp\{x^2 + b\,x\}\,dx$$

Here, b is either a constant or at least not a function of x. Without the b x term, the exponent would be a perfect square and we would have the answer from the prior section.

The strategy of *completing the square* is to convert the exponent into a perfect square. We do this as follows:

$$x^2 + b\,x = x^2 + b\,x + b^2/4 - b^2/4$$

$$x^2 + b\,x = (x + b/2)^2 - b^2/4$$

The first term on the right is a perfect square and the second is a constant. We can therefore write:

$$Q = \exp\{-b^2/4\} \int \exp\{\ (x + b/2)^2\ \}\,dx$$

Now define u such that:

$$(x + b/2) = u / \sqrt{2}$$

$$dx = du / \sqrt{2}$$

With this substitution the integral becomes:

$$Q = \exp\{-b^2/4\} \int \exp\{\ u^2 / 2\ \}\,du / \sqrt{2}$$

$$Q = \exp\{-b^2/4\} \sqrt{(2\pi)} / \sqrt{2}$$

$$Q = \exp\{-b^2/4\} \sqrt{\pi}$$

Chapter 14
More Calculus

4A§14.1 Path & Loop Integrals

Another type of integral is the *path integral*, also called a *line integral*. The integration path or line need not be a straight line. Figure 14-1 shows a path Γ of length L that starts at point A and ends at point B. The path may be arbitrarily complicated. Path Γ may be a *closed* loop, in which case A = B, or it may be *open*, as shown in the figure.

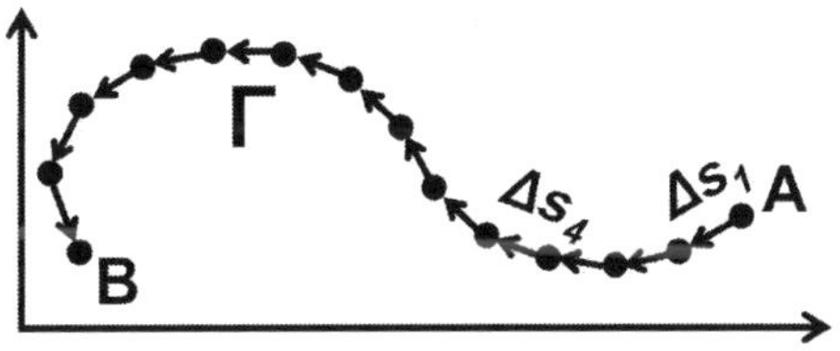

Figure 14-1 A to B Path Integral

Like other integrals, a path integral is the sum of N steps, each of length L / N, in the limit that N goes to infinity and the step size L / N goes to zero. Each step can be represented by vector $\boldsymbol{\Delta s}$, which we number $\boldsymbol{\Delta s_1}$ through $\boldsymbol{\Delta s_N}$; $\boldsymbol{\Delta s_1}$ and $\boldsymbol{\Delta s_4}$ are labeled in Figure 14-1. Consider the integral Q:

$$Q = \int_\Gamma ds = \text{limit } N \rightarrow \infty \ \{\Sigma_k \Delta s_k\}$$

Here, the subscript Γ on the integral sign denotes the path, and Σ_k is the sum over steps, from k = 1 to k = N. In the limit that N goes to ∞, Σ_k becomes an infinite sum of infinitesimal step lengths, which is the total length L of path Γ.

Figure 14-2 demonstrates how path integrals can vary with the chosen path.

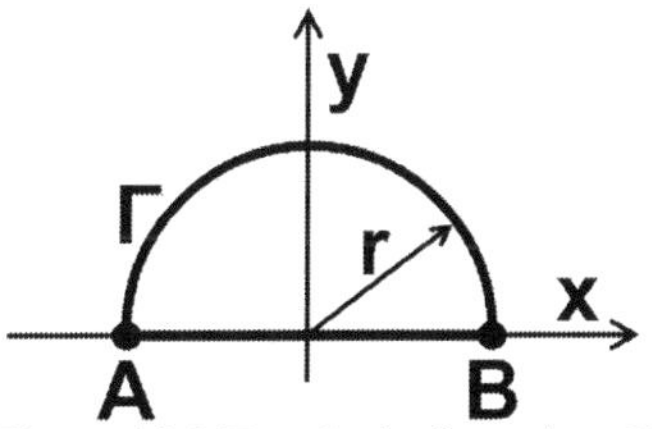

Figure 14-2 Two Paths from A to B

Let's evaluate two path integrals from point A to point B, one along the semi-circular path Γ and the other along the straight line AB. The integrand in both cases is:

$$r = \sqrt{(x^2 + y^2)}$$

Along Γ, the integrand has a constant value r, while y is always zero along AB.

$$\int_\Gamma r \, ds = r\,(\pi\, r) = \pi\, r^2$$

$$\int_{AB} r \, ds = \int_{-r}^{+r} |x| \, dx = 2\,(x^2 / 2)\,|_0^{+r} = r^2$$

Clearly, we must ensure we pick the correct path for each specific situation.

A *loop integral* is a path integral in which the path forms a closed loop, as shown in Figure 14-3.

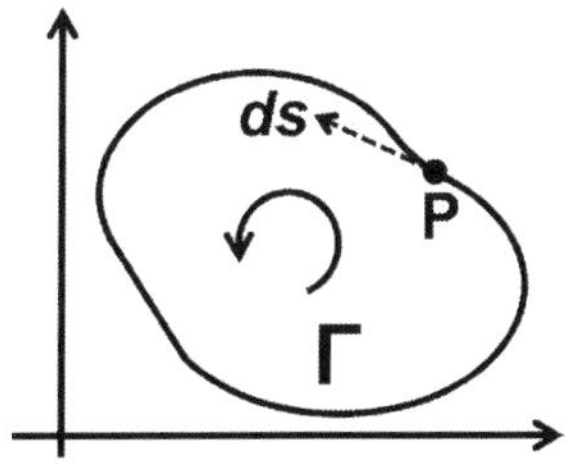

Figure 14-3 Integral Around Loop Γ

The normal sign convention for a loop integral is to go around the loop counterclockwise. This is another instance of the right hand rule (see 4A§11.2): with your right thumb pointing toward you (out of the page), the fingers of your right hand curl in the counterclockwise direction. If you integrates in the clockwise direction, multiply the result by –1.

4A§14.2 Area & Volume Integrals

Multidimensional integrals are employed to integrate quantities over a specified area or volume. In N-dimensions, we need to integrate over N variables, typically coordinate axes. We evaluate multiple integrals sequentially: we first integrate over one variable, then integrate that result over a second variable, and continue until all integrals are done. The order of integration is arbitrary; regardless of which variables are integrated first or last, one obtains the same ultimate result.

For simplicity of discussion, I will use the word "volume" here to describe any defined region with any number of dimensions, whether it is a 2-D area, a 3-D volume, or an n-D hyper-volume.

In rectilinear coordinate systems, integrating over an n-dimensional volume simply requires doing n one-dimensional integrals. For example, consider an integral over x, over y, and over z with these limits:

$$x: 0 \text{ to } X$$

$$y: 0 \text{ to } Y$$

$$z: 0 \text{ to } Z$$

This integral is over the entire volume of a cuboid whose opposite corners are at (0, 0, 0) and (X, Y, Z). To make this as simple as possible, let the integrand f be a constant.

$$Q = \int\int\int f \, dx \, dy \, dz$$

$$Q = \int\int (f \,|_0^X) \, dy \, dz$$

$$Q = X \int\int f \, dy \, dz$$

$$Q = X \int (f \,|_0^Y) \, dz$$

$$Q = X\,Y \int f \, dz$$

$$Q = X\,Y\,(f \,|_0^Z)$$

$$Q = f\,X\,Y\,Z = f\,V$$

Here, V = X Y Z is the cuboid's volume.

If we now let f be a function of the coordinates, Q becomes:

$$Q = \langle f \rangle V$$

Here, $\langle f \rangle$ represents the average value of f throughout the cuboid.

This procedure works for any rectilinear coordinate system with any number of dimensions.

Let's next examine the 2-D integral of f across a disk of radius R that is centered at the origin. Assume for now that f is constant. Figure 14-4 shows how we will do this. for each y value, from y = –R to y = +R, we sum a rectangle of height dy and width 2(xlim), where:

$$xlim = \sqrt{(R^2 - y^2)}$$

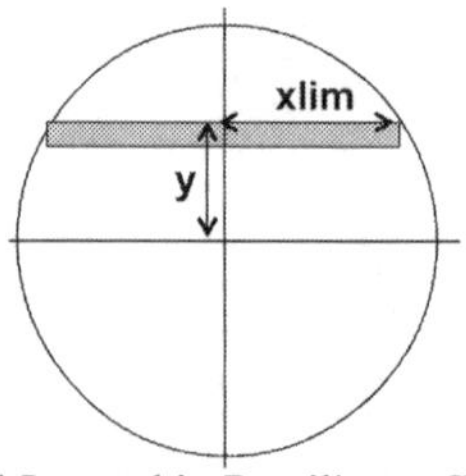

Figure 14-4 Integral in Rectilinear Coordinates

The integral is:

$$Q = \int\int f\, dx\, dy$$

$$Q = f \int_{-R}^{+R} x|_{-xlim}^{+xlim}\ dy$$

$$Q = f \int 2 \sqrt{(R^2 - y^2)}\, dy$$

From tables, this integral is:

$$Q = f \{\ y \sqrt{(R^2 - y^2)} + R^2 \arcsin(y / R)\ \} |_{-R}^{+R}$$

$$Q = f \{\ 0 - 0 + R^2 [\pi/2 - (-\pi/2)]\ \}$$

$$Q = f\, \pi\, R^2$$

The result Q is exactly what we should expect: the integral equals f multiplied by the area within the circle of radius R.

4A§14.3 Volume Elements

Let's try the previous integral again, but now using polar coordinates. With r going from 0 to R, and θ going from 0 to 2π, the integral seems to be:

$$Q = \int_0^{2\pi} \int_0^R f\, dr\, d\theta$$

$$Q = f \int_0^{2\pi} R\, d\theta$$

$$Q = f\, 2\pi\, R, \text{ WRONG}$$

This calculation is wrong because dr dθ is not the area enclosed when the radius sweeps from r to r + dr and the angle sweeps from θ to θ + dθ. In fact dr dθ does not even have the units of area; its units are distance × radians. The problem is illustrated in Figure 14-5.

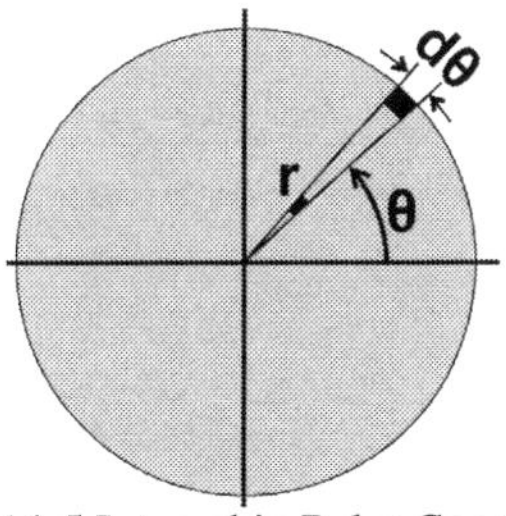

Figure 14-5 Integral in Polar Coordinates

Here, two black enclosed regions have the same width dθ and the same length dr, but clearly have very different areas. The proper width is not dθ, but rather r dθ, and the incremental area is:

for 2-D polar coordinates: r dr dθ

This expression is one example of a *volume element*, the measure of the space enclosed by infinitesimal changes in each coordinate axis. This name is commonly used regardless of the dimensionality of that space. Every coordinate system has a volume element; for rectilinear coordinates, the volume element is the product of the coordinate differentials, with no additional factors.

Let's redo the prior integral with the proper volume element:

$$Q = \int_0^{2\pi} \int_0^R \mathbf{r\, dr\, d\theta} = (R^2/2)\,(2\pi) = \pi\, R^2$$

This is the correct result.

Here are the volume elements for all coordinate systems that physicists commonly use.

2-D polar: r dr dθ

2-D rectilinear: dx dy

3-D rectilinear: dx dy dz

3-D cylindrical: r dr dø dz

3-D spherical: r^2 sinθ dr dθ dø

4-D spacetime: c Δt Δx Δy Δz

We now know how to integrate in two other non-rectilinear coordinate systems: 3-D cylindrical and 3-D spherical, denoted C and S respectively.

$$C = \int_0^Z \int_0^{2\pi} \int_0^R r \, dr \, d\phi \, dz = (R^2 / 2) (2\pi) (Z)$$

$$C = \pi R^2 Z$$

$$S = \int_0^{2\pi} \int_0^{\pi} \int_0^R r^2 \sin\theta \, dr \, d\theta \, d\phi$$

$$S = (R^3/3) (2) (2\pi) = 4\pi R^3 /3$$

The above quantities Q, C, and S are respectively: the area of a disk of radius R; the volume of a cylinder of radius R and length Z; and the volume of a ball of radius R. If we multiply each integrand by f, the integrals would become:

$$< f > \times \text{(volume integrated over)}$$

This is exactly what integrals are supposed to do: sum an expression over a defined space, or equivalently, find an expression's average value multiplied by the volume of that space.

For some other exotic coordinate system W, the volume element equals the determinant of the *Jacobian* matrix J_{kn} that is defined by:

$$J_{kn} = \partial W_k / \partial X_n$$

Here W_k are the W coordinates, and X_n are rectilinear coordinates. We discuss matrices and determinants in Chapter 16.

4A§14.4 Variational Calculus

In many situations, we seek the value of x at which a function f(x) reaches a maximum or minimum (an *extremum*). We find the answer by setting $\partial f/\partial x = 0$, and solving this equation for x.

Variational calculus solves more complex problems, those in which the entity that varies is more than simply a single quantity x. A typical example is: what path from point A to point B minimizes some function Q? We cannot simply take the derivative of Q in one direction; we must instead be able to compare alternative paths in multiple dimensions.

To do this, we need a more powerful tool: the *calculus of variations*.

Here is how variational calculus works. Imagine two paths that start at the same x and t, and end at the same x and t, as shown in Figure 14-6. The true path (what we seek) is represented by the solid curve, and one alternate path is represented by the dashed curve.

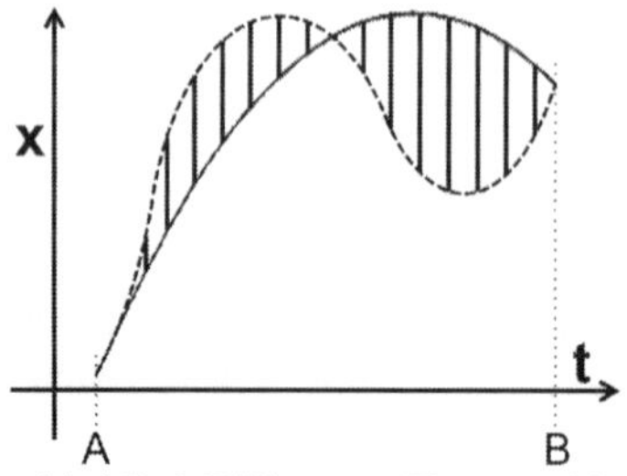

Figure 14-6 Path Difference: True vs. Alternate

Let's define x(t) and w(t) such that:

true path: x(t)

alternate: x(t) + w(t)

Thus w(t) is the difference between the true path and the alternate path. The length of each vertical line in Figure 14-6 represents the value of w(t) at selected times. In our analysis, we will consider alternate paths that deviate only slightly from the true path, which means:

$$| w(t) | << | x(t) | \text{ for all } t$$

Let's consider the example of minimizing the *action* S, which equals an object's kinetic energy minus its potential energy U, as given by:

$$S = mv^2 / 2 - U(x)$$

Along the true path, the integral is:

$$S_{true} = \int_A^B \{ (dx/dt)^2\, m/2 - U(x) \}\, dt$$

Along the alternate path, the integral is:

$$S_{alt} = \int_A^B \{ (d[x + w]/dt)^2\, m/2 - U(x + w) \}\, dt$$

Since the true path has the least action, S_{alt} must be greater than or equal to S_{true}. We define the *variation in S*, δS, to be:

$$\delta S = S_{alt} - S_{true} >= 0$$

$$\delta S = \int_A^B \{ [(d[x + w]/dt)^2 - (dx/dt)^2]\, (m/2) - U(x + w) + U(x) \}\, dt$$

Let's simplify this piece by piece, beginning with the U terms. Since w is small, we will drop terms proportional to w^2 and higher powers. From the definition of a derivative, we have:

$$U(x+w) = U(x) + (dU/dx)\, w$$

$$- U(x+w) + U(x) = - (dU/dx)\, w$$

Now, let's simplify the difference of the second derivatives in the δS equation.

$$(d\,[x + w]\,/dt)^2 - (dx/dt)^2$$

$$= (dx/dt + dw/dt)^2 - (dx/dt)^2$$

$$= 2\, (dx/dt)\, (dw/dt)$$

We then have:

$$\delta S = \int_A^B \{m\, (dx/dt)\, (dw/dt) - (dU/dx)\, w\}\, dt$$

Let's ignore the clutter for a moment and look at the "big picture". The integrand has this form:

$$P\, dw/dt - Q\, w$$

If it were P w – Q w, we would immediately have a solution. So, the way forward is to turn P dw/dt into R w, for some R. But how?

Here is the **first key step:** *integration by parts*. Recall that for any two functions u and v:

$$\int_A^B u\, (dv/dt)\, dt = u\, v\, |_A^B - \int_A^B v\, (du/dt)\, dt$$

In the present case, set:

$$v = w$$

$$u = dx/dt$$

$$dv/dt = dw$$

$$du/dt = d^2x/dt^2$$

Integration by parts yields:

$$\int_A^B (dx/dt)\, (dw/dt)\, dt = (dx/dt)\, w\, |_A^B - \int_A^B w\, (d^2x/dt^2)\, dt$$

Here is **the second key step**: w(A) = w(B) = 0. The alternate path and the true path both start at the same x and t and end at the same x and t. We vary the path *between* the endpoints, but not *at* the endpoints. The δS equation reduces to:

$$\delta S = - \int_A^B \{ m\, (d^2x/dt^2) + (dU/dx) \}\, w\, dt$$

As we know, functions change very slowly near their extrema. When the alternate path is the same as the true path, when w = 0 everywhere, δS = 0. When the alternate path is close to the true path, δS will be very close to zero, with δS deviating from zero only in the second order.

Our problem boils down to finding the x(t) for which δS = 0 for any small path deviation w(t).

The third key step: along the true path the term in { }'s *is zero everywhere*. Why? Consider a function w(t) that is nonzero only between t* and t* + Δt. If Δt is small enough, we can approximate w(t) as constant, and get:

$$\delta S = - \{m\, [d^2x(t^*)/dt^2] + [dU(t^*)/dx] \}\, w(t^*)\, \Delta t$$

If the term in { }'s is not zero at t*, we can make S smaller by choosing some nonzero w(t*). But that contradicts the definition that x(t) is the true path, which requires δS >= 0 for any nonzero w(t*).

Since this requirement applies to every value of t*, the true path is defined by:

$$\text{for all t: } 0 = \{ m\, [d^2x(t)/dt^2] + [dU(t)/dx] \}$$

We therefore have our solution:

$$\text{for all t: } m\, d^2x/dt^2 = -\, dU/dx$$

In one dimension, F = – dU/dx, resulting in Newton's familiar equation:

$$m\, a = F$$

With variational calculus we proved that, for any conservative force F, objects move according to Newton's second law.

Feynman provides sage advice on this variational method:

> "It turns out that the whole trick of the calculus of variations consists of writing down the variation of S and then integrating by parts so that the derivatives of [w] disappear. It is always the same in every problem in which derivatives appear. … [Next] comes something which always happens—the integrated part disappears."

4A§14.5 Project, Divide & Conquer

Another approach to solving complicated equations is to divide them into multiple simpler equations.

One such technique employs projecting a vector equation onto any selected vector using the dot product. In ***Feynman Simplified 2D***, Chapter 44, Feynman derives the following frightening equation for the steady flow of a non-viscous, incompressible fluid:

$$0 = (\check{\boldsymbol{D}} \times \boldsymbol{v}) \times \boldsymbol{v} + \check{\boldsymbol{D}}\, (\boldsymbol{v} \bullet \boldsymbol{v}) / 2 + \check{\boldsymbol{D}}\, P / \mu + \check{\boldsymbol{D}}\, \phi$$

Here, $\boldsymbol{v}$ is the fluid velocity, μ is its mass density, ϕ is its potential energy per unit mass and P its pressure. We can learn something about fluid flow from the dot product of this equation with $\boldsymbol{v}$.

$$0 = \boldsymbol{v} \bullet \{ (\check{\boldsymbol{D}} \times \boldsymbol{v}) \times \boldsymbol{v} + \check{\boldsymbol{D}}\, (\boldsymbol{v} \bullet \boldsymbol{v}) / 2 + \check{\boldsymbol{D}}\, P / \mu + \check{\boldsymbol{D}}\, \phi \}$$

The first term is zero since (anything) × $\boldsymbol{v}$ is always orthogonal to $\boldsymbol{v}$. This eliminates the most challenging term in the equation, and provides Bernoulli's theorem:

$$0 = \boldsymbol{v} \bullet \check{\boldsymbol{D}}\, \{ \boldsymbol{v} \bullet \boldsymbol{v} / 2 + P / \mu + \phi \}$$

This says the quantity in { }'s, which is the fluid's energy per unit mass, is constant along every fluid streamline.

We can also project a vector equation onto the surface perpendicular to any vector $\boldsymbol{n}$. For simplicity, assume $\boldsymbol{n}$ is a unit vector. We take the dot product of the equation with $\boldsymbol{n}$, and then subtract the result from the initial equation. We can write all this as the vector operator:

$$(1 - \boldsymbol{n} \bullet)$$

Here is a trivial example, with $\boldsymbol{n} = (1, 0, 0)$ operating on vector (A, B, C).

$$(1 - \boldsymbol{n} \bullet)\, (A, B, C) = (A, B, C) - (A, 0, 0) = (0, B, C)$$

The final result is a vector within the yz-plane that is perpendicular to n.

Both projection procedures can be generalized to tensor equations.

A related technique divides a vector field $\boldsymbol{u}$ into two parts: vector field $\boldsymbol{u}_{zd}$ that has zero divergence; and vector field $\boldsymbol{u}_{zc}$ that has zero curl.

The equation for a disturbed solid, perhaps our planet after an earthquake, is:

$$\varrho\, \partial^2\boldsymbol{u}/\partial t^2 - \mu\, \check{D}^2\boldsymbol{u} = (\lambda + \mu)\, \check{\boldsymbol{D}}\, (\check{\boldsymbol{D}} \bullet \boldsymbol{u})$$

Here, ϱ is the mass density, λ and μ are elasticity constants, and $\boldsymbol{u}$ is the displacement vector field. This looks like a harmonic equation, except for the term on the right.

Using $\boldsymbol{u} = \boldsymbol{u}_{zd} + \boldsymbol{u}_{zc}$, we obtain an equation that I will label ZDZC:

$$\text{(ZDZC): } (\lambda + \mu)\, \check{\boldsymbol{D}}\, (\check{\boldsymbol{D}} \bullet \boldsymbol{u}_{zc}) = \varrho\, \partial^2(\boldsymbol{u}_{zd} + \boldsymbol{u}_{zc})/\partial t^2 - \mu \check{D}^2(\boldsymbol{u}_{zd} + \boldsymbol{u}_{zc})$$

Taking the divergence of (ZDZC) eliminates $\boldsymbol{u}_{zd}$.

$$0 = \varrho\, \check{\boldsymbol{D}} \bullet \partial^2(\boldsymbol{u}_{zc})/\partial t^2 - (\lambda + 2\mu)\, \check{D}^2 (\check{\boldsymbol{D}} \bullet \boldsymbol{u}_{zc})$$

$$0 = \check{\boldsymbol{D}} \bullet \{\, \varrho\, \partial^2\boldsymbol{u}_{zc}/\partial t^2 - (\lambda + 2\mu)\, \check{D}^2\, \boldsymbol{u}_{zc} \,\}$$

Since $\boldsymbol{u}_{zc}$ has zero curl, the expression in { }'s has both zero divergence and zero curl everywhere. A solution that satisfies these requirements is the harmonic equation:

$$0 = \varrho\, \partial^2\boldsymbol{u}_{zc}/\partial t^2 - (\lambda+2\mu)\, \check{D}^2\boldsymbol{u}_{zc}$$

Taking the curl of (ZDZC) eliminates $\boldsymbol{u}_{zc}$.

$$0 = \varrho\, \check{\boldsymbol{D}} \times \partial^2(\boldsymbol{u}_{zd})/\partial t^2 - \mu\, \check{\boldsymbol{D}} \times \check{D}^2\, \boldsymbol{u}_{zd}$$

Again, we have a field with zero divergence and zero curl everywhere. A solution is the harmonic equation:

$$0 = \varrho\, \partial^2\boldsymbol{u}_{zd}/\partial t^2 - \mu\, \check{D}^2\, \boldsymbol{u}_{zd}$$

Thus, $\boldsymbol{u}_{zd}$ corresponds to *transverse* or *shear* waves with velocity c_{trans}, and $\boldsymbol{u}_{zc}$ corresponds to *compression* or *longitudinal* waves with the greater velocity $c_{long} = \sqrt{[(\lambda + 2\mu) / \varrho]}$, where:

transverse waves velocity: $c_{trans} = \sqrt{[\mu / \varrho]}$

longitudinal wave speed: $c_{long} = \sqrt{[(\lambda + 2\mu) / \varrho]}$

4A§14.6 One More Trick

I want to share another fine example of Feynman's mathematical trickery. As shown in ***Feynman Simplified 1B***, Chapter 20, the intensity of Planck black body radiation is proportional to:

$$I(z) = z^3 / (\exp\{z\} - 1)$$

with $z = \hbar\omega / kT$

Feynman shows us how to integrate I(z) from z = 0 to z = ∞. We can rewrite Planck's equation as:

$$[z^3 / \exp\{z)]\ [1/(1 - \exp\{-z\}]$$

Since $\exp\{-z\} < 1$, for $z > 0$, and $I(z{=}0) = 0$, we can employ the infinite series:

$$1 / (1 - x) = 1 + x + x^2 + x^3 + \ldots$$

We then have:

$$I(z) = [z^3 \exp\{-z\}]\ [1 + \exp\{-z\} + \exp\{-2z\} + \ldots\ ]$$

$$I(z) = z^3\ [\ \exp\{-z\} + \exp\{-2z\} + \exp\{-3z\} + \ldots\ ]$$

$$I(z) = z^3\, \Sigma_n \exp\{-nz\},\ \text{ sum from } n = 1 \text{ to } \infty$$

Now, for each n, we need to calculate:

$$I_n = \int z^3 \exp\{-nz\}\, dz$$

Let's start with a simpler integral:

$$\int_0^\infty \exp\{-nz\}\, dz = (-1 / n) \exp\{-nz\}\ |_0^\infty = 1 / n$$

Next comes a great trick. Take the *third derivative* of this equation *with respect to n*, yielding:

$$\int_0^\infty [\ d^3 \exp\{-nz\} /dn^3\ ]\, dz = d^3\, (n^{-1}) /dn^3$$

$$\int_0^\infty (-z)\,(-z)\,(-z) \exp\{-nz\}\, dz = (-1)\,(-2)\,(-3)\, n^{-4}$$

$$\int_0^\infty z^3 \exp\{-nz\}\, dz = 6\, n^{-4}$$

Hence:

$$\int I(z)\, dz = \int z^3\, \Sigma_n \exp\{-nz\}\, dz$$

$$\int I(z)\, dz = 6\, \Sigma_n\, n^{-4} = 6\, (1 + 1/2^4 + 1/3^4 + \ldots)$$

In Appendix 2, the sum in ( )'s in the prior equation is given as $\pi^4 / 90$, which makes our integral:

$$\int I(z)\, dz = \pi^4 / 15$$

Chapter 15
Differential Equations

Almost all physics equations are differential equations, equations that define relationships among derivatives of quantities of interest. There is no one simple procedure for solving differential equations, because there is no one simple procedure for solving integrals.

We learn how to solve differential equations on a case-by-case basis, and by experience. For theoretical physicists, solving differential equations is a survival art.

4A§15.1 Linear Differential Equations

Let's begin with linear differential equations because: (1) we can actually solve these equations; and (2) many fundamental laws of physics are linear, or approximately so. Perhaps the most common and most important differential equation describes **harmonic motion**:

$$d^2x/dt^2 + \omega^2 x = 0$$

The solution to this equation can be written in two different forms:

$$x(t) = A \cos(\omega t + \phi)$$

$$x(t) = \text{Re} [A \exp\{i\omega t\}]$$

In both solutions, the system oscillates with amplitude A and frequency ω. In the first solution, ø is a constant phase shift, and A is a real constant. In the second solution, A may be complex (thus providing a phase shift), and x is the real part of the complex function in [ ]'s.

The general class of linear differential equations with constant coefficients has the form:

$$a_n d^nx/dt^n + \ldots + a_1 dx/dt + a_0 x = f(t)$$

If j is the largest index for which a_j is nonzero, we say the equation is jth *order*. In physics, we generally deal with second order differential equations, because forces are linked to accelerations, which are the second order derivatives of position. For any order, we can define an *operator* that encompasses in one symbol all the derivative operations that are the guts of our equation. We define Λ so that:

$$\Lambda(x) = a_n d^nx/dt^n + \ldots + a_1 dx/dt + a_0 x$$

All the physics is encoded into Λ, in the number of coefficients and their values. All that remains is plugging in whatever function x we wish.

The most important feature of linear systems is *linear superposition*: the ability to combine solutions by simple addition. Any linear system governed by Λ, with two solutions x and y, and any two constants a and b, has another solution (a x + b y), as shown here:

$$\Lambda(a x + b y) = a \Lambda(x) + b \Lambda(y)$$

Furthermore, let u be a force-free solution of this system, a solution when all forces are zero. This means Λ(u) = 0. Now, if x is a solution when an external driving force is applied, then:

$$\Lambda(x + u) = \Lambda(x) + \Lambda(u)$$

$$\Lambda(x + u) = \Lambda(x) + 0$$

$$\Lambda(x + u) = \Lambda(x)$$

Hence, we can simply add any *transient solution* u to any forced solution x and get another solution. Transient solutions describe situations in which a driving force ceases, or when a driving force begins. An initially stationary system does not instantaneously transition from zero motion to periodic motion. At the start, the motion is a combination of periodic and transient motions, with the latter diminishing exponentially.

Thus we can linearly sum all types of solutions to solve a wide range of problems. This is true even when the driving force F is not a simple harmonic function.

4A§15.2 Linear System Example

Let's consider an example: a mass m on a spring. For small displacements x(t), we can approximate the spring force F with Hooke's law, $F = - k x$, making this a linear system. In isolation, this system is a simple harmonic oscillator, whose differential equation and solution were presented at the start of this chapter:

$$m\, d^2x/dt^2 + k\, x = 0$$

$$x(t) = A \cos(\omega t+\phi)$$

$$\text{with } \omega = \sqrt{(k / m)}$$

If that mass is now subject to an external driving force f cos(ßt), the system's differential equation becomes:

$$m\, d^2x/dt^2 = -k\, x + f \cos(\beta t)$$

A trial solution $x(t) = D \cos(\beta t)$ yields:

$$-\beta^2\, m\, D \cos(\beta t) = -k\, D \cos(\beta t) + f \cos(\beta t)$$

$$(k / m - \beta^2)\, m\, D = f$$

$$D = (f / m) / (\omega^2 - \beta^2)$$

Let's now add a damping force F, such as friction, that is proportional to the velocity of the mass, according to:

$$\text{damping force: } F = - \mu\, m\, dx/dt$$

The system's differential equation becomes:

$$m\, d^2x/dt^2 = - k\, x + f \cos(\beta t) - \mu\, m\, dx/dt$$

We will now employ complex numbers and variables to more conveniently describe this motion, and use the trial solution $x(t) = \text{Re}\,(D \exp\{i\beta t\})$ yielding:

$$- \beta^2\, m\, x = - k\, x + f \cos(\beta t) - i\beta\, \mu\, m\, x$$

$$(k / m - \beta^2 + i\beta\, \mu)\, m\, D = f$$

$$D = (f / m) / (\omega^2 - \beta^2 + i\mu\, \beta)$$

In this case, we find D is complex, with amplitude r and lag angle θ, according to:

$$r = (f / m) / \sqrt{[\, (\omega^2 - \beta^2)^2 + (\mu\, \beta)^2\,]}$$

$$\sin\theta = m\, r\, \mu\, \beta / f$$

At resonance, where $\beta = \omega$:

$$r = f / \mu\, m\, \beta$$

$$\theta = \pi / 2$$

$$Q = \omega / \mu$$

Here, Q is the peak-to-width ratio.

If an oscillator's driving force stops at $t = 0$, one of three behaviors results depending on the values of ω and μ. Figure 15-1 shows the three possible outcomes: curve O is the overdamped response, curve U is the underdamped response, and curve C is the critically damped response.

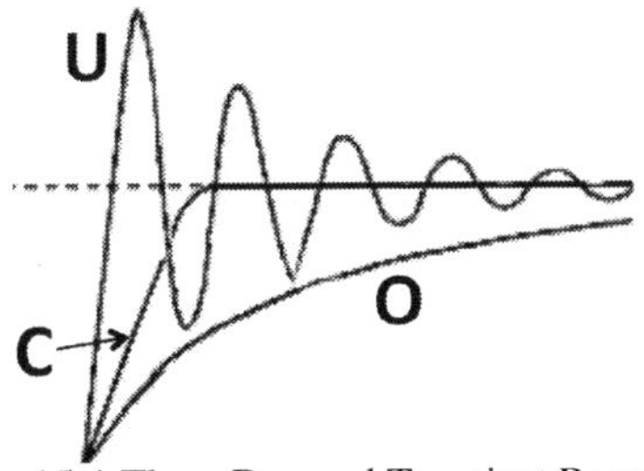

Figure 15-1 Three Damped Transient Responses

We characterize these three behaviors by defining: $\Omega^2 = \omega^2 - \mu^2 / 4$

If $\Omega^2 > 0$, the system is **underdamped**:

$$x = A \exp\{-t\mu / 2\} \cos(a + \Omega t)$$

motion oscillates while decreasing slowly

If $\Omega^2 = 0$, system is **critically damped**:

$$x = A \exp\{-t\mu / 2\}$$

motion decreases rapidly

If $\Omega^2 < 0$, system is **overdamped**, define $\beta^2 = -4\,\Omega^2 = \mu^2 - 4\omega^2 > 0$

$$x = A \exp\{-t(\mu + \beta) / 2\} + B \exp\{-t(\mu - \beta) / 2\}$$

motion decreases slowly

4A§15.3 Quasi-Linear System Example

So far, we have analyzed systems that were assumed to be linear. This is an idealization that may be only approximately correct for many real systems. While the most general nonlinear systems cannot be analyzed simply, we can gain insight into the interesting behaviors of *quasi-linear systems*, those that are only slightly nonlinear.

Consider an electronic device with an input x(t) that produces an output y(t). In a nonlinear device, the relationship may have the form:

$$y(t) = K [\, x(t) + \varepsilon\, x^2(t)\,]$$

If ε is small enough, the nonlinear term $\varepsilon\, x^2(t)$ is small compared with x(t). Now examine a sinusoidal input x, and the output y, described by:

$$x(t) = \cos(\omega t)$$

$$y(t) = K [\cos(\omega t) + \varepsilon \cos^2(\omega t)]$$

If ε were zero, the system would be governed by a linear differential equation, resulting in the thin, sinusoidal curve in Figure 15-2. If ε is slightly greater than zero, the response is slightly nonlinear, resulting in the bold curve that is more complex than the sinusoidal curve.

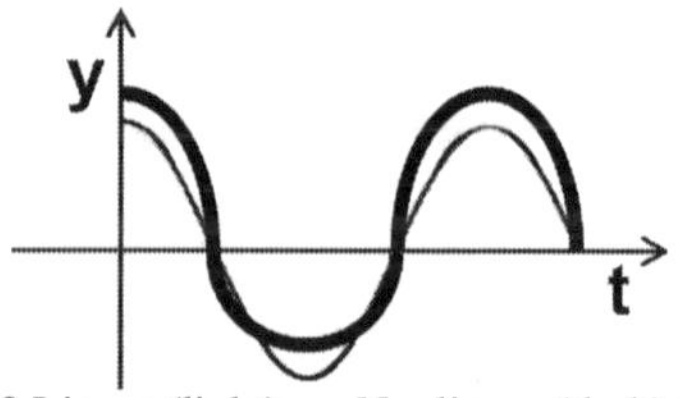

Figure 15-2 Linear (light) vs. Nonlinear (dark) Responses

Recalling that:

$$\cos(2A) = \cos^2 A - \sin^2 A = 2\cos^2 A - 1$$

we can rewrite the y(t) equation as:

$$y(t) = K \cos(\omega t) + K\varepsilon / 2 + (K\varepsilon / 2) \cos(2\omega t)$$

The first of the three terms on the right side of this equation is the normal linear response at the same frequency ω as the input x. The second term adds a constant offset to y(t), shifting its average value. This shifting of the entire response is called *rectification*. The third term adds a higher frequency harmonic to y(t). A nonlinearity proportional to x^2 results in a *second harmonic* at frequency 2ω. Nonlinearities proportional to x^3 or x^4 would add third or fourth harmonics at frequencies 3ω or 4ω, respectively. The most general nonlinearity would introduce an entire spectrum of harmonics. Adding harmonics is called *modulation*.

Now consider the response of a quasi-linear device to an input x(t) with two components of different frequency and amplitude that produces an output y(t).

$$x(t) = A \cos(\omega t) + B \cos(\Omega t)$$

$$y(t) = K [x(t) + \varepsilon\, x^2(t)]$$

Expanding y(t) yields:

$$y(t) = K\ x(t) + K\ \varepsilon\ [A \cos(\omega t) + B \cos(\Omega t)]^2$$

$$y(t) = K\ x(t) + K\ \varepsilon\ [A^2 \cos^2(\omega t) + B^2 \cos^2(\Omega t)] + 2K\ \varepsilon\ A\ B \cos(\omega t) \cos(\Omega t)$$

The term in [ ]'s is the same as the prior example; it produces second harmonics. The last term, containing the product of cosines of different frequencies, produces sidebands with cosines of the sum and difference of ω and Ω.

We can rewrite this equation as:

$$y(t) = K\ A \cos(\omega t) + K\ B \cos(\Omega t) + (K\ \varepsilon / 2)\ [A^2 + B^2]$$

$$- (K\ \varepsilon / 2)\ [A^2 \cos(2\omega t) + B^2 \cos(2\Omega t)]$$

$$+ K\ \varepsilon\ A\ B\ \{\cos([\omega + \Omega]\ t) + \cos([\omega - \Omega]\ t)\}$$

We have here a combination of interesting effects:

linear response: $K\ A \cos(\omega t) + K\ B \cos(\Omega t)$

rectification: $(K\ \varepsilon / 2)\ [A^2 + B^2]$

harmonics: $A^2 \cos(2\omega t) + B^2 \cos(2\Omega t)$

and sidebands: $\cos([\omega + \Omega]\ t) + \cos([\omega - \Omega]\ t)$

If ω is nearly equal to Ω, the sidebands include a term with a frequency of about 2ω and another at $\omega - \Omega$. If $\omega >> \Omega$, the two sidebands are at nearly the same frequency. An entirely equivalent way of looking at this term comes from considering its prior form: $\cos(\omega t) \cos(\Omega t)$. If ω is nearly equal to Ω, this term produces *beats*. If $\omega >> \Omega$, y(t) oscillates rapidly at frequency ω, while slowly modulating at frequency Ω. Both descriptions are perfectly correct.

Note that all the nonlinear effects are proportional to the second power of amplitudes: A^2, B^2, or A B. This means nonlinear effects are more important for larger inputs.

These nonlinear effects — rectification, harmonics, modulation, sum and difference frequencies — have many practical implications. It is believed that the human ear is somewhat nonlinear. Very loud sounds give us the sensation of harmonics, and sum and difference frequencies, even when the input is monotonic.

4A§15.4 Separating Coupled Differential Equations

We sometimes encounter problems described by two coupled differential equations, such as the following equations for the voltage V and current J in a transmission line whose inductance and capacitance per unit length are L and C.

$$\partial V/\partial x = - L\ \partial J/\partial t$$

$$- \partial J/\partial x = C\ \partial V/\partial t$$

We separate V and J by differentiating the first equation with respect to x and the second with respect to t, so that both contain the term $\partial^2 J/\partial t \partial x$.

$$\partial^2 V/\partial x^2 / L = - \partial^2 J/\partial t \partial x$$

$$- \partial^2 J/\partial t \partial x = C\ \partial^2 V/\partial t^2$$

$$\partial^2 V/\partial x^2 - L\ C\ \partial^2 V/\partial t^2 = 0$$

The same trick, done the other way, yields two equations containing $\partial^2 V/\partial t \partial x$.

$$\partial^2 V/\partial x \partial t = - L\ \partial^2 J/\partial t^2$$

$$- \partial^2 J/\partial x^2 / C = \partial^2 V/\partial x \partial t$$

$$\partial^2 J/\partial x^2 - C\ L\ \partial^2 J/\partial t^2 = 0$$

We see that both V and J satisfy the 1-D wave equation with $v^2 = 1 / (L\ C)$:

$$\partial^2 \psi/\partial x^2 - \partial^2 \psi/\partial t^2 / v^2 = 0$$

4A§15.5 Separation of Variables By Axes

Another approach to divide and conquer is called *separation of variables*. An example is separating a complex function of multiple coordinates into simpler functions, each describing the dependence along a different coordinate axis. Consider the harmonic equation:

$$\check{D}^2 \psi = c^{-2}\, \partial^2\psi/\partial t^2$$

Since this is a linear differential equation, finding a complete basis set of solutions is sufficient. All other solutions will be linear combinations of those basis set solutions. To search for solutions in polar coordinates, we begin with a trial solution that is the product of two new functions $X(\boldsymbol{r})$ and $T(t)$, as follows:

$$\psi(\boldsymbol{r},t) = X(\boldsymbol{r})\, T(t)$$

We insert this into the prior differential equation, then divide by ψ, yielding:

$$[1 / X(\boldsymbol{r})]\, \check{D}^2 X(\boldsymbol{r}) = [1 / c^2\, T(t)]\, \partial^2 T/\partial t^2$$

The left side is a function of $\boldsymbol{r}$ only, while the right side is a function of t only. The two sides can be equal only if both are equal to some constant that we define to be $-k^2$. This trial solution has thus divided a complex differential equation into two simpler equations:

$$\check{D}^2 X(\boldsymbol{r}) = -\, k^2\, X(\boldsymbol{r})$$

$$\partial^2 T(t)/\partial t^2 = -\, k^2\, c^2\, T(t)$$

The T equation has simple sinusoidal solutions. We next expand the X equation in spherical coordinates using the expression for $\check{D}^2$ given in Appendix 5, which yields:

$$0 = k^2 X + (1/r^2)\, \partial(\, r\, \partial X/\partial r\,)/\partial r$$
$$+ (1/r^2 \sin\theta)\, \partial(\, \sin\theta\, \partial X/\partial\theta\,)/\partial\theta$$
$$+ (1/r^2 \sin^2\theta)\, \partial^2 X/\partial\phi^2$$

To attack this intimidating equation, we repeat the separation of variables strategy by defining a trial solution that is the product of three new functions $R(r)$, $\Theta(\theta)$, and $\Phi(\phi)$, as follows:

$$X(\boldsymbol{r}) = R(r)\, \Theta(\theta)\, \Phi(\phi)$$

We insert this into the prior equation, then multiply by r^2/X, yielding:

$$0 = k^2 r^2 + (1/R)\, \partial(\, r\, \partial R/\partial r\,)/\partial r$$
$$+ (1/\Theta \sin\theta\,)\, \partial(\, \sin\theta\, \partial\Theta/\partial\theta\,)/\partial\theta$$
$$+ (1/\Phi \sin^2\theta)\, \partial^2\Phi/\partial\phi^2$$

If we were to multiply everything by $\sin^2\theta$, the last term would be a function of ϕ only, while the other terms would be independent of ϕ. This is possible only if:

$$\partial^2\Phi/\partial\phi^2 / \Phi = -\, m^2$$

Here, m is some constant. Hence, Φ also has simple sinusoidal solutions.

We insert the last result into the prior equation, yielding:

$$0 = k^2 r^2 + (1/R)\, \partial(\, r\, \partial R/\partial r\,)/\partial r$$
$$+ (1/\Theta \sin\theta\,)\, \partial(\, \sin\theta\, \partial\Theta/\partial\theta\,)/\partial\theta$$
$$-\, m^2 / \sin^2\theta$$

This separates into two equations, for some constant A.

$$+\, A = k^2 r^2 + (1/R)\, \partial(\, r\, \partial R/\partial r\,)/\partial r$$

$$-\, A = (1/\Theta \sin\theta\,)\, \partial(\, \sin\theta\, \partial\Theta/\partial\theta\,)/\partial\theta - m^2 / \sin^2\theta$$

The top line is a function of r only, while the lower line is a function of θ only.

This process converts one very complex equation into two very simple equations and two solvable equations.

4A§15.6 Separation of Variables By Scale

Another separation of variables approach addresses complex functions that have both small scale and large scale structure. Here, we separate the complex function into simpler functions, each describing the dependence on a different scale. Consider a difficult differential equation: Schrödinger's equation for the orbit of an electron in an atom.

$$\partial^2(\varrho\, \psi)/\partial \varrho^2 = -\, (\varepsilon + 2 / \varrho)\, \varrho\, \psi$$

Here, ε is the electron energy, ϱ is the radial coordinate, and ψ is the electron wave amplitude.

At small distances from the nucleus, ψ oscillates, but at large distances, ψ must decrease exponentially with ϱ. This is because, at large distances, a bound electron's potential energy is greater than its total energy. With a negative kinetic energy, wave number k becomes imaginary and the normal oscillatory exp{ikϱ} term becomes exp{–Kϱ}, where K is real and K= k / i.

We separate the large-scale exponential from the small-scale oscillations by making this substitution:

$$\varrho\, \psi = g(\varrho)\, \exp\{-\, \beta\, \varrho\}$$

Here, β is an arbitrary constant, and g(ϱ) is the unknown small-scale function of distance. After a lot of math, our differential equation becomes:

$$\partial^2 g/\partial \varrho^2 - 2\beta\, \partial g/\partial \varrho + (\beta^2 + \varepsilon + 2 / \varrho)\, g = 0$$

Believe it or not, this is progress.

Let's choose $\beta^2 = -\varepsilon$, reducing our equation to:

$$\partial^2 g/\partial \varrho^2 - 2\beta\, \partial g/\partial \varrho + 2g / \varrho = 0$$

If not for the ϱ in the denominator of the third term, this would be a simple equation. We can solve this with a Taylor series.

$$\text{Let: } g(\varrho) = \Sigma_k\, a_k\, \varrho^k$$

This technique will work if the coefficients a_k approach zero for large k. Putting the Taylor series into our equation yields:

$$0 = \Sigma_n \{\, (n+1)\, n\, a_{n+1} - 2\beta\, n\, a_n + 2a_n \,\}\, \varrho^{n-1}$$

Here the sum is from n = 1 to n = N. Feynman shows that N must be a finite integer to prevent ψ(ϱ) from increasing exponentially at large distance from the nucleus. The above equation is valid for all values of ϱ. This can only be true if the coefficient of each power of ϱ is zero. This is an **important general rule for polynomials** that is well worth remembering. (Proof: for any finite polynomial of order J with coefficients b_j, j = 0 to J, differentiate the polynomial J times. Thie result is: $0 = b_J\, J!$; hence $b_J = 0$. Next, differentiate it J – 1 times, proving $b_{J-1} = 0$. Continue until proving $b_0 = 0$.)

After rearranging, we obtain for all n > 0:

$$a_{n+1} = a_n\, \{\, 2\, (\beta\, n - 1) / n\, (n+1) \,\}$$

With any choice of a_1, we can recursively calculate a_2, a_3, … in terms of β, which is related to the electron's energy. For the electron to be bound to the nucleus, Feynman shows that β must equal 1/n for some integer n.

This is one of the most important discoveries of science: **electrons in atoms have quantized energies**.

This is the basis of the Periodic Table, chemistry, biology, solid state physics, digital electronics, and everything else we know about atoms. And without this, we would not know that the universe is expanding.

4A§15.7 Solving Laplace's 2-D Equation

Using functions of complex variables, we can solve many interesting physical phenomena that are governed by the two-dimensional Laplace equation, which is:

$$\partial^2\phi/\partial x^2 + \partial^2\phi/\partial y^2 = 0$$

All the usual mathematical functions of real variables can be extended to become functions of complex variables. We have done this before, for example, in analyzing harmonic phenomena with exponentials with complex exponents. Any function F(β) of a complex variable β = x + iy can be expressed as the sum of its real and imaginary parts.

Recall this example of exponentials and complex numbers:

$$\exp\{\beta\} = \exp\{x + iy\} = \exp\{x\} (\cos y + i \sin y)$$

Similarly, for two real functions U and V, let:

$$F(\beta) = U(x,y) + i\, V(x,y)$$

For example:

$$F(\beta) = \beta^2 = x^2 + 2ixy - y^2$$

$$U(x,y) = x^2 - y^2$$

$$V(x,y) = 2xy$$

Feynman says: "… a miraculous mathematical theorem [proves that] U and V *automatically* satisfy the relations:"

$$\partial U/\partial x = + \partial V/\partial y$$

$$\partial V/\partial x = - \partial U/\partial y$$

We can confirm that our example satisfies these relations.

$$\partial U/\partial x = + \partial V/\partial y = +2x$$

$$\partial V/\partial x = - \partial U/\partial y = +2y$$

Taking the second order partial derivatives of the prior pair of equations yields:

$$\partial^2 U/\partial x^2 = \partial^2 V/\partial x \partial y = - \partial^2 U/\partial y^2$$

$$\partial^2 U/\partial x^2 + \partial^2 U/\partial y^2 = 0$$

$$\partial^2 V/\partial x^2 = - \partial^2 U/\partial x \partial y = - \partial^2 V/\partial y^2$$

$$\partial^2 V/\partial x^2 + \partial^2 V/\partial y^2 = 0$$

Thus, we can pick any function F(β) and immediately have two solutions, U and V, to the Laplace equation. Feynman says: "We can write down as many solutions as we wish—by just making up functions—then we just have to find the *problem* that goes with each solution. It may sound backwards, but it's a possible approach." Let's take an example.

$$F(\beta) = \beta^2 = (x^2 + iy)^2$$

$$U(x,y) = x^2 - y^2 = \text{some constant A}$$

$$V(x,y) = 2xy = \text{some constant B}$$

Both the U and V equations are solved by hyperbolae, with the U solutions everywhere orthogonal to the V solutions. They describe the equipotentials of a quadrupole lens, in two different configurations.

4A§15.8 Cylindrical Harmonics

Bessel functions often arise in problems involving cylindrical symmetry. *Bessel functions of the first kind* are denoted $J_n(x)$, and solve the differential equation:

$$x^2\, d^2y/dx^2 + x\, dy/dx + (x^2 - n^2)\, y = 0$$

These are also called *cylindrical harmonics*.

Chapter 16
Tensors & Matrices

The mathematics of matrices and tensors is essential in analyzing complex systems, particularly for quantum mechanics, general relativity, and multi-dimensional situations.

4A§16.1 What is a Matrix?

Matrices are rectangular arrays of components laid out in rows and columns. Each component may be an ordinary number, a complex number, a function, or a quantum mechanical amplitude. An n × m ("n-by-m") matrix has n rows, m columns, and n times m components.

In physics, the most common matrices have the same number of rows and columns; these are called **square matrices**.

Let's consider the example of a 3×3 matrix M. The components of M are denoted M_{ij}, where the first index i denotes the component's row number and the second index j denotes the column number. We often use M to denote the entire matrix, but sometimes we write M_{ij} to emphasize that it is a matrix. The layout of matrix M is shown below.

$$M_{ij} = \begin{pmatrix} M_{11} & M_{12} & M_{13} \\ M_{21} & M_{22} & M_{23} \\ M_{31} & M_{32} & M_{33} \end{pmatrix}$$

It is essential to remember that in matrix mathematics, M_{ij} is **not** equal to M_{ji} in general.

A **diagonal matrix** is a square matrix in which the only nonzero components are those with equal row and column numbers ($M_{ij} = 0$ if i does not equal j).

Multiplication by a scalar is the simplest matrix operation. A *scalar* is a single entity that might be a number, a function, or an amplitude. You can think of a scalar as being a 1×1 matrix. Multiplying matrix M by scalar s just means multiplying each component of M by s.

$$s\,M = \begin{pmatrix} sM_{11} & sM_{12} & sM_{13} \\ sM_{21} & sM_{22} & sM_{23} \\ sM_{31} & sM_{32} & sM_{33} \end{pmatrix}$$

Matrix addition is defined only for matrices with the same number of rows and the same number of columns. If matrix A = matrix M plus matrix N, then $A_{ij} = M_{ij} + N_{ij}$, for all i and j, as shown here:

$$M+N = \begin{pmatrix} M_{11}+N_{11} & M_{12}+N_{12} & M_{13}+N_{13} \\ M_{21}+N_{21} & M_{22}+N_{22} & M_{23}+N_{23} \\ M_{31}+N_{31} & M_{32}+N_{32} & M_{33}+N_{33} \end{pmatrix}$$

Matrix subtraction is simply multiplying one matrix by the scalar –1, followed by matrix addition.

Matrix multiplication is a bit trickier. If matrix C equals the product of matrix A times matrix B:

$$C_{ij} = \Sigma_k\, A_{ik}\, B_{kj}$$

Here we sum over all values of k for each combination of i and j. This product is defined only when the number of columns in A equals the number of rows in B, which ensures that each term in the above sum over k is well defined. The product matrix C has the same number of rows as A and the same number of columns as B.

The calculation of C_{32} is schematically illustrated below: the summation runs across row 3 of A and down column 2 of B.

$$\begin{pmatrix} A_{11} & A_{12} & A_{13} \\ A_{21} & A_{22} & A_{23} \\ A_{31} & A_{32} & A_{33} \end{pmatrix} \begin{pmatrix} B_{11} & B_{12} & B_{13} \\ B_{21} & B_{22} & B_{23} \\ B_{31} & B_{32} & B_{33} \end{pmatrix}$$

$$= \begin{pmatrix} C_{11} & C_{12} & C_{13} \\ C_{21} & C_{22} & C_{23} \\ C_{31} & \boxed{C_{32}} & C_{33} \end{pmatrix}$$

$$\Sigma_k \, A_{3k} \, B_{k2} = C_{32}$$

Let's try an example: multiply a 2×4 matrix by a 4×3 matrix to produce a 2×3 matrix, as shown below.

$$\begin{pmatrix} 1 & 4 & 5 & 3 \\ 3 & 5 & 6 & 9 \end{pmatrix} \begin{pmatrix} 2 & 7 & 8 \\ 4 & 1 & 2 \\ 3 & 2 & 4 \\ 5 & 9 & 1 \end{pmatrix}$$

$$= \begin{pmatrix} 48 & 48 & 39 \\ 89 & 119 & 67 \end{pmatrix}$$

You might check this yourself to see if I got the right answer. The row 2, column 2 component is:

$$3\times7 + 5\times1 + 6\times2 + 9\times9 = 21 + 5 + 12 + 81 = 119$$

Multiplication order is critical: A times B is **not** equal to B times A in general. When several matrices are multiplied together, we always multiply from right to left. To evaluate the matrix product ABC, multiply B times C, then multiply that result by A.

Vectors with n components can be considered either n×1 matrices or 1×n matrices. Vectors and matrices can be multiplied together. For example, consider a 3×3 matrix R and two 3-component vectors ø and Ψ:

$$\Psi_i = \Sigma_j \, R_{ij} \, ø_j$$

Here Ψ and ø could be spin state vectors of a spin 1 particle, and R could be the rotation operator that transforms states from one coordinate basis to another. The particle's state in one basis, ø, is transformed into its state in another basis, Ψ, by multiplying ø by R.

The ***identity matrix***, also called the ***unit matrix***, is denoted by the *Kronecker delta* δ_{ij}.

$$\delta_{ij} = \begin{pmatrix} 1 & 0 & 0 \\ 0 & 1 & 0 \\ 0 & 0 & 1 \end{pmatrix}$$

This may seem trivial, but the unit matrix is just as useful as the number 1.

Matrix equations can be valid only if every term is a matrix with the same number of rows and the same number of columns. The equation $M = 0*\delta_{ij}$ means that every component of matrix M is zero. In such equations, δ_{ij} is sometimes not explicitly shown, but is implicitly assumed.

Matrix components can be complex, particularly in quantum mechanics. They can be complex both in the sense of being complicated, and in the sense of having real and imaginary parts (x + iy). Matrices are **Hermitian** if $M_{ij} = M_{ji}^*$ for each ij, where M_{ji}^* is the complex conjugate of M_{ji}. In general the sum of two Hermitian matrices *is* Hermitian, but their product *is not*.

Matrix division is not defined. However, most matrices of interest in physics have inverses. If A^{-1} is the ***inverse*** of matrix A, then:

$$A\,A^{-1} = A^{-1}\,A = \delta_{ij}$$

Matrix A has an inverse if and only if it is a *non-singular,* n × n matrix. In this case, A^{-1} is also a non-singular, n × n matrix. A matrix is singular if any of its rows equals a linear combination of its other rows, or if any of its columns equals a linear combination of its other columns. Conversely, if the rows of a matrix are all linearly independent, and if its columns are also all linearly independent, the matrix is non-singular and has an inverse.

Simple examples of singular matrices are those whose components are all zero across an entire row or an entire column.

Each n × n matrix has a *determinant*. The determinant of a nonsingular matrix is nonzero, while the determinant of a singular matrix is zero.

4A§16.2 Matrix Determinants

Determinants are defined only for N × N matrices.

Determinants have a geometric significance. Consider each column (or each row) to be an independent vector. The determinant of a 2x2 matrix equals ± the area of the parallelogram whose two sides are the two column vectors, with the sign determined by the order of the column vectors. The determinant of a 3x3 matrix equals ± the volume of the parallelepiped whose three sides are the three column vectors. The determinant of a N × N matrix equals ± the measure of the N-dimensional space enclosed by the N column vectors.

The simplest way to calculate the determinant of a matrix is by iteration.

The determinant of a 1×1 matrix simply equals its sole component.

$$\text{for } M = (M_{11}), \text{ Det } | M | = M_{11}$$

The determinant of a 2x2 matrix M equals the product of the upper-left and lower-right components minus the product of the other two components, as shown below:

$$\textbf{Det } |M_{ij}| = \begin{vmatrix} a & b \\ c & d \end{vmatrix}$$

$$\text{Det } | M_{ij} | = a\,d - c\,b$$

The determinant of a 3x3 matrix M has three contributions. Begin by picking any row or any column; all choices yield the same determinant. In the example below, we chose row 1. We proceed to step across the row, sequentially selecting each component and evaluating its contribution. The first contribution equals the first component of row 1, M_{11} = a, multiplied by the determinant of the *minor* of M_{11}. The minor of M_{11} is the 2x2 matrix formed by deleting M_{11}'s row and column. In the example below, the minor for each contribution is shown in bold type.

$$\textbf{Det } |M_{ij}| = \begin{vmatrix} a & b & c \\ d & e & f \\ g & h & j \end{vmatrix}$$

$$= a\begin{vmatrix} a & b & c \\ d & \mathbf{e} & \mathbf{f} \\ g & \mathbf{h} & \mathbf{j} \end{vmatrix} - b\begin{vmatrix} a & b & c \\ \mathbf{d} & e & \mathbf{f} \\ \mathbf{g} & h & \mathbf{j} \end{vmatrix} + c\begin{vmatrix} a & b & c \\ \mathbf{d} & \mathbf{e} & f \\ \mathbf{g} & \mathbf{h} & j \end{vmatrix}$$

$$\text{Det } | M_{ij} | = + a\,(e\,j - h\,f) - b\,(d\,j - g\,f) + c\,(d\,h - g\,e)$$

In this example, the minor of M_{11} = a is the matrix formed by components e, f, h, and j. The second contribution equals b (row 1, column 2) multiplied by the determinant of its minor, which is the matrix formed by d, f, g, and j. The third contribution equals c (row 1, column 3) multiplied by the determinant of its minor, the matrix d, e, g, h.

Each of the three contributions must be summed with the proper sign: $(-1)^{r+c}$, where r is the row number and c is the column number of the selected component. For example, for the first contribution we selected row 1 column 1, which has sign $(-1)^{1+1} = +1$. For the second contribution we selected row 1 column 2, which has sign $(-1)^{1+2} = -1$. The signs alternate as one proceeds across the chosen row or column.

For larger matrices, repeat this procedure iteratively. Pick any row (or column). Multiply each component in that row (or column) by the determinant of its minor and by its proper sign. Sum the contributions across the entire row (or column).

In any N × N matrix, the minor of component M_{ij} is the (N–1) × (N–1) matrix obtained by eliminating the ith row and the jth column of M_{ij}.

An equivalent equation for the determinant of an N × N matrix M is:

$$\text{Det} \mid M_{ij} \mid = \Sigma \, \text{Sign}(abc\ldots) \bullet M_{1a} \bullet M_{2b} \bullet M_{3c} \bullet \ldots$$

Here, abc... is a permutation of the integers 1 through N. Each term in the sum is the product of N components, with one selected from each row and one selected from each column. The sum extends over all permutations of the integers 1 through N, and Sign(abc..) equals +1 for even permutations and –1 for odd permutations. A permutation is even (or odd) if it is obtained from the sequence 1, 2, 3, ... N by an even (or odd) number of swaps of adjacent integers. For example, 1243 is obtained from 1234 with one swap, so Sign(1243) = –1, while 1423 requires two swaps, so Sign(1423) = +1.

4A§16.3 Matrix Inverses

Calculating the inverse of a matrix M is only a bit more work than calculating its determinant, Det(M). The equation is:

$$M^{-1}_{ij} = (-1)^{i+j} \, \text{Det}(\text{minor of } M_{ij}) \, / \, \text{Det}(M)$$

For the prior 3×3 matrix M, the inverse M^{-1} is:

$$M_{ij} = \begin{pmatrix} a & b & c \\ d & e & f \\ g & h & j \end{pmatrix}$$

$$M^{-1}_{ij} = \begin{pmatrix} ej-hf & gf-dj & dh-ge \\ hc-bj & aj-gc & gb-ah \\ bf-ec & dc-af & ae-db \end{pmatrix} / \, \text{Det}(M)$$

4A§16.4 Eigenvalues & Eigenvectors

For an n×n matrix M and an n-component vector ***V***, it sometimes occurs that:

$$M \, \boldsymbol{V} = E \, \boldsymbol{V}$$

for some constant E. In component notation this is:

$$\Sigma_k \, M_{ik} \, V_k = E \, V_i$$

This means multiplication by M may change the length and polarity of ***V*** but not its orientation.

In this case, ***V*** is called an *eigenvector* and E is the corresponding *eigenvalue* of M. The terms are derived from the German word "eigen" that denotes an intrinsic characteristic. Since these are linear equations, if ***V*** is a solution to the above equation so must be 10***V***. Therefore, we generally normalize each eigenvector so that it has unit length ($\boldsymbol{V} \bullet \boldsymbol{V} = 1$).

An n × n matrix M may have up to n linearly independent eigenvectors; it will have n eigenvalues, some of which may be the same.

If matrix M has n linearly independent eigenvectors V_j, j=1 to n, each with a corresponding eigenvalue E_j, we can construct an n × n matrix Q whose jth column is M's jth eigenvector. This means:

$$Q_{ij} = (V_i)_j \text{ for each value of i.}$$

Here, $(V_i)_j$ means the ith component of the jth eigenvector. We can rewrite the prior eigenvector equation to encompass all n eigenvectors.

$$\Sigma_k \, M_{ik} \, Q_{km} = E_m \, V_{im}$$

Note this equation is valid for each value of i and m; we are not summing over i or m.

Since the eigenvectors are linearly independent, Q has an inverse Q^{-1}. Multiplying the prior equation on the left by Q^{-1} yields:

$$\Sigma_i Q^{-1}_{ni} \Sigma_k M_{ik} Q_{km} = E_m \Sigma_i Q^{-1}_{ni} V_{im}$$

$$\Sigma_{ik} Q^{-1}_{ni} M_{ik} Q_{km} = \Sigma_i E_m Q^{-1}_{ni} Q_{im}$$

$$(Q^{-1} M Q)_{nm} = E_m \delta_{nm}$$

$$(Q^{-1} M Q)_{nm} = E_n \text{ if } n = m, \text{ else} = 0$$

This means the matrix $(Q^{-1} M Q)$ is a diagonal matrix whose diagonal components are the eigenvalues of M. In physics, we interpret this as: the phenomenon described by M has its most natural representation in a coordinate system whose axes are M's eigenvectors. In that system, the action along each axis is independent of the actions along the other axes.

For example, in a crystalline solid, an external electric field displaces electrons in the crystal's atoms from their equilibrium positions. For a given electric field strength, the amount of displacement may be different in different directions because of the types of atoms and their locations within the crystal structure. In an arbitrary xyz coordinate system, an electric field in the x-direction may cause electron displacements along all three axes, complicating the mathematics.

We can represent that complexity with a 3×3 polarization matrix Π that may have 9 nonzero components. However, there is always a coordinate system whose axes align with the crystal's *principal axes*, the directions in which an electric field produces displacements that are entirely parallel to that field. In this coordinate system, Π is a diagonal matrix with only 3 nonzero components.

Whenever possible, complicated matrix problems are greatly simplified by finding the eigenvalues and eigenvectors of a matrix, and then diagonalizing that matrix.

4A§16.5 Characteristic Polynomial

Here is one approach for finding the eigenvalues and eigenvectors of a non-singular, square matrix. For an N × N matrix H, an N-component vector ***V***, and a constant E, consider the equation:

$$0 = (H_{jk} - \delta_{jk} E) \boldsymbol{V}$$

The left side is a matrix with all zero components, and hence its determinant equals zero. For the equation to be valid, the right side must also be a matrix with all zero components and zero determinant.

We can write the equation for the determinant of $(H_{jk} - \delta_{jk} E)$ and set it equal to zero. This results in the **characteristic polynomial** of $(H_{jk} - \delta_{jk} E)$, with E as the independent variable. In general, this polynomial contains all powers of E from the zeroth power to the Nth power. As we know from the algebra of complex numbers, such equations have N roots, some of which might be equal. Each root is an eigenvalue of the matrix H.

With the N eigenvalues denoted E_n, n=1...N, we then have N equations of the form:

$$H_{jk} \boldsymbol{V}_n = \delta_{jk} E_n \boldsymbol{V}_n$$

Here, $\boldsymbol{V}_n$ is the eigenvector corresponding to eigenvalue E_n. These equations can be solved to yield the N eigenvectors. With the eigenvectors, we can diagonalize the matrix H. However, the required effort is proportional to N^2.

4A§16.6 Solving a Sample Problem

Let's see how these ideas are employed to solve a simple matrix problem: the stationary states of a two-state system. This is a common problem in quantum mechanics. It also arises in macroscopic problems, such as two pendulums joined with a spring.

In the latter case, if one pendulum is set in motion while the other is stationary, the first exerts a force through the spring on the second. The spring transfers energy from the first to the second. Eventually, the second swings and the first stops. The process then repeats in reverse. Over time, the first pendulum swings, then stops, then swings again, while the second does the opposite.

What we seek are the *stationary states* of this system: the states of motion that repeat at a single frequency with a definite energy. We know the solutions for the motion of a single pendulum: oscillation at frequency $\omega = \sqrt{(g / L)}$, with ω in radians per second, g being Earth's gravitational acceleration, and L being the pendulum length.

For two identical pendulums, we define two *basis states*:

|1> = pendulum #1 swinging at frequency ω

|2> = pendulum #2 swinging at frequency ω

Like two coordinate axes, any motion of two pendulums is a linear combination of |1> and |2>, which we can write in vector form as:

$$(p, q) = p\,|1> + q\,|2>$$

This is logically equivalent to:

$$(x, y) = x\,|e_x> + y\,|e_y>$$

with $|e_x>$ and $|e_y>$ being unit vectors in the x- and y-directions. The equation of motion can then be written in matrix form as:

$$\begin{pmatrix} H_{11} & H_{12} \\ H_{21} & H_{22} \end{pmatrix} \begin{pmatrix} p \\ q \end{pmatrix} = E \begin{pmatrix} p \\ q \end{pmatrix}$$

Here, E is the energy of motion, and the matrix H is called the *Hamiltonian*. We calculate the eigenvalues by setting the determinant of $(H_{jk}-\delta_{jk}\,E)$ equal to zero.

$$M_{jk} = H_{jk} - \delta_{jk}E = \begin{pmatrix} H_{11}-E & H_{12} \\ H_{21} & H_{22}-E \end{pmatrix}$$

$$0 = \text{Det}\,|\,M\,| = (H_{11} - E) \bullet (H_{22} - E) - H_{21} \bullet H_{12}$$

$$0 = H_{11} \bullet H_{22} - E \bullet (H_{11} + H_{22}) + E^2 - H_{21} \bullet H_{12}$$

$$E = E_0 \pm E^*$$

Here the two values of E are the eigenvalues, and E_0 and E^* are given by:

$$E_0 = (\, H_{11} + H_{22}\,) / 2$$

$$E^* = \sqrt{\{\, E_0^{\,2} - H_{11}\, H_{22} + H_{21}\, H_{12}\,\}}$$

Now let's calculate the eigenvectors. For energy E, the equation of motion produces two equations, one from each row. We show both, but we actually only need one.

$$H_{11}\, p + H_{12}\, q = E\, p$$

$$H_{21}\, p + H_{22}\, q = E\, q$$

$$H_{12}\, q = (E - H_{11})\, p$$

$$p / q = H_{12} / (E - H_{11})$$

For each of the two eigenvalues, the ratios of eigenvector components are:

for $E = E_0 + E^*$: $p / q = H_{12} / (\, E_0 + E^* - H_{11}\,)$

for $E = E_0 - E^*$: $p / q = H_{12} / (\, E_0 - E^* - H_{11}\,)$

In each case, we normalize p and q such that $|p|^2 + |q|^2 = 1$.

Let's examine the simplest case, where $H_{11} = H_{22} = E_0$ and $H_{21} = H_{12} = A$.

$$E^* = \sqrt{\{\, E_0^{\,2} - E_0^{\,2} + A^2\,\}} = A$$

for $E = E_0 + A$: $p / q = -A / (\, E_0 + A - E_0) = -1$

for $E = E_0 - A$: $p / q = -A / (\, E_0 - A - E_0) = +1$

for $E_0 + A$, $p = -\,q$; pendulums swing oppositely

for $E_0 - A$, $p = +\,q$; pendulums swing together

For each state of motion, the pendulums swing with the same amplitude and frequency. In the state with higher energy ($E = E_0 + A$), the pendulums always swing in opposite directions. In the state with lower energy ($E = E_0 - A$), the pendulums always swing in the same direction.

4A§16.7 Rotations as Matrices

Recall from Chapter 10 the equations for the rotations of coordinate axes. For a rotation of the xy-axes by angle θ about the z-axis, the transformation equations that relate the coordinates of a point P in the xyz system to P's coordinates in the x*y*z* system are:

$$x^* = +x\cos\theta + y\sin\theta$$

$$y^* = -x\sin\theta + y\cos\theta$$

$$z^* = z$$

We can write this rotation transformation as a matrix R(θ), whose components are:

$$\begin{array}{cc} & \begin{array}{ccc} k=x & y & z \end{array} \\ \begin{array}{c} j=x^* \\ y^* \\ z^* \end{array} & \begin{pmatrix} +\cos\theta & +\sin\theta & 0 \\ -\sin\theta & +\cos\theta & 0 \\ 0 & 0 & 1 \end{pmatrix} \end{array}$$

Note that $R^{-1}(\theta)$, the inverse of R(θ), is simply R(–θ), which is the *transpose* of R:

$$R_{jk}^{-1} = R_{kj}$$

The vector Q transforms into vector Q* when the coordinate axes rotate by angle θ about the z-axis. In matrix notation, this is written:

$$Q^*_j = \Sigma_k Z_{jk} Q_k$$

In Chapter 10, we said that we can determine if some vector ***s*** is a proper vector by taking its dot product with a known proper vector ***r***. If, and only if, ***s*** is a proper vector will ***s • r*** be an invariant scalar that is unchanged by any rotation. Here is the proof.

Take any rotation matrix R that transforms ***r*** into ***r**** and ***s*** into ***s****, and let $R^{-1}_{jk} = R_{kj}$ be R's inverse matrix, the matrix that transforms vectors back to the original coordinate system. Let's compare ***s • r*** in the rotated and original systems.

$$s^* \bullet r^* = \Sigma_k s^*_k r^*_k$$

Now replace ***r**** by R ***r***.

$$s^* \bullet r^* = \Sigma_k s^*_k (\Sigma_j R_{kj} r_j)$$

Next, rearrange the sums.

$$s^* \bullet r^* = \Sigma_{kj} R_{kj} s^*_k r_j$$

And, finally, we replace R_{kj} with R^{-1}_{jk}.

$$s^* \bullet r^* = \Sigma_j r_j (\Sigma_k R^{-1}_{jk} s^*_k)$$

The right hand side equals ***s • r*** if and only if:

$$\Sigma_k (R^{-1}_{jk} s^*_k) = s_j$$

This equation is valid if and only if ***s*** is a proper vector.

4A§16.8 What is a Tensor?

Tensors are a generalization of vectors and matrices. Tensor calculus is a beautiful branch of mathematics that empowers us to elegantly and effectively describe many complex, multi-dimensional phenomena.

Tensors are essential in general relativity, where 4-D spacetime curves, twists, and stretches differently at every point, at every instant, and in every direction.

Tensors are also employed in 3-D analyses of mechanics and wave propagation in anisotropic materials, those whose properties are different in different directions.

The most important thing to know about tensors is that any tensor equation that is valid in one coordinate system is automatically valid without any modifications in all coordinate systems, regardless of their rotation or motion relative to the original coordinate system.

That generality is one reason that the mathematics of general relativity is so challenging, but it is also one of the most powerful tools in solving problems. If we can identify a coordinate system in which we can solve a complex problem with a tensor equation, we have immediately solved the problem in all coordinate systems. General relativity is the only major branch of physics in which tensor equations are universally employed.

Tensors are arrays of components that transform properly between coordinate systems. In 3-D, they transform according to Euclidian coordinate rotations. In 4-D spacetime, they also transform according to the Lorentz transformation.

Tensors can have one component or millions of components, each being a different function of all coordinates.

Let's consider some quantities that are not tensors. Temperature is a simple quantity that changes with time and location, making it a function of the four coordinates of spacetime. Its values are different in different coordinate systems, but these values do not change according to the rotation matrices or the Lorentz transformation. Hence, temperature is not a tensor. Similarly, energy by itself is not a tensor. But, the proper combination of energy and momentum — (E/c, p_x, p_y, p_z) — is a tensor because its components do transform properly.

Tensors are characterized by their *rank* and by the dimensionality of the space in which they are defined. In physics, the most common spaces are Euclidian 3-D, and 4-D spacetime. The most common tensors have rank 0, 1, 2, or 4. The largest meaningful tensor I know is a rank 10 tensor with 1,048,576 components; don't worry — I will not share this tensor with you in this book.

The simplest tensors are scalars; these are *rank 0* tensors. These include π, 7, 0, and your age — all numbers that have the same values in all coordinate systems.

We are also very familiar with *rank 1* tensors: vectors. Every proper 3-vector is a 3-D rank 1 tensor, and every proper 4-vector is a 4-D rank 1 tensor.

An example of a 4-D *rank 2* tensor is the Faraday tensor shown below.

$F_{\mu\sigma}$

	σ = t	x	y	z
μ = t	0	$-E_x/c$	$-E_y/c$	$-E_z/c$
x	E_x/c	0	$-B_z$	$+B_y$
y	E_y/c	$+B_z$	0	$-B_x$
z	E_z/c	$-B_y$	$+B_x$	0

When tensors are shown as arrays of components, the components are generally enclosed in square brackets [ ], as above.

The Faraday tensor has the special property of being *antisymmetric*. This means $F_{\mu\sigma} = -F_{\sigma\mu}$ for all combinations of indices μ and σ. As a result, all the diagonal components are zero, and components on opposite sides of the diagonal are the same, but with the opposite sign. With 4 zero components and 6 redundant components, the number of independent components in the Faraday tensor is only $16 - 4 - 6 = 6$. These are the 3 components of ***E*** and the 3 components of ***B***.

Tensors of rank 2 and greater can have interesting symmetry properties. Some tensors are *symmetric*, meaning that $G_{\mu\sigma} = +G_{\sigma\mu}$ for all combinations of μ and σ. A rank 2, symmetric tensor has $6 + 4 = 10$ independent components, and 6 others that are redundant.

4A§16.9 Tensor Ranks & Indices

Since tensors can have so many components, we use indices to avoid writing them all out individually. Above, we used two indices to identify the components of a rank 2 tensor. Let's now discuss tensor indices in general.

A rank N tensor has N indices that each range over the same set of allowed values. A rank 2 tensor, for example, must have the same number of rows and columns, unlike a matrix that may have a different number of rows and columns. A rank N tensor has 4^N components in 4-D, and 3^N components in 3-D.

The sums and differences of tensors with the same rank and indices are also tensors. The product of two tensors is a tensor, but the quotient of two tensors is not generally a tensor.

In 3-D and flat (non-curved) 4-D spacetime, tensor indices are written as subscripts. For example, consider two alternative notations for the components of the rank 1, position tensor:

$$r_1 = r_x = x$$

$$r_2 = r_y = y$$

$$r_3 = r_z = z$$
$$r_0 = r_t = ct$$
$$r_\sigma = (ct, x, y, z)$$

Here, $\sigma = 0, 1, 2, 3$, or if you prefer, $\sigma = t, x, y, z$.

These equations demonstrate a critical difference between the *free index* σ, which can have any value in the allowed range, and the *fixed indices* 0, 1, 2, 3, x, y, z, and t. The latter refer to specific components, whereas the former refers to any component corresponding to any possible value of σ.

Free indices are just labels. The specific letters we choose have no significance mathematically or physically: x_σ and x_μ mean exactly the same thing, as do $A_{\mu\sigma}$ and $A_{\beta\alpha}$. The significance of free indices lies in the relationships they establish. For example: $A_{\mu\sigma} = B_{\sigma\mu}$ means that every component of B equals the component of A on the opposite side of the diagonal. Just like a vector equation, this tensor equation would mean exactly the same thing for any letters we might substitute for μ and σ.

4A§16.10 Tensor Algebra

A common tensor equation is:

$$L_\mu = 0$$

This establishes the same equation for all values of the free index μ. In this case, it means each component of L equals zero. Every tensor equation is valid for all values of every *unmatched* free index. By "unmatched", we mean a free index that appears no more than once in each product term. Hence:

$$A_{\sigma\mu} = 1$$

means every component of tensor A equals 1.

We said above that the product of two tensors is itself a tensor. For example, if A_μ and B_σ are two position 4-vectors (rank 1 tensors), their product $A_\mu B_\sigma$ is a rank 2 tensor $(AB)_{\mu\sigma}$. The product of a rank N tensor with a rank M tensor is always a tensor of rank N+M.

When a free index appears twice in one term, the *Einstein summation convention* directs us to sum over all values of the repeated index. This is similar to a vector dot product, and is called a *tensor contraction* (contraction in the sense that the number of indices and the tensor's rank decrease). For example:

$$A_\mu B_\mu = A_0 B_0 + A_1 B_1 + A_2 B_2 + A_3 B_3$$

Here μ appears twice in the product term $A_\mu B_\mu$, which directs us to sum over all values of μ. All quantities on the right side are components, hence their sum is a scalar with no free indices. The tensor contraction of a rank N tensor with a rank M tensor has rank N+M–2.

In a proper tensor equation, each nonzero term must have the same set of unmatched free indices. For example:

$$\text{valid for all } \sigma\text{: } A_{\sigma\mu}\, x_\mu + C_\sigma = B_\sigma$$

$$\text{invalid: } A_{\sigma\mu}\, x_\mu + C_\beta = B_\mu$$

In the lower line, σ, μ, and β are not in every term.

The *unit tensor* [1] is as important in tensor calculus as 1 is in arithmetic. The components of the unit tensor are 1 when all indices are the same, and 0 whenever any two indices are different. The rank 2, 4-D unit tensor is defined by:

$$\delta_{\mu\sigma} = 1 \text{ if } \mu = \sigma \text{ and zero otherwise.}$$

In 4-D, the tensor contraction $\delta_{\mu\mu}$ equals:

$$\delta_{\mu\mu} = \delta_{tt} + \delta_{xx} + \delta_{yy} + \delta_{zz} = 4$$

You will recognize $\delta_{\mu\sigma}$, the *Kronecker delta*. In tensor calculus, we can extend the *Kronecker delta* to any number of indices, such as the rank 4 tensor $\delta_{\mu\sigma\beta\gamma}$.

Like matrices, all non-singular tensors have inverses: $A^{-1}{}_{\mu\sigma}$ is the inverse of $A_{\mu\sigma}$. The contraction of a tensor with its inverse always equals the unit tensor.

$$A^{-1}{}_{\sigma\mu}\, A_{\mu\beta} = A_{\sigma\mu}\, A^{-1}{}_{\mu\beta} = [1] = \delta_{\sigma\beta}$$

This is as close as tensor calculus gets to dividing by a tensor. For tensors A and C, and a non-singular tensor B:

$$\text{If } A\,B = C, \text{ then } A = C\,B^{-1}$$

We define tensors as arrays of components that transform properly. In 3-D, that means they transform according to the rules of Euclidean rotations. As discussed above, each 3-D rotation R has a corresponding 3×3 matrix R_{jk}. Rotating coordinate axes transforms vector Q into vector Q* according to:

$$Q^*_j = R_{jk}\, Q_k$$

The requirement that a tensor transform properly under any rotation R can be written:

$$\text{rank 1: } A^*_j = R_{jk}\, A_k$$
$$\text{rank 2: } A^*_{jk} = R_{jn}\, R_{km}\, A_{nm}$$
$$\text{rank 3: } A^*_{jk\mu} = R_{jn}\, R_{km}\, R_{\mu\sigma}\, A_{nm\sigma}$$

In each line, A* is the transformed tensor of the original tensor A. In 3-D for rank 1, we sum over all three values of k. For rank 2, we sum over all 3×3 = 9 values of n and m. For rank 3, we sum over all 3×3×3 = 27 values of n, m, and σ. For rank 2, the above transform can also be written:

$$A^*_{jk} = R_{jn}\, R_{km}\, A_{nm}$$
$$A^*_{jk} = R_{jn}\, R^{-1}_{mk}\, A_{nm}$$
$$A^* = R\,A\,R^{-1}$$

Note that since tensor operations are fully defined by paired free indices, the order of product terms is irrelevant, unlike matrix multiplication.

We can form tensors by combining any proper vectors. For example, if A_j and B_k are proper 3-vectors, then:

$$C_{jk} = A_j\, B_k \text{ is a proper rank 2, 3-D tensor.}$$

Let's show that C transforms properly for any rotation R by transforming each vector.

$$(A^*_n)\,(B^*_m) = (R_{nj}\, A_j)\,(R_{mk}\, B_k)$$
$$(A^*_n)\,(B^*_m) = R_{nj}\, R_{mk}\, A_j\, B_k$$
$$(A^*_n)\,(B^*_m) = R_{nj}\, R_{mk}\, C_{jk} = C^*_{nm}$$

Hence, C transforms properly and is therefore a tensor. The same logic applies to tensors of any rank. For example:

$$F_{\mu\sigma}\, \Lambda_{\alpha\beta}\, R_{\delta\varepsilon}$$

is a proper rank 6 tensor, although it has no physical meaning as far as I know.

To understand the ***Feynman Lectures***, that is as much as you need to know about tensors.

Those who wish a glimpse of general relativistic tensor calculus can enjoy the next section (there's no exam), while others can skip to the following section.

4A§16.11 Tensor Calculus in Curved Spacetime

In 4-D curved spacetime, the dot product cannot be simply the sum of the products of corresponding components. This is because coordinate axes may change directions and rulers may change lengths, and all that can happen differently at every location and instant in time.

We therefore need a *metric* $g_{\mu\sigma}$ to reveal the geometry at each *event* (ct, x, y, z) in spacetime. That metric specifies the *invariant interval*, the "true distance", between any two nearby events. It turns out that knowing the interval between all nearby events completely determines the geometry everywhere. In curved space, the dot product of two 4-vectors A_μ and B_μ is:

$$A_\mu\, B^\mu = A^\mu\, B_\mu = g_{\mu\sigma}\, A^\mu\, B^\sigma = g_{\mu\sigma}\, A^\sigma\, B^\mu =$$
$$+ g_{tt}\, A^t\, B^t + g_{tx}\, A^t\, B^x + g_{ty}\, A^t\, B^y + g_{tz}\, A^t\, B^z$$
$$+ g_{xt}\, A^x\, B^t + g_{xx}\, A^x\, B^x + g_{xy}\, A^x\, B^y + g_{xz}\, A^x\, B^z$$
$$+ g_{yt}\, A^y\, B^t + g_{yx}\, A^y\, B^x + g_{yy}\, A^y\, B^y + g_{yz}\, A^y\, B^z$$
$$+ g_{zt}\, A^z\, B^t + g_{zx}\, A^z\, B^x + g_{zy}\, A^z\, B^y + g_{zz}\, A^z\, B^z$$

If ds_μ is the separation 4-vector between two nearby events, the invariant interval ds^2 between those points is:

$$ds_\mu = (cdt, dx, dy, dz)$$
$$ds^2 = ds_\mu\, ds^\mu = ds^\mu\, ds_\mu = g_{\mu\sigma}\, ds^\mu\, ds^\sigma$$

Here, ds^2 is an invariant scalar — it measures the separation between nearby events, and has the same value in any coordinate system. In Feynman's sign convention, ds^2 equals $c^2\, d\tau^2$ where τ is *proper time*, the time measured by an ideal clock moving between these nearby events. Note that some indices are subscripted while others are superscripted. The former are called *covariant indices*, while the latter are called *contravariant indices*. Since superscripts look exactly like exponents, we try to avoid using exponents in tensor calculus whenever confusion might arise: x^2 always means the second component of x^μ, while $(x)^2$ means x-squared. Because the square of coordinate differentials occur so frequently, an exception to this rule is dx^2, which always means $(dx)^2$.

In 4-D curved spacetime, we only sum repeated free indices if one is covariant and the other is contravariant. The difference between the two is demonstrated by the covariant and contravariant position 4-vectors (in Feynman's sign convention).

$x^\alpha = (ct, -x, -y, -z)$ is contravariant

$x_\alpha = (ct, +x, +y, +z)$ is covariant

The metric in flat spacetime, in Feynman's sign convention, is:

$$g_{\mu\sigma}$$

	σ = t	x	y	z
μ = t	1	0	0	0
x	0	–1	0	0
y	0	0	–1	0
z	0	0	0	–1

In polar coordinates, the metric near a black hole in Feynman's sign convention is:

$$g_{\mu\sigma}$$

	σ = t	r	θ	φ
μ = t	Ω	0	0	0
r	0	$-1/\Omega$	0	0
θ	0	0	$-r^2$	0
φ	0	0	0	$-r^2\sin^2\theta$

Here, the $g_{\theta\theta}$ and $g_{\phi\phi}$ components are the normal polar coordinate factors that are unaffected by gravity. Gravity dilates time and stretches space through the factor $\Omega = 1 - 2\,G\,M / c^2\, r$, where G is Newton's gravitational constant, M is the black hole's mass, and r is the distance from its center. Note that odd things happens when $\Omega = 0$ at $r = 2\,G\,M / c^2$, the location of the black hole's event horizon. One interesting effect is that the event horizon is timeless — the passage of time has no effect whatsoever on the event horizon, because g_{tt} is zero at that radius.

In the most common modern notation, the metric $g_{\mu\sigma}$ has a minus sign on the time component and plus signs on the three spatial components, the opposite of Feynman's convention.

The Lorentz transformation tensor is:

$$\Lambda^\mu{}_\sigma$$

	σ = t	x	y	z
μ = t	γ	$-\beta\gamma$	0	0
x	$-\beta\gamma$	γ	0	0
y	0	0	1	0
z	0	0	0	1

Here, $\beta = v / c$ and $\gamma = 1 / \sqrt{(1 - \beta^2)}$. Some typical index operations are:

(1) Lowering an Index: $x_\mu = g_{\mu\sigma}\, x^\sigma$

(2) Raising an Index: $x^\mu = g^{\mu\sigma}\, x_\sigma$

(3) Lorentz Transform: $X_\sigma = \Lambda^\beta{}_\sigma\, x_\beta$

Like other square matrices, the metric tensor for most geometries can be *diagonalized*, meaning all non-diagonal components can be made zero with suitable transformations. The invariant interval is then reduced to 4 terms, and the inverse metric is simply $g^{\alpha\alpha} = 1 / g_{\alpha\alpha}$.

Diagonalizing the metric tensor can mix the coordinates in surprising ways. For example, the time coordinate t and radial distance coordinate r might be replaced by the coordinates u = ct + r and w = ct – r, leaving nothing that represents pure time. But since the tensor calculus of general relativity works in any coordinate system, such mixing is mathematically valid; it can simplify our calculations even when it defies our intuition.

When one becomes comfortable with tensor notation, it is possible to drop the indices altogether, as we do in vector algebra. We can then write equation (3) as:

$$X = \Lambda\, x$$

4A§16.12 Einstein's Field Equations

The ultimate equation of general relativity, and Einstein's greatest contribution to mankind, are his *field equations* that are written:

$$G = 8\pi\, T$$

Here, G represents the geometry of spacetime, and T represents all forms of energy, including mass and momentum. G is now called the *Einstein tensor*, and T is called the *mass-energy-stress tensor*. Both are symmetric, rank 2 tensors. We say equations, the plural, because G = 8π T represents 16 component equations: 4 describe the conservation of energy and momentum; 6 relate energy density to spacetime curvature; and the remaining 6 are redundant.

John Archibald Wheeler said the meaning of G = 8π T is:

"The geometry of spacetime tells mass and energy how to move,

while mass and energy tell space how to curve."

Brian Greene describes Einstein's field equations as the choreography of the cosmic ballet of the universe. It is a duet in which both parties lead one another.

I hope you found this brief taste of general relativity intriguing. For a thorough yet accessible explanation of the most profound theory of science see my ebook ***General Relativity 1: Newton & Einstein***.

4A§16.13 Cross Product as a Tensor

The cross product of two vectors should properly be considered a tensor. For example, the equation for torque is:

$$\tau = r \times F$$

We can write this as:

$$\tau_{jk} = r_j\, F_k - r_k\, F_j$$

Here we see that τ_{jk} is formed by two pairs of proper polar 3-vectors. As we discussed in the **Tensor Algebra** section, each product term on the right side is a proper rank 2 tensor, and so therefore is their difference. Clearly, τ_{jk} is antisymmetric: $\tau_{jk} = -\tau_{kj}$. In 3 D, this means τ_{jk} has only three independent nonzero components. Feynman says it is "almost by accident" that these three components form a proper axial 3-vector. He says "accident" because this is true only in 3-D. In 4-D, for example, antisymmetric tensors have 6 nonzero components, which clearly cannot make a 4-vector.

The same logic applies to any vector cross product; each can be written as a rank 2, antisymmetric tensor.

Recall that vectors are divided into two classes, polar and axial, depending on how they change in reflection (inverting the polarity of one coordinate axis). Recall also that almost all axial vectors are defined by cross products. Similarly, tensors are divided into two classes, normal tensors and pseudo-tensors, depending on how they change in reflection. The cross product is a pseudo-tensor.

Chapter 17 Numerical Integration

4A§17.1 Using Computers

Numerical integration is necessary when we can compute y(x) for any x, but cannot obtain an analytical solution to the inetgral $\int y(x)dx$.

The standard procedure is to divide the total range of x into N subintervals, each of width Δx. For each subinterval, define x_j to be its x-value and $y_j = y(x_j)$. The integral is then estimated to be the sum of the areas of N rectangles, with the jth rectangle having area $y_j \Delta x$. In the limit that Δx goes to zero and N goes to ∞, we obtain an exact integral.

While $N = \infty$ is impossible, integration precision improves as N increases, encouraging us to choose the largest affordable N. Since most simple functions are analytically integrable, numerical integration is typically employed for quite complicated functions, where the primary cost is in calculating the y_j values.

As an example of standard numerical integration, consider integrating $y(x) = x^2$ from x = 0 to x = 8. I deliberately chose a function that is easy to integrate analytically:

$$S = \int_0^8 x^2\, dx = x^3 / 3 \,|_0^8 = 512 / 3 = 170.666...$$

Let's see how well the standard numerical integration procedure works. We being with $\Delta x = 2$. Figure 17-1 shows three rectangles of width $\Delta x = 2$ and heights:

$$y(0) = 0$$
$$y(2) = 4$$
$$y(4) = 16$$
$$y(6) = 36$$

The rectangle of height y(0) = 0 is invisible.

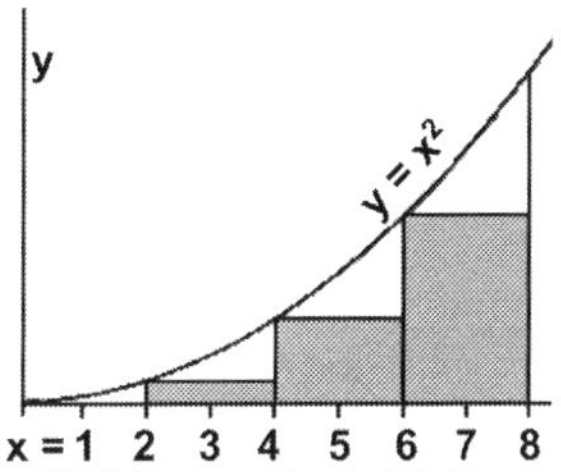

Figure 17-1 Integrating with Rectangles

The areas of these four rectangles sum to:

$$S_4 = 2 \times (0 + 4 + 16 + 36) = 112$$

This is 30% less than the correct value. The problem is obvious from the figure: the rectangles do not cover the entire area under the curve. For better precision, we must add more rectangles.

A good strategy is to double the number of rectangles. This way, we can reuse the y-values that we already computed. To go from 4 to 8 rectangles, we only need to compute 4 more y-values.

If we do not know the true value of the integral, how will we know when to stop doubling the number of rectangles? Generally, one stops when the change in the final result from the last doubling is acceptably "small enough". Clearly, "small enough" depends on the cost versus value of greater precision for the specific circumstances.

For this integral, the results with N rectangles, for various values of N, are:

$$N = 4: \quad S_4 = 112$$
$$N = 8: \quad S_8 = 140$$
$$N = 16: \quad S_{16} = 155$$
$$N = 32: \quad S_{32} = 162.75$$

Each time N doubles, the error decreases by about a factor of 2. But even at N = 32, the standard procedure is still nearly 5% below the true value of 170.666....

4A§17.2 Trapezoidal Integration

A simple way to improve on the standard procedure is to employ trapezoids rather than rectangles, as shown in Figure 17-2.

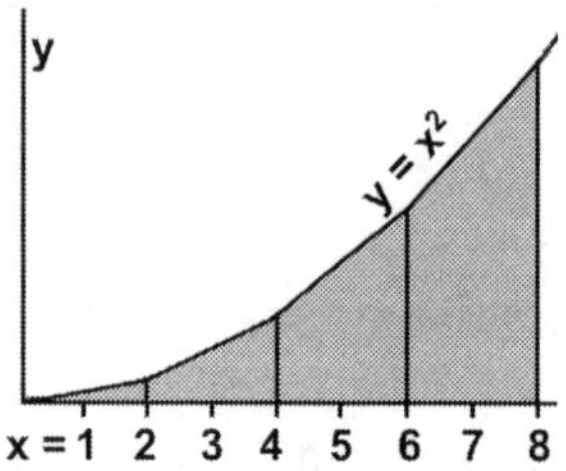

Figure 17-2 Trapezoids Replacing Rectangles

The area of a trapezoid equals its base multiplied by its average height, so the area of the largest trapezoid in the figure is:

$$\Delta x\ [y(6) + y(8)] / 2$$

The equation for the total area of the 4 trapezoids in Figure 17-2 is:

$$\begin{aligned} S_4 = &\ \Delta x\ [y(0) + y(2)] /2 \\ &+ \Delta x\ [y(2) + y(4)] /2 \\ &+ \Delta x\ [y(4) + y(6)] /2 \\ &+ \Delta x\ [y(6) + y(8)] /2 \end{aligned}$$

$$S_4 = \Delta x\ [y(0) + 2y(2) + 2y(4) + 2y(6) + y(8)] /2$$

The endpoints, y(0) and y(8), each have half the weighting of the interior points. The difference between summing rectangles and summing trapezoids is that the rectangle sum includes only one endpoint, at full weight, while the trapezoid sum includes both endpoints at half-weight. We can rewrite this above sum as:

$$\begin{aligned} S_4 = &\ \Delta x\ [y(0) + y(2) + y(4) + y(6)] \\ &+ \Delta x\ [y(8) - y(0)] /2 \end{aligned}$$

The sum in the upper line is the area of 4 rectangles, exactly matching our prior procedure. Switching this numerical integral from rectangles to trapezoids simply adds the expression in the lower line.

In general, for any value of N, switching from rectangles to trapezoids requires one more y-value and increases the integral by:

$$\Delta x\ [y(x\text{-max}) - y(x\text{-min})] /2$$

Here, x-min and x-max are the integration endpoints.

Recalculating the integral of $y = x^2$ for N trapezoids, for the same values of N, yields:

$$\begin{aligned} N = 4&: \quad S_4 = 176 \\ N = 8&: \quad S_8 = 172 \\ N = 16&: \quad S_{16} = 171 \\ N = 32&: \quad S_{32} = 170.75 \end{aligned}$$

For N = 32, trapezoids reduce the error from 5% to 0.05%. The improvement is not always so dramatic. Trapezoids provide exact integrals for linear functions, which are trivial to integrate in any case. The example we choose here is a quadratic function of x, whose deviation from linearity decreases with the square of Δx, playing to trapezoids' strength. For more complicated functions, the improvement may be less significant. But the cost is minor: only 1 more y-value than the standard rectangle procedure.

4A§17.3 Romberg Integration

Romberg integration improves the precision and minimizes the cost of numerical integration.

Let's consider an example: integrating the function y(x) = sin(x) from x = 0 to π/2. This function is also easy to integrate analytically:

$$S = \int_0^{\pi/2} \sin(x)\,dx = -\cos(x)\,|_0^{\pi/2} = 0 - (-1) = 1$$

Let's see how well various methods can match the correct value of 1. Figure 17-3 shows an example of the standard procedure: numerically summing four rectangles, each of width π/8, with heights y(0), y(π/8), y(2π/8), and y(3π/8). The first rectangle is invisible since it has zero height.

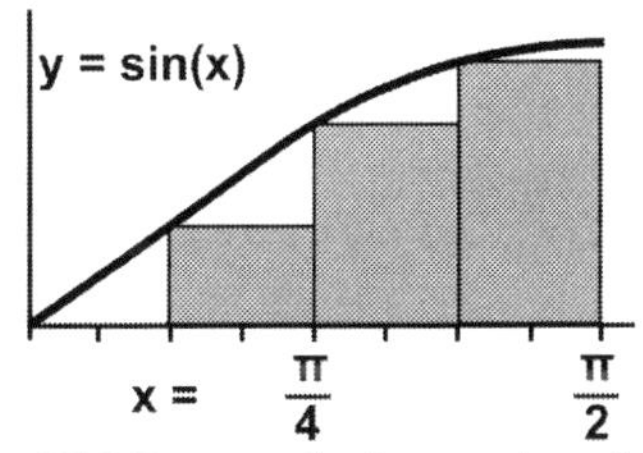

Figure 17-3 Rectangular Integration of sin(x)

The sum of the areas of these 4 rectangles is:

$$S_4 = (\pi/8) \times (0 + 0.38268 + 0.70711 + 0.92388)$$
$$S_4 = 0.79077$$

Continuing with the standard rectangle approach, we try more and more, narrower and narrower rectangles. Here are the results obtained with N rectangles, for various N:

$$N = 2{:}\ \Delta x = \pi/4{:}\ S_2 = 0.55536$$
$$N = 4{:}\ \Delta x = \pi/8{:}\ S_4 = 0.79077$$
$$N = 8{:}\ \Delta x = \pi/16{:}\ S_8 = 0.89861$$
$$N = 16{:}\ \Delta x = \pi/32{:}\ S_{16} = 0.95011$$
$$N = 32{:}\ \Delta x = \pi/64{:}\ S_{32} = 0.97526$$

Even with 32 values of sin(x), the numerical integral is off by 2.5%.

Using trapezoids instead of rectangles substantially improves these results, yielding:

$$N = 2{:}\ \Delta x = \pi/4{:}\ S_2 = 0.94806$$
$$N = 4{:}\ \Delta x = \pi/8{:}\ S_4 = 0.98712$$
$$N = 8{:}\ \Delta x = \pi/16{:}\ S_8 = 0.99679$$
$$N = 16{:}\ \Delta x = \pi/32{:}\ S_{16} = 0.99920$$
$$N = 32{:}\ \Delta x = \pi/64{:}\ S_{32} = 0.99980$$

For N = 32, trapezoids are 100 better times than rectangles.

But Romberg integration is even better.

The Romberg scheme is to fit a curve to S(Δx), and extrapolate that curve to Δx = 0. In Figure 17-4 , the fitted curve passes through 3 of the 5 plotted S-values, with an arrow pointing to its intercept at Δx = 0.

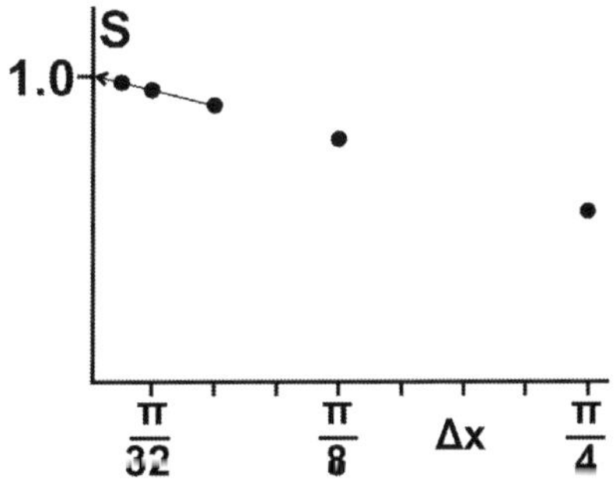

Figure 17-4 Extrapolating S to Δx = 0

In this case, a quadratic curve is an excellent choice. Curve fitting is an art onto itself, as we will discuss in subsequent chapters. Fitting a quadratic to three data points is straightforward. As I prove below, a quadratic that passes through points f(z=1), f(z=2), and f(z=4) has a value at z = 0 given by:

$$f(0) = \{ 8\, f(1) - 6\, f(2) + f(4) \} / 3$$

Fitting a quadratic and extrapolating to $\Delta x = 0$ yields results that are extremely close to the correct answer, as these data show:

Set (2, 4, 8): Romberg = 1 – 118 ppm

Set (4, 8, 16): Romberg = 1 – 7.3 ppm

Set (8, 16, 32): Romberg = 1 – 0.5 ppm

Here, the label Set (J, K, L) means fitting a quadratic to the numerical integrals with J, K, and L rectangles. Also, "ppm" means parts per million, so 1 – 118 ppm = 0.999,882.

The power of the Romberg method is evident: it provides greater precision with less calculation.

With 8 values of the integrand, the standard procedure has an error of 10%. But with the Romberg method, the error is nearly 900 times smaller. Romberg with 8 rectangles is 200 times better than the standard rectangle method with 32 rectangles, and 2 times better than 32 trapezoids.

Note that in the Romberg scheme, there is no advantage to substituting trapezoids for rectangles. Since trapezoids increase each S-value by an amount that is linearly proportional to Δx, using them would change the slope of the Romberg curve, but not its intercept at $\Delta x = 0$.

Next, I will show you how to fit a quadratic to three points spaced in the ratio 1 : 2 : 4. If you wish, you can skip the derivation, and just remember the result.

4A§17.4 Fitting a Quadratic to Three Points

You can use these quadratic fit coefficients on any function, if you have its values at the right spacings. You can also use this procedure to calculate other coefficients for different spacings.

Define function $f(z) = A + B\, z + C\, z^2$. Our objective is to calculate A, the value of f at z = 0.

$$f(1) = A + B + C$$
$$f(2) = A + 2B + 4C$$
$$f(4) = A + 4B + 16C$$

Subtract the upper and lower pairs.

$$f(2) - f(1) = B + 3C$$
$$f(4) - f(2) = 2B + 12C$$

Solve for C, then B, and finally A.

$$\{f(4) - f(2)\} - 2\, \{f(2) - f(1)\} = 6C$$
$$C = \{ f(4) - 3\, f(2) + 2\, f(1) \} / 6$$
$$B = \{ f(2) - f(1) \} - 3C$$
$$B = \{ f(2) - f(1) \} - \{ f(4) - 3\, f(2) + 2\, f(1) \} / 2$$
$$B = \{ 5\, f(2) - f(4) - 4\, f(1) \} / 2$$
$$A = f(1) - B - C$$
$$6A = 6\, f(1) - 15\, f(2) + 3\, f(4) + 12\, f(1) - f(4) + 3\, f(2) - 2\, f(1)$$
$$A = \{ 16\, f(1) - 12\, f(2) + 2\, f(4) \} / 6$$
$$A = \{ 8\, f(1) - 6\, f(2) + f(4) \} / 3$$

QED

Chapter 18
Data Fitting

4A§18.1 Curve Fitting

Imagine that we wish to measure the luminosity of a remote star as a function of time. We might be interested in the star's intrinsic properties — its variability, the frequency of flares, or the intensity of star-spots. We might also be searching for periodic luminosity dips that reveal orbiting planets.

Ideally, our measurements would yield L(t), the luminosity at all times. But real measurements actually yield $L_j \pm \Delta L_j$, luminosities with instrumental uncertainties, taken at a set of discrete times t_j.

To properly interpret real measurements, physicists often attempt to represent their data with continuous analytic functions. Called *curve fitting*, this craft is a blend of science and art. We sometimes fit data to theoretical functions, and other times to empirical functions, such as Taylor series.

Let's consider some examples.

If N competent physicists measure the mass of individual oxygen atoms, we might sum their results and divide by N to obtain an average value. In doing so, we have chosen to represent this data by a single constant, the simplest possible function. This seems entirely reasonable since oxygen atoms should have the same mass in everyone's laboratory ... unless those atoms are moving at relativistic velocities. We would then be more successful using a more complex function that included each atom's velocity.

Next, consider measurements of the distance x_j that an airplane has traveled by time t_j, as it flies from Los Angeles to London. We might fit that data with a linear function: $x(t) = A + B\,t$. Or we might try a more complicated function that addresses a slow climb followed by increasing speed as spent fuel decreases weight.

How do we know which function to use, and whether or not we have a "good" fit?

Fit quality is generally evaluated with the ratio: χ^2 / #df, chi-square per degree of freedom (also discussed in Chapter 9). For N data points, x_j for j = 1 to N, and corresponding values of a fitting function f_j, χ^2 is given by:

$$\chi^2 = \Sigma_j \{ x_j - f_j \}^2 w_j / u_j^2$$

Here, we sum over j = 1 to N. The w_j are optional weighting factors that are entirely subjective; you can employ them to stress data that you deem more important or more valuable than others. Most commonly, we treat all data points equally and set all weighting factors to 1.

The u_j's are the one standard deviation uncertainties in $\{x_j - f_j\}$. If the f-values have zero uncertainty, and the x_j's are the numbers of rare events, $u_j = \sqrt{x_j}$, if x_j is not zero. If some x_j's are zero, you can remove those j's from the summation, or set u_j to 0 or 1, depending on the circumstances.

When the collected data has gaps, those zeros should not be included in computing χ^2. If the x_j's are the numbers of rare events, and one measurement, say x_k is zero, you should probably set $u_k = 1$. Conversely, if you are absolutely sure that f_k must be precisely zero for some k, a nonzero x_k is definitive proof of an erroneous fit, and a zero or tiny value of u_k is appropriate. This decision can be as much art as science.

Next, let's discuss #df, the *number of degrees of freedom*, which equals the number of terms in the summation of χ^2 minus the number of *adjustable parameters* in the fit. If we are fitting our data to a straight line, there are two adjustable parameters: the line's slope and intercept. If we are fitting our data to a fifth-order polynomial, there are six adjustable parameters: A through F in:

$$f(t) = A + B\,t + C\,t^2 + D\,t^3 + E\,t^4 + F\,t^5$$

In almost all fits, there is at least one adjustable parameter. We must scale the fitting function to match the total number of events in the data, which is determined by budgets and manpower rather than by the laws of nature. We therefore use one adjustable parameter to ensure:

$$\Sigma_j f_j = \Sigma_j x_j$$

Let's now examine how we use χ^2 / #df to fit functions to data, and later discuss what χ^2 / #df says about the acceptability of that fit.

Imagine that we wish to experimentally determine a function x(t). We first measure data points x_j at times t_j, each with uncertainty u_j, for j =1 to N. We next try to fit that data with a quadratic polynomial:

$$f(t) = A + B\,t + C\,t^2$$

This fit has N – 3 degrees of freedom.

The equation for χ^2 is:

$$\chi^2 = \Sigma_j \{ x_j - A - B\,t_j - C\,t_j^2 \}^2 / u_j^2$$

If χ^2 were zero, the function f would represent the data perfectly. Evidently, the best fit corresponds to the values of parameters A, B, and C that minimize χ^2. We therefore compute the partial derivatives of χ^2 with respect to each parameter, set those derivatives equal to zero, and solve for A, B, and C.

$$0 = \partial\chi^2/\partial A = -2\,\Sigma_j \{ x_j - A - B\,t_j - C\,t_j^2 \} / u_j^2$$

$$0 = \partial\chi^2/\partial B = -2\,\Sigma_j \{ x_j - A - B\,t_j - C\,t_j^2 \}\, t_j / u_j^2$$

$$0 = \partial\chi^2/\partial C = -2\,\Sigma_j \{ x_j - A - B\,t_j - C\,t_j^2 \}\, t_j^2 / u_j^2$$

We can rearrange these equations to read:

$$\Sigma_j A / u_j^2 + \Sigma_j B\,t_j / u_j^2 + \Sigma_j C\,t_j^2 / u_j^2 = \Sigma_j x_j / u_j^2$$

$$\Sigma_j A\,t_j / u_j^2 + \Sigma_j B\,t_j^2 / u_j^2 + \Sigma_j C\,t_j^3 / u_j^2 = \Sigma_j x_j\,t_j / u_j^2$$

$$\Sigma_j A\,t_j^2 / u_j^2 + \Sigma_j B\,t_j^3 / u_j^2 + \Sigma_j C\,t_j^4 / u_j^2 = \Sigma_j x_j\,t_j^2 / u_j^2$$

To reduce clutter, define:

$$T_k = \{ \Sigma_j t_j^k u_j^2 \} / \{ \Sigma_j 1 / u_j^2 \}$$

$$X_k = \{ \Sigma_j x_j t_j^k / u_j^2 \} / \{ \Sigma_j 1 / u_j^2 \}$$

With those definitions, the prior equations become:

$$A\,T_0 + B\,T_1 + C\,T_2 = X_0$$

$$A\,T_1 + B\,T_2 + C\,T_3 = X_1$$

$$A\,T_2 + B\,T_3 + C\,T_4 = X_2$$

These equations can be written in matrix form as follows:

$$(A\ \ B\ \ C)\begin{pmatrix} T_0 & T_1 & T_2 \\ T_1 & T_2 & T_3 \\ T_2 & T_3 & T_4 \end{pmatrix} = \begin{pmatrix} X_0 \\ X_1 \\ X_2 \end{pmatrix}$$

If the matrix, call it M, is singular (if its determinant is zero), the equations are under-constrained; they do not provide enough information to solve for A, B, and C. But if M is non-singular, it has an inverse matrix M^{-1}. We can then multiply both sides of this equation on their right ends by M^{-1}. Since $M\,M^{-1}$ equals the unit matrix, the left side reduces to the vector (A, B, C), while the right side becomes a vector whose components are the best fit values of A, B, and C.

This procedure is valid for any number N of data points and any nth order polynomial, provided $n < N$. The procedure can also be straightforwardly generalized for fitting functions other than polynomials.

Let's try an example: fit five data points with the quadratic $f(t) = A + B\,t + C\,t^2$.

Let N = 5, all $u_j = 1$, and $t_j = (-2, -1, 0, +1, +2)$. These messy equations then reduce to:

$$T_0 = (+1\ +1\ +1\ +1\ +1) / 5 = 1$$

$$T_1 = (-2\ -1\ +0\ +1\ +2) / 5 = 0$$

$$T_2 = (+4\ +1\ +0\ +1\ +4) / 5 = 2$$

$$T_3 = (-8\ -1\ +0\ +1\ +8) / 5 = 0$$

$$T_4 = (+16\ +1\ +0\ +1\ +16) / 5 = 6.8$$

$$X_0 = (+x_1 + x_2 + x_3 + x_4 + x_5) / 5$$
$$X_1 = (-2x_1 - x_2 + x_4 + 2x_5) / 5$$
$$X_2 = (+4x_1 - x_2 + x_4 + 4x_5) / 5$$

The matrix M and its inverse M^{-1} are:

$$\underset{\mathbf{M}}{\begin{pmatrix} 1 & 0 & 2 \\ 0 & 2 & 0 \\ 2 & 0 & 6.8 \end{pmatrix}} \quad \underset{\mathbf{5.6\times M^{-1}}}{\begin{pmatrix} 13.6 & 0 & -4 \\ 0 & 2.8 & 0 \\ -4 & 0 & 2 \end{pmatrix}}$$

Note the determinant of M:

$$\text{Det } M = 1 (2 \times 6.8 - 0) - 0 + 2 (0 - 2 \times 2)$$
$$\text{Det } M = 13.6 - 8 = 5.6$$

Multiplying the vector (X_0, X_1, X_2) by M^{-1} yields:

$$A = \{ 13.6 (+x_1 + x_2 + x_3 + x_4 + x_5) + 0 - 4 (+4x_1 - x_2 + x_4 + 4x_5) \} / (5 \times 5.6)$$
$$B = \{ 0 + 2.8 (-2x_1 - x_2 + x_4 + 2x_5) + 0 \} / 28$$
$$C = \{ -4 (+ x_1 + x_2 + x_3 + x_4 + x_5) + 0 + 2 (+4x_1 - x_2 + x_4 + 4x_5) \} / 28$$

These reduce to:

$$A = (-6x_1 + 24x_2 + 34x_3 + 24x_4 - 6x_5) / 70$$
$$B = \{-2x_1 - x_2 + x_4 + 2x_5 \} / 10$$
$$C = (+2x_1 - x_2 - 2x_3 - x_4 + 2x_5) / 14$$

Before we continue, note that the endpoints, x_1 and x_5, dominate the equations for B and C. This is generally true for all polynomial fits and also for some other functions. Statistical fluctuations or systematic errors in endpoint data can dramatically alter the fit coefficients.

After all that math, it is worthwhile checking some simple cases.

If the x_j are all equal ($x_j = Q$), the fit coefficients are:

$$A = 28Q / 28 = Q$$
$$B = (-3Q + 3Q) / 10 = 0$$
$$C = (+8Q - 8Q) / 28 = 0$$

If the x_j vary linearly ($x_j = Q\, t_j$), the fit coefficients are:

$$A = (+4.8Q -9.6Q +9.6Q -4.8Q) / 28 = 0$$
$$B = (+4Q +Q +Q +4Q) / 10 = Q$$
$$C = (-8Q +2Q -2Q +8Q) / 28 = 0$$

If the x_j vary quadratically ($x_j = Q\, t_j^2$), the fit coefficients are:

$$A = (-9.6Q +9.6Q +9.6Q -9.6Q) / 28 = 0$$
$$B = (-8Q -Q +Q +8Q) / 10 = 0$$
$$C = (+16Q -2Q -2Q +16Q) / 28 = Q$$

All three simple test cases yield the proper results. This means we probably have not made a big mistake.

Here is a practice problem if you wish: fit a quadratic to three data points: x_1, x_2, and x_3 measured at times –1, 0, and +1, respectively. Try doing this without looking at the preceding example. The answers are in Section 4A§18.3.

Having derived the fitting function, we can then calculate the figure of merit χ^2 / #df, unless #df = 0. In the latter case, we have no means of judging the quality of fit. Any three data points can be fit with a quadratic with $\chi^2 = 0$; but because #df = 0, we have no basis for believing that the true function is well-represented by that quadratic.

In general, a "good" fit has χ^2 / #df less than 1, or not "too much" more than 1. If χ^2 / #df is much greater than 1, the fitted function is a bad match to the data. As #df increases, the expected χ^2 becomes closer and closer to 1, and what constitutes "too much" decreases rapidly. Appendix 4 contains a table quantifying the statistical expectations for χ^2 for various numbers of degrees of freedom.

4A§18.2 Curve Fitting Cautions

The results of curve fitting should be considered suggestive rather than definitive; the results must be judiciously interpreted.

In searching for a "good" fit, one generally starts with simple functions, and then adds more complex terms as needed. If χ^2 / #df = 6 for a quadratic polynomial, one might try a third-order polynomial. One might keep adding higher-order terms until χ^2 / #df becomes "good". But do not add more terms than necessary.

The coefficients in polynomial fitting are not stable. This means, the coefficient of t^3 in a fourth-order fit can be very different from the coefficient of t^3 in a fifth-order fit: the fit coefficients *do not* converge asymptotically to "true" values. Lower-order polynomial fits are generally more stable than higher-order fits.

As discussed above, beware of endpoint issues. Any systematic problems that impact the endpoints will be magnified in a polynomial fit. Even unavoidable statistical fluctuations in endpoint values can substantially alter the fit. It is often wise to try alternative fits with the endpoints removed; if the fit function changes dramatically, you are on thin ice and may wish to reexamine the endpoint data.

Fits are much less sensitive to inner data points, but these too deserve attention. In addition to evaluating the sum of χ^2 over all points, look at the χ^2 contribution from each data point individually. χ^2 is the square of (the actual deviation divided by one standard deviation). The statistical expectation for individual χ^2 contributions is that:

$$\chi^2 > 1 \text{ in } 32\% \text{ of data points}$$

$$\chi^2 > 4 \text{ in } 5\% \text{ of data points}$$

$$\chi^2 > 9 \text{ in } 0.3\% \text{ of data points}$$

Data points with χ^2 well beyond these statistical expectations merit judicious scrutiny. The objective is to avoid biased data; biases can be statistical, systematic, or human (people have a tendency to get rid of "ugly" data).

Generally, one should throw out data only if it is almost certainly wrong. In an attempt to minimize sensitivity to statistical fluctuations, some researchers always throw out the maximum and minimum measured values in each data bin. If you do adopt that or any similar policy, you must be consistent to avoid unintentional biases.

Before publishing results, you might test your fits for stability. If you are organizing data into bins, check to see if you get a similar fit with more or fewer bins. The entire fitting process may well be erroneous if the fit changes substantially depending on arbitrary analysis choices, such as bin size.

An Australian physicist suggests calculating medians instead of averages, because a median (the value that half the data are greater than and half are less than) is less impacted by statistical fluctuations than is an average or mean (the sum of all data values divided by their number). To my knowledge, that interesting idea never caught on.

4A§18.3 Answer to Quadratic Fit

The challenge is to fit x(t) = A + B t + C t^2 to three data points: x_1, x_2, and x_3 that are measured at t = –1, 0, and +1, respectively.

$$T_0 = (+1 +1 +1) / 3 = 1$$

$$T_1 = (-1 +0 +1) / 3 = 0$$

$$T_2 = (+1 +0 +1) / 3 = 2 / 3$$

$$T_3 = (-1 +0 +1) / 3 = 0$$

$$T_4 = (+1 +0 +1) / 3 = 2 / 3$$

$$X_0 = (+x_1 + x_2 + x_3) / 3$$

$$X_1 = (-x_1 + x_3) / 3$$

$$X_2 = (+x_1 + x_3) / 3$$

With these quantities, the matrix M and its inverse M^{-1} are:

$$\underset{3M}{\begin{pmatrix} 3 & 0 & 2 \\ 0 & 2 & 0 \\ 2 & 0 & 2 \end{pmatrix}} \quad \underset{M^{-1}}{\begin{pmatrix} +3 & 0 & -3 \\ 0 & 3/2 & 0 \\ -3 & 0 & 9/2 \end{pmatrix}}$$

The determinant of M equals 4 / 27.

Multiplying the vector (X_0, X_1, X_2) by M^{-1} yields:

$$A = 3(+x_1 + x_2 + x_3) / 3 + 0 - 3 (+x_1 + x_3) / 3$$

$$B = 0 + (3/2) (-x_1 + x_3) / 3 + 0$$

$$C = -3 (+x_1 + x_2 + x_3) / 3 + 0 + (9/2) (+x_1 + x_3) / 3$$

These reduce to:

$$A = x_2$$

$$B = (-x_1 + x_3) / 2$$

$$C = (x_1 - 2x_2 + x_3) / 2$$

QED

Chapter 19
Transforms & Fourier Series

In the prior chapter, we fit polynomials to represent data. In this chapter, we will discover how to represent functions with sinusoids, transforms, and spherical harmonics (sinusoids in two dimensions).

4A§19.1 Fourier Series

Any physically realistic waveform, however complex, can be represented by a linear sum of sinusoidal functions. Let's now discover the mathematical procedure by which this is accomplished: the Fourier series.

Consider a violin string, vibrating between two fixed points, the neck stop and the violinist's finger. Define those fixed points to be x = 0 and x = L. The violin bow excites a transverse oscillation that must have zero amplitude at each fixed point. The equation for string displacement f(x,t) is:

$$f(x, t) = A \sin(\omega t + \phi) \sin(n \pi x / L)$$

$$\lambda_n = 2L / n$$

$$\omega_n = 2\pi v / \lambda_n$$

Here, ω_n is the oscillation frequency, A is the oscillation amplitude, ϕ is the initial phase, and λ_n is the wave length. The wave velocity v is determined by the string's mass density and tension, each of which has a constant value (after tuning). Frequency ω_n is determined by v and λ_n according to a standard wave equation.

The right-most sine function in f(x, t) ensures that f(0, t) = 0. But we must also require that the oscillation *mode number* n be an integer so that f(L, t) = 0. Since the sine function is zero twice in each wavelength, there must be an integral number of half-wavelengths in length L:

$$L = n \lambda_n /2$$

Each value of n yields a valid solution to the string's wave equation.

Since this is a linear system, any linear sum of such solutions is also a solution. Hence, the most general wave solution is:

$$f(x, t) = \Sigma_n A_n \sin(\omega_n t + \phi_n) \sin(n \pi x / L)$$

$$\text{with } \omega_n = n v \pi / L, \text{ for any integer } n > 0$$

Let's now define $\omega = v \pi / L$, the frequency at which all modes repeat, which means:

$$\omega_n = n \omega = n v \pi / L$$

We can also eliminate the phase angles ϕ_n by using:

$$\sin(n \omega t + \phi_n) = \sin(n \omega t) \cos(\phi_n) + \sin(\phi_n) \cos(n \omega t)$$

Let's next consider the wave form at a fixed value of x, and define:

$$a_n = A_n \cos(\phi_n) \sin(n \pi x / L)$$

$$b_n = A_n \sin(\phi_n) \sin(n \pi x / L)$$

With these definitions, we have another expression for the most general waveform at x:

$$f(t) = \Sigma_n \{ a_n \cos(n \omega t) + b_n \sin(n \omega t) \}$$

The above expression is called the *Fourier series* for f(t), a function that repeats with frequency ω.

In musical terms, the n = 1 term is the *first harmonic*, n = 2 is the *second harmonic*, etc.

Waves are almost always expressed with an average value of zero, such as sin(x), but the above sums can accommodate waves with offsets, such as 1 + sin(x), by including an n = 0 term, which is simply the constant a_0.

Figure 19-1 shows how increasing the number of terms in a Fourier series better approximates a true square wave shown in the upper third of this figure.

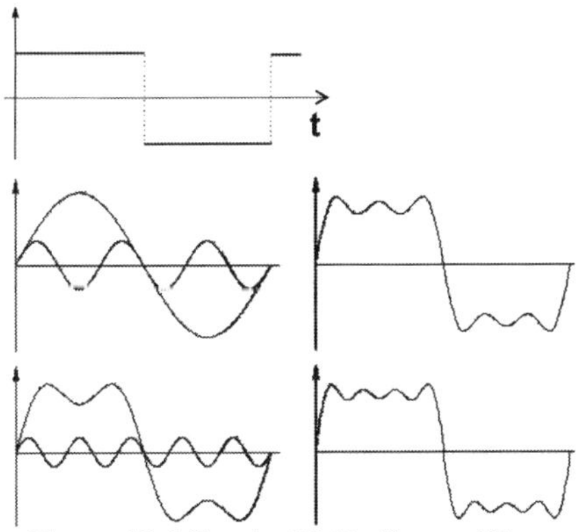

Figure 19-1 Fourier Fit To Square Wave

On the left side of Figure 19-1, the middle image plots the $n = 1$ and $n = 3$ modes [$\sin(\omega t)$ and $\sin(3\omega t)$], while the lower image plots the best Fourier fit with modes $n = 0$ to 3. On the right side, are the best fits with modes $n = 0$ to 5 in the middle image, and $n = 0$ to 7 in the lower image. Adding more modes clearly improves the fit.

4A§19.2 Musical Quality & Consonance

Fourier series help us describe the musical concepts of *quality* and *consonance*.

Even when the tones from a violin and an oboe have the same pitch (the same ω), the differing characteristics of these instruments produce sounds with different *tone qualities*, due to differing coefficients of higher frequencies in their Fourier series representations. Sounds that contain only one harmonic are called *pure tones*. Sounds composed of many strong harmonics are called *rich tones*.

4A§19.3 Calculating Fourier Coefficients

Fourier analysis is a powerful tool for linear systems because any periodic function f is a sum of sine and cosine functions, with some set of Fourier coefficients. We can often solve difficult equations for the special case of simple sinusoidal functions. The solution for the complex function f is then simply the sum of the sinusoidal solutions, with the appropriate Fourier coefficients.

All we need is a procedure to compute the Fourier coefficients. Not surprisingly, the person who developed this procedure was Jean-Baptiste Joseph Fourier (1768-1830). Consider again the Fourier series:

$$f(t) = \Sigma_n \{ a_n \cos(n\,\omega t) + b_n \sin(n\,\omega t) \}$$

Define T to be one full repetition period: $T = 2\pi / \omega$.

The integral of any sinusoid over a full cycle is zero. Hence only the cosine part of the $n = 0$ term remains on the right hand side after that integration.

$$\int_T f(t)\,dt = \int_T a_0 \cos(0)\,dt = T\,a_0$$

$$a_0 = (1 / T) \int_T f(t)\,dt$$

Fourier discovered that calculating the other coefficients is not much harder. Multiply the Fourier series by $\cos(j\,\omega t)$ for some integer $j > 0$, and then integrate over T:

$$\int_T f(t) \cos(j\omega t)\,dt = \Sigma_n \{ \int_T a_n \cos(n\,\omega t) \cos(j\,\omega t)\,dt \} + \Sigma_n \{ \int_T b_n \sin(n\,\omega t) \cos(j\,\omega t)\,dt \}$$

Note that:

$$2 \cos(n\,\omega t) \cos(j\,\omega t) = \cos([n + j]\,\omega t) + \cos([n - j]\,\omega t)$$

$$2 \sin(n\,\omega t) \cos(j\,\omega t) = \sin([n + j]\,\omega t) + \sin([n - j]\,\omega t)$$

$$2 \sin(n\,\omega t) \sin(j\,\omega t) = \cos([n - j]\,\omega t) - \cos([n + j]\,\omega t)$$

Each term on the right side of each above equation is a sinusoidal function of frequency m ω, for m either n + j or n – j. If m is nonzero, the integral over period T is zero. The right side of the middle equation is zero for any combination of n and j. Therefore, we have shown that after integrating from t = 0 to t = T = 2π / ω:

$$\int \cos(n\omega t)\cos(j\omega t)\,dt = T/2 \text{ if } n = j, \text{ else } = 0$$

$$\int \sin(n\omega t)\cos(j\omega t)\,dt = 0 \qquad \text{for any n and j}$$

$$\int \sin(n\omega t)\sin(j\omega t)\,dt = T/2 \text{ if } n = j, \text{ else } = 0$$

The prior integral becomes:

$$\int_T f(t)\cos(j\,\omega t)\,dt = \int_T a_j \cos(j\,\omega t)\cos(j\,\omega t) = a_j\,T/2$$

$$a_j = (2/T)\ \int_T f(t)\cos(j\,\omega t)\,dt$$

We can repeat this logic multiplying f(t) by sin(jωt), and integrating over T. The result is:

$$b_j = (2/T)\ \int_T f(t)\sin(j\,\omega t)\,dt$$

We have been very successful analyzing many repetitive motion problems using exponentials with complex exponents. We can employ that technique here as well, and rewrite the Fourier series equations as:

$$f(t) = \text{Real part of } \Sigma_n \{\, z_n \exp(in\omega t)\,\}$$

$$\text{with } z_n = a_n + ib_n$$

$$z_n = (2/T)\int_T f(t)\exp(-in\,\omega t)\,dt, \text{ for } n > 0$$

$$a_0 = (1/T)\int_T f(t)\,dt, \text{ and } b_0 = 0$$

4A§19.4 Evaluating a Fit

We showed above how to calculate the Fourier coefficients a_n and b_n. But how do we know when to stop calculating — how many terms in the Fourier series do we need?

Generally, one calculates Fourier coefficients for larger and larger values of n until their values become "small" — until a_N and b_N are much smaller in absolute value than the a_n and b_n for all n < N. Then evaluate the fit by calculating the *rms* (*root mean square*) deviation:

$$(\text{rms})^2 = (1/T)\int_T \{\, f(t) - F(t)\,\}^2\,dt$$

$$\text{with } F(t) = \Sigma_n \{\, a_n \cos(n\,\omega t) + b_n \sin(n\,\omega t)\,\}$$

Here, the sum is from n = 0 to N. We cannot use the χ^2 / #df tables to evaluate rms, because f and F have no statistical fluctuations, But rms does provide a quantitative measure of the fit precision. Ideally, rms is zero, indicating a perfect fit. With a nonzero rms, you can decide whether the benefit of potentially greater precision is worth the cost and effort of calculating more coefficients.

4A§19.5 Fourier Series of Square Wave

With these equations we can calculate the Fourier series for a square wave. Let the square wave be a periodic function, shown in Figure 19-2, repeating with period T and defined by:

$$f(t) = +1 \text{ for } 0 =< t < T/2$$

$$f(t) = -1 \text{ for } T/2 < t < T$$

$$f(t + T) = f(t) \text{ for any t}$$

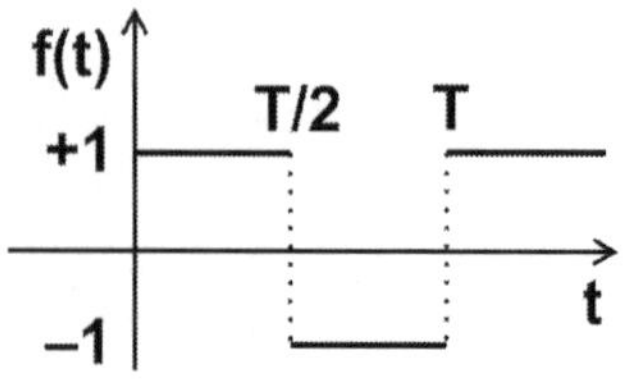

Figure 19-2. Square Wave Function

Clearly a_0, the average value of f(t), is zero. To compute the other coefficients, we must separate each integral into two parts: (1) the integral from t = 0 to T / 2, where f(t) = 1; and (2) the integral from t = T/2 to T, where f(t) = –1. Recall the relation ωT = 2π.

$$a_j = (2 / T) \{ \int_{(1)} \cos(j\ \omega t)\ dt - \int_{(2)} \cos(j\ \omega t)\ dt \}$$

$$a_j = (2 / T\ j\ \omega) \{ \sin(j\ \omega\ T/2) - 0\ \ - \sin(j\ \omega\ T) + \sin(j\ \omega\ T/2) \}$$

$$a_j = (1 / j\pi) \{ 2\sin(j\ \pi) - \sin(j\ 2\pi) \}$$

$$a_j = (1 / j\pi) \{ 0 - 0 \}$$

$$b_j = (2 / T) \{ \int_{(1)} \sin(j\ \omega t)\ dt - \int_{(2)} \sin(j\ \omega t)\ dt \}$$

$$b_j = (2 / T\ j\ \omega) \{ -\cos(j\ \omega\ T/2) + 1\ \ + \cos(j\ \omega T) - \cos(j\ \omega\ T/2) \}$$

$$b_j = (1 / j\pi) \{ 1 - 2\cos(j\ \pi)\ + \cos(j\ 2\pi) \}$$

$$b_j = (1 / j\pi) \{ 2 - 2\cos(j\ \pi) \}$$

$a_j = 0$ for all j

$b_j = 0$ for j even

$b_j = (4 / j\ \pi)$ for j odd

$$f(t) = (4 / \pi) \{\sin(\omega t) + \sin(3\omega t)/3 + \sin(5\omega t)/5 + \ldots \}$$

Since all sinusoids are continuous functions, the Fourier series of a square wave cannot exactly match the square wave at its discontinuity (t = T / 2 in this case). Here we find:

$$f(T/2) = (4 / \pi) \{\sin(\pi) + \sin(3\pi)/3 + \sin(5\pi)/5 + \ldots \} = 0$$

The Fourier series yields the value half way between the square wave's value at t < T / 2 and at t > T / 2. This seems reasonable. Natural phenomena are almost never discontinuous. Any physically realistic function that goes from +1 to –1 must pass through zero.

4A§19.6 Fourier Transform

Fourier series are appropriate for waves confined to a finite space. Let's now extend the Fourier methodology to an infinite space.

We showed above that the equations for the Fourier series representation of f(t) can be written:

$$f(t) = \text{Real part of } \Sigma_n \{ z_n \exp(in\omega t) \}$$

with $z_n = a_n + ib_n$

$$z_n = (2 / T) \int_T f(t) \exp(-in\ \omega t)\ dt, \text{ for } n > 0$$

$$a_0 = (1 / T) \int_T f(t)\ dt, \text{ and } b_0 = 0$$

Here the Fourier series is the sum of a set, perhaps an infinite set, of terms with discrete frequencies. For a space of length L, this discrete set includes all frequencies that are multiples of ω = π v / L.

We can generalize this by extending the Fourier sum to a continuous sum, an integral of terms with a continuous range of frequencies.

For any physically realistic function s(t), the **Fourier transform** S(f) and its inverse transform are:

$$S(f) = \int s(t) \exp\{ -i\ 2\pi\ f\ t\}\ dt$$

$$s(t) = \int S(f) \exp\{ +i\ 2\pi\ f\ t\}\ df$$

The last equation shows that the Fourier transform has an inverse operation that brings us back to s(t). We need to specify the range of the above integrals: both integrals range from –∞ to +∞, if:

$$\int_{-\infty}^{+\infty} | s(t) |\ dt \text{ is finite}$$

Note that s(t) is a function of time, whereas S(f) is a function of frequency. Roughly speaking, S(f) is the amount of frequency f in the sum of complex exponentials that equals s(t).

For example, consider a function s(t) and its Fourier transform S(f):

$$s(t) = [\sin(t) + \cos(3t)]\ D$$

$$\text{with } D = \exp\{ -t^2 / 10000\}$$

We included the factor D to make the integral of s(t) finite, while very slowly reducing the amplitude of s(t). In this way, D does not change significantly within one oscillation of either sin(t) or cos(3t).

$$S(f) = \int s(t) \exp\{-i\ 2\pi\ f\ t\}\ dt$$

$$S(f) = \int s(t)\ [\cos(2\pi\ f\ t) - i \sin(2\pi\ f\ t)]\ dt$$

$$S(f) = \int D \sin(t) \cos(2\pi\ f\ t)\ dt + \int D \cos(3t) \cos(2\pi\ f\ t)\ dt$$

$$- i \int D \sin(t) \sin(2\pi\ f\ t)\ dt - i \int D \cos(3t) \sin(2\pi\ f\ t)\ dt$$

We can make these approximations:

$$\int D \sin(t) \cos(2\pi\ f\ t)\ dt = 0$$

$$\int D \cos(3t) \sin(2\pi\ f\ t)\ dt = 0$$

$$\int D \cos(3t) \cos(2\pi\ f\ t)\ dt = 0 \text{ unless } 2\pi\ f = 3$$

$$\int D \sin(t) \sin(2\pi\ f\ t)\ dt = 0 \text{ unless } 2\pi\ f = 1$$

For $2\pi\ f = 1$:

$$\int D \sin(t) \sin(2\pi\ f\ t)\ dt = \int D \sin^2(t)\ dt = \int (1/2) \exp\{-t^2 / 10000\}\ dt = \sqrt{(\pi / 2)}$$

For $2\pi\ f = 3$:

$$\int D \cos(3t) \cos(2\pi\ f\ t)\ dt = \int D \cos^2(3t)\ dt = \int (1/2) \exp\{-t^2 / 10000\}\ dt = \sqrt{(\pi / 2)}$$

To this approximation, S(f) is nonzero at only two values of f, because only two sinusoidal functions need be summed to reproduce s(t).

Since we know how to solve linear differential equations for sinusoidal forces, we know how to solve them for any series, even an infinite one, of sinusoids. Thus, with Fourier series and transforms, we can solve linear differential equations for any realistic function.

4A§19.7 Fourier Transform of a Gaussian

Now let's find the Fourier representation of a Gaussian distribution. Gaussians are very important because many natural phenomena follow such distributions. The equation for a Gaussian distribution, G(x), with mean zero and standard deviation σ is:

$$G(x) = \exp\{ -x^2 / 2\sigma^2 \} / \sqrt{(2\pi\ \sigma^2)}$$

The Fourier transform of function G(x) is:

$$S(k) = \int_{-\infty}^{+\infty} \exp\{ -x^2 / 2\sigma^2 - ik\ x\}\ dx / (2\pi\ \sigma)$$

This is not the prettiest exponent, but it is integrable with a neat trick from Chapter 13 called *completing the square*. We can make the exponent of the integrand a perfect square by adding the right constant A to the exponent of G(x).

$$-(x / \sigma\sqrt{2} + A)^2 = -x^2 / 2\sigma^2 - 2x\ A / \sigma\sqrt{2} - A^2$$

To complete the square, choose A such that:

$$2x\ A / \sigma\sqrt{2} = ik\ x$$

$$A = ik\ \sigma / \sqrt{2}$$

We can then rewrite the exponent of the integrand as:

$$-x^2 / 2\sigma^2 - ik\ x = -(x / \sigma + ik\ \sigma)^2 / 2 - k^2\ \sigma^2 / 2$$

The exponential then becomes:

$$\exp\{-(x/\sigma + ik\sigma)^2/2\}\exp\{-k^2\sigma^2/2\}$$

To reduce clutter, define $u = x/\sigma + ik\sigma$, which means $du = dx/\sigma$. The integral then becomes:

$$S(k) = \exp\{-k^2\sigma^2/2\} \int \exp\{-u^2/2\}\, \sigma\, du / (2\pi\sigma)$$

$$S(k) = \exp\{-k^2\sigma^2/2\} \sqrt{(2\pi)} / (2\pi)$$

$$S(k) = \exp\{-k^2\sigma^2/2\} / \sqrt{(2\pi)}$$

We see that S(k), the Fourier transform of the Gaussian G(x), is also a Gaussian. Also note that the standard deviation of G(x) is σ, while the standard deviation of S(k) is $1/\sigma$.

For any Gaussian distribution, about 50% of the population is contained within $1/\sqrt{2}$ standard deviations of the mean. In the context of a wave packet, we can view $1/\sqrt{2}$ standard deviations as being the uncertainties Δx and Δk in the values of x and k, respectively.

The product of these two uncertainties is:

$$\Delta x\, \Delta k = \{\sigma/\sqrt{2}\}\{(1/\sigma)/\sqrt{2} = 1/2$$

This analysis proves that reducing Δx increases Δk, and vice versa, demonstrating the unavoidable tradeoff between the uncertainty of a wave packet's location and the uncertainty of its wave number. Since quantum mechanics equates momentum p with $\hbar k$, we have proven:

$$\Delta x\, \Delta p = \hbar/2, \text{ for Gaussian distributions}$$

which is the Heisenberg Uncertainty principle of quantum mechanics.

4A§19.8 Green's Function

Let's now examine the Fourier transform of a strange but particularly useful function. The *Dirac delta function* $\delta(x)$ equals zero for all x except $x = 0$, but is, loosely speaking, infinite at $x = 0$. To be more precise, for any function f(x):

$$\int_{-\infty}^{+\infty} \delta(x)\, f(x)\, dx = f(0)$$

The Dirac delta function effectively picks out the single value of f at $x = 0$. The Dirac delta function can represent an instantaneous force: a sudden impact that transfers energy and momentum but whose duration is infinitesimal. This is an idealization that is never fully realized in practice, but is very useful mathematically and conceptually.

The Fourier transform of the Dirac delta function is simple:

$$S(f) = \int_{-\infty}^{+\infty} \delta(t) \exp\{-i\, 2\pi f t\}\, dt$$

$$S(f) = e^{-0} = 1$$

This means the Fourier transform of the Dirac delta function includes all frequencies equally. Since we can solve linear differential equations for sinusoids, we can solve them for an instantaneous impulsive force, as represented by the Dirac delta function.

We can now take that another step forward. As sketched in Figure 19-3, any function F(t) can be broken down into a series of thin rectangles, each having width dt. The rectangle at time t has height F(t).

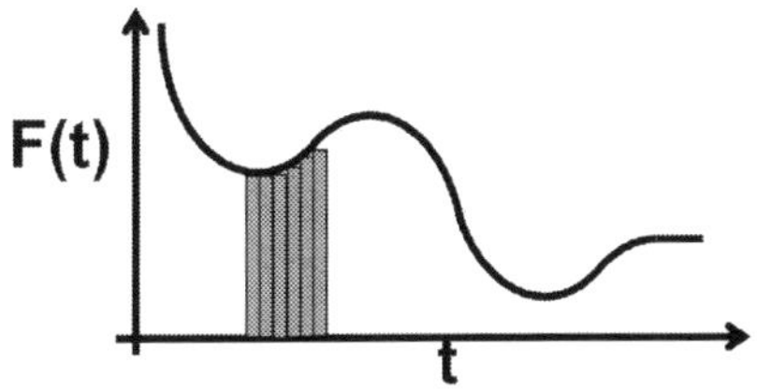

Figure 19-3 Function F(t) Decomposed Into Rectangles

In the limit that dt goes to zero, the rectangles become a series of instantaneous impulses that we now know we can solve. This approach is called the *Green's function method*, named after George Green (1793–1841).

4A§19.9 Spherical Harmonics

Spherical harmonics occur frequently in the analysis of both atomic scale and cosmic scale phenomena when using 3-D spherical coordinates. Spherical harmonics are typically written:

$$Y_{j,m}(\theta, \phi)$$

Here, θ is the polar angle, and ϕ is the azimuthal angle (see Figure 2-6). In addition, j is the total angular momentum of this distribution, and m is the component of j in the $\theta = 0$ direction.

Another common representation of spherical harmonics uses the *associated Legendre functions* $P_j^m(\cos\theta)$:

$$P_j^m(\cos\theta) \exp\{im\,\phi\} = Y_{j,m}(\theta, \phi)$$

It is helpful to know both representations, since both are often referenced and tabulated in many books and online.

The spherical harmonics $Y_{j,m}(\theta,\phi)$ are mutually orthogonal and form a complete set. This means any distribution on the surface of a sphere can be represented by a linear sum of spherical harmonics.

For example, let Q be a function of angles θ and ϕ, and express Q as a linear superposition of spherical harmonics. This is:

$$Q(\theta, \phi) = \Sigma_{jm} A_{jm} Y_{j,m}(\theta, \phi)$$

Now multiply this equation by $Y^*_{J,M}(\theta, \phi)$, the complex conjugate of any selected harmonic $Y_{J,M}(\theta, \phi)$, and integrate over all values of the angles ($\theta = 0$ to π; $\phi = 0$ to 2π).

$$\int Q(\theta, \phi)\, Y^*_{J,M}(\theta, \phi)\, d\theta\, d\phi = \int \Sigma_{jm} A_{jm} Y_{j,m}(\theta, \phi)\, Y^*_{J,M}(\theta, \phi)\, d\theta\, d\phi$$

Since the Y's are mutually orthogonal, each j m term in the summation vanishes when integrated, except the one term in which $j = J$ and $m = M$. Therefore, we have:

$$A_{JM} = \int Q(\theta, \phi)\, Y^*_{J,M}(\theta, \phi)\, d\theta\, d\phi \,/ \int Y_{J,M}(\theta, \phi)\, Y^*_{J,M}(\theta, \phi)\, d\theta\, d\phi$$

Repeating this process for all required values of J and M yields the complete set of A's needed to represent Q. One can evaluate the quality of fit, and determine how many spherical harmonics are required, in the same manner that we described above for evaluating a Fourier series fit.

The $Y_{j,m}$ functions for j = 0, 1, and 2 are tabulated here.

	P	j	m	$Y_{j,m}(\theta, \phi)$
s	+	0	0	1
p	–	1	+1	$-\sin\theta\, e^{+i\phi} / \sqrt{2}$
			0	$\cos\theta$
			–1	$+\sin\theta\, e^{-i\phi} / \sqrt{2}$
d	+	2	+2	$+\sin^2\theta\, e^{+2i\phi} \sqrt{(3/8)}$
			+1	$+\sin\theta \cos\theta\, e^{+i\phi} \sqrt{(3/2)}$
			0	$(3\cos^2\theta - 1) / 2$
			–1	$-\sin\theta \cos\theta\, e^{-i\phi} \sqrt{(3/2)}$
			–2	$-\sin^2\theta\, e^{+2i\phi} \sqrt{(3/8)}$

The unlabeled left-most column above shows the historical symbol assigned to each value of j. These symbols originated well before anyone understood the underlying physics. While somewhat archaic, they remain in frequent use. The column labeled P is the *parity* of Y, which equals $(-1)^j$. Upon mirror reflection, an even or positive parity harmonic remains unchanged, while an odd or negative parity harmonic flips polarity. The above $Y_{j,m}(\theta, \phi)$ are not normalized. This is adequate for most purposes, because spherical harmonics are often combined with functions of radius and time, with normalization occurring thereafter.

Chapter 20
Advanced Data Analysis

4A§20.1 Is it Really an Elephant?

A notable physicist once said: "With 8 parameters, I can fit an elephant." While exaggerated for dramatic effect, the message is: with enough data manipulation, almost any conclusion is possible.

Data manipulation is a case of less is better than more.

Clever data analysis drives science forward, but it is a hazardous road. It is best to exercise great care and remain skeptical.

An interesting example is a 2013 paper on exoplanets. The paper admits that, among 42,000 Sun-like stars, they actually detected ZERO Earth-like planets in Earth-like orbits. Yet, after extensive analysis, they claim to prove that 12% of Sun-like stars have an Earth-like planet in the habitable zone. Is it really possible to prove that more than 5000 Earths exist where none are found? To evaluate this claim, you as a scientist must understand the powers and pitfalls of many different data analysis techniques.

When doing your own data analysis, be the devil's advocate. Ask yourself: How could my analysis be wrong? Do not stop asking until you have solid answers to refute every imaginable objection. Be equally zealous in questioning results that you "know" are true as you would results that you find disturbing. It is best to be the first to ask penetrating questions — you certainly will not be the last interrogator, and not all of those will be friendly.

In this chapter, we explore some cutting-edge data analysis tools, and how to avoid leaving your blood on their sharp edges.

4A§20.2 Monte Carlo Methods

Many very complicated problems in physics are solved using *Monte Carlo* methods. The most common application is estimating the detection efficiency of experimental apparatus.

Monte Carlo calculations have similarities to numerical integration, but are much more versatile.

Let's start with a simple example: finding the area A of the gray region in Figure 20-1. This is possible for any A on two conditions: (1) for any x and y, we must be able to determine whether or not the point (x, y) is within A; and (2) A must be contained within known limits.

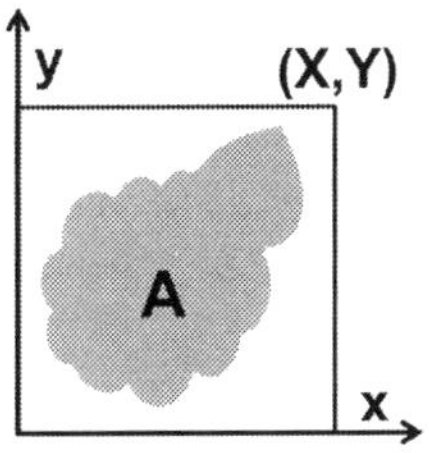

Figure 20-1 Finding Area A

In this case, the known limits of A are 0 to X horizontally and 0 to Y vertically.

The Monte Carlo procedure seems incredibly simple: randomly select N points within the known limits, and determine the number n that fall within A. The area of A is then:

$$A = (n / N) XY$$

Here, XY is the total area within the known limits, and n / N is the fraction of selected points within A. This is equivalent to randomly throwing darts at the total area XY and counting the fraction that hit A.

What could be simpler?

4A§20.3 A Real Monte Carlo Example

Now consider a more challenging real-world situation: radiation cancer therapy. A beam of electrons with a known energy distribution and cross-section enters a patient whose life is threatened by a malignant tumor. The beam enters the patient at a known location and angle. These electrons lose energy and scatter as they pass through tissues of various densities and shapes. While each individual scattering event is random, physicists have carefully measured the probability distributions of scattering angles and energy loss.

We need to calculate how much energy is deposited in the tumor, and how much in the patient's vital organs. While the challenge varies with the type of tumor and type of healthy tissue, the general rule is: if the radiation dose is 2% less than optimal there is a 50% chance of not killing the tumor, which thereby kills the patient indirectly; if the dose is 2% more than optimal there is a 50% chance of the treatment killing the patient directly. Our calculations must be very precise.

With extreme care, such precision is possible, but only with Monte Carlo methods and extensive particle physics measurements.

Electrons lose energy and also scatter as they penetrate matter. Most of their energy is deposited where they come to a stop. Both energy loss and scattering depend on the amount and type of matter traversed. For most human tissue, energy loss and scattering are proportional to distance traveled multiplied by the density of matter; this product has units of grams/cm². Energy loss and scattering rates increase in matter with a high atomic number, such as bones rich in calcium.

An electron's probability of not interacting within range R is given by:

$$p(R) = \exp\{-R/\lambda\}$$

Here, λ is the mean free path that varies with electron energy and material type. Both R and λ are measured in grams/cm². If one randomly picks a probability p between 0 and 1, the range R with the correct distribution is given by:

$$R = -\lambda \ln\{p\}$$

Deriving an optimal treatment plan begins with a CT-scan that produces a high-resolution cross-sectional map of the mass density within the patient. This map is typically hundreds of pixels wide by hundreds of pixels tall. An example is shown in Figure 20-2.

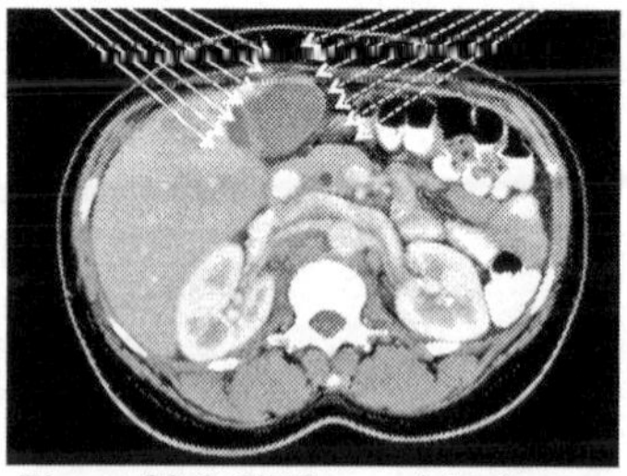

Figure 20-2 CT Scan of Abdomen

The white arrows indicate two possible beam orientations aimed at a tumor in the patient's liver. We wish to maximize the radiation delivered to the tumor while minimizing radiation delivered to vital organs, such as the spinal cord. Also seen in the image are both kidneys, the descending aorta, muscles of the back, intestines partially filled with fluid, six ribs in cross-section, and an ample layer of fat under the skin.

We seek to calculate a dose map of the energy deposited in each pixel from an exposure to N electrons with average energy E and entry angle θ. The key steps in such calculations are:

(1) Set number of simulated electrons n = 0. Zero the dose map.

(2) Add 1 to n. For electron #n, randomly select energy E, position within beam profile, and angle according to their distributions. Project electron #n onto the patient's skin.

(3) Randomly select interaction depth R.

(4) Project electron #n forward 1 pixel. If electron has exited patient, go to (8). Set ΔR = mass density × pixel width. Calculate energy loss ΔE in ΔR.

(5) Add ΔE to dose map; subtract ΔE from E. If E <= 0 go to (8).

(6) Subtract ΔR from R; if R > 0 go to (4)

(7) Randomly scatter electron #n: Compute new electron direction and scattering energy loss ΔE. Add ΔE to dose map; subtract ΔE from E. If $E > 0$ go to (3).

(8) Electron #n completed. If $n < N$ go to (1).

(9) Dose map is complete.

Having calculated the dose map for one beam energy and angle, we repeat this procedure to obtain dose maps for several other beam conditions. We then search for the optimum treatment plan: the combination of exposure intensity at each beam condition that maximizes tumor dose and minimizes vital organ dose. We will discuss search procedures shortly.

Clearly, this application of physical principles and techniques is quite different from typical particle physics research. But both employ the same analysis methods that must overcome the same pitfalls.

This is a good time to ask: How do we estimate the precision of a Monte Carlo analysis? When is N, the number of simulated electrons, large enough?

Most likely, K, the number of electrons passing through any particular pixel, has a Gaussian distribution (most things do). For rare events, one standard deviation of K equals $\sqrt{K}$. But, since Monte Carlo calculations are often Big Black Boxes, it may be worth verifying that the results truly do have a Gaussian distribution.

One simple test is comparing separate trials. Imagine that we decide to calculate a dose map with N = 100,000. We can pause after every 10,000 simulated electrons, and save the dose map for that group. When we have ten maps, we can average them to create a master map, and then plot the deviations of each of the ten maps from the master. If those deviations have a Gaussian distribution, we can calculate the rms (root mean square) differences between the ten maps and the master map. The estimated standard deviation of the master dose map is that rms divided by $\sqrt{9}$. It's 9 because we have 10 maps, but removed one degree of freedom in calculating the master. If the deviations are not Gaussian, there may be a systematic effect that should be identified.

You can then decide if the benefit of halving the standard deviation justifies the cost of quadrupling N.

In oncology, we want to drive the dose uncertainties well below the ±2% figure quoted earlier.

In a physics experiment, your final result might be n, the number of detected particles, divided by ε, the Monte Carlo-computed detection efficiency. The uncertainty in your final result will be the square root of the sum of the squares of the fractional uncertainties in n and ε. Since physics experiments are typically much more expensive than computer simulations, you will want the fractional uncertainty in ε to be well below the statistical fractional uncertainty in n.

One potential pitfall of Monte Carlo methods is human: unjustified confidence. Because computers do arithmetic perfectly, it is easy to think that computers are always correct. Never assume that. Remember the programmer's adage: GIGO — Garbage In, Garbage Out.

Another potential pitfall of Monte Carlo methods is their reliance on computer-generated random trial values of key variables. Existing computers typically provide sequences of random numbers between 0 and 1. The problem is computer-generated random numbers are only *pseudo-random*.

True random numbers are *independent* samples from a *fixed* uniform distribution, with an equal probability of having any value between 0 and 1. Pseudo-random numbers have the proper average distribution, but they are not entirely independent: they may be *correlated* with one another in various ways.

The essence of current computer technology is absolute determinism: computers invariably and precisely execute specified instructions. This makes it impossible for any existing computer to produce a truly random result. That capability may come with quantum computers.

In some applications, each trial, such as simulating a coin flip, might require only one random number, so correlations might not be a problem. But most Monte Carlo applications use many random numbers to simulate each trial. In the oncology example, we might need more than 10 random numbers to track a single electron through a patient's body. In such cases, random number correlations can lead to serious biases.

Consider a particle physics example: a Monte Carlo simulation of the decay:

$$K^+ \to \pi^+ + \pi^+ + \pi^-$$

Here, a kaon decays into three charged pions. In the kaon's rest frame, the three pions share 75 MeV of kinetic energy. This means there are two degrees of freedom for how that energy is distributed: the kinetic energy of pion #1, and the kinetic energy of pion #2. Having selected those, the kinetic energy of pion #3 is fixed. There are additional degrees of freedom associated with angular orientations. We can examine the distribution of pion momenta using a *Dalitz plot*.

The upper portion of Figure 20-3 is a Dalitz plot of real experimental measurements reported by Brian Lindquist. Each dot represents one decay. We see that the dots are distributed uniformly, as they should be.

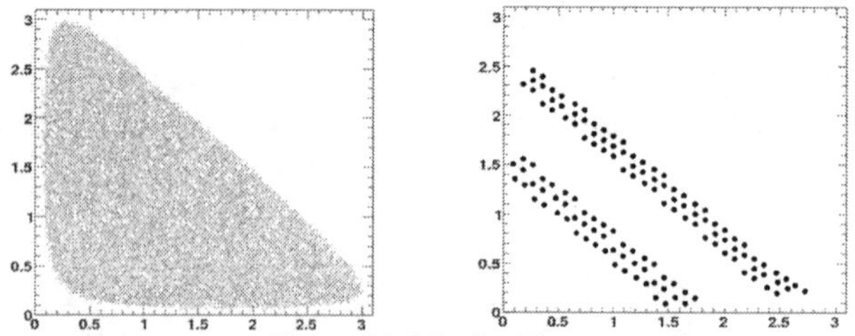

Figure 20-3 Dalitz Plots

The lower image is a possible Dalitz plot of a Monte Carlo simulation of this decay, where pseudo-random number correlations are obvious. Pairs of consecutive computer-generated random numbers have related values — they are not independent. The "random" pairs do not fill the space uniformly, causing subtle errors in the simulation.

Without explicitly looking for this effect, you might well never notice this systematic error. This actually happened in my thesis experiment, but fortunately a wise colleague found it early on. His solution was to get a first pseudo-random number and use it to determine how many of the following pseudo-random numbers to discard before actually using the next one. That brute force fix eliminated any detectable correlations.

Monte Carlo analyses typically require an enormous amount of input data that computers accept without question. If the input data has an error, the output will be wrong. If your programming instructions have an error, the output will be wrong. If you forget an important physical effect, the output will be wrong.

Monte Carlo analysis is a powerful tool that must be used with great skill and scrutiny.

4A§20.4 Searching for Optima

The oncology example leads us to another analysis challenge: searching for optima. In cancer therapy, we seek the greatest tumor-to-vital-organ dose ratio. In calculating electron distributions in complex molecules, we seek the lowest energy levels. In designing physics experiments, we use Monte Carlo simulations to achieve the greatest detection efficiency for a fixed budget.

In these cases and many more, if there are N adjustable parameters (variables), we must search an N-dimensional space. This can become enormously expensive as N increases.

Let's consider a simple example: maximizing a function f within some range of two variables x_1 and x_2. (If you need to minimize a function g, just maximize $f = -g$.)

Figure 20-4 shows an example, with the largest values of f shown in the darkest shades of gray. Such plots are often called *landscapes*, and the optimization procedure is call *hill climbing*.

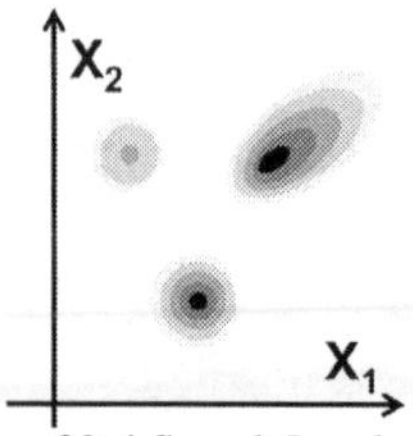

Figure 20-4 Search Landscape

Before discussing how to find the maximum, examine three challenges that are illustrated in the figure: (1) *local maxima*; (2) *plateaus*; and (3) *skewed ridges*.

The modest hill in the upper left is a local maximum but not a global maximum. Its peak is a better solution than nearby alternatives, but it is not the best solution in the landscape.

A second challenge is the plateau: most of the landscape is flat. At any randomly selected point, neighboring points have a very similar f. This means there is no clear guidance as to where to search next.

The third challenge is in the upper right: a ridge that is much wider in one direction than in another. The problem this ridge poses is that it is skewed, extended in a direction quite different from either axis. As hill climbing algorithms typically step along one axis at a time, it will take a great many small steps to zigzag back and forth along this ridge.

Let's discover how these algorithms typically work.

Hill climbing algorithms start at one set of values of each variable. They cycle through the variables, and sequentially evaluate f at larger and smaller values of each variable. Whenever they find a larger value of f, they restart their search at that set of variable values. The most basic algorithm is:

(1) Choose an initial step size Δ_j for each variable X_j, j = 1 to N.
Step sizes are typically some percentage of X_j's allowed range.
Set each X_j to a starting value that may be random, or may be the center of X_j's range.
Calculate $F = f(X_1, \ldots, X_N)$ and set MAX = F.
Set END to whatever number of search loops you choose, and set LOOP = 0.

(2) Set $Y_j = X_j$ for all j, and set J = 1.

(3) Add Δ_J to Y_J. Calculate $F = f(Y_1, \ldots, Y_N)$.
If F > MAX, go to (3), else subtract Δ_J from Y_J, and go to (4).

(4) Set MAX = F. Set $X_J = Y_J$. Go to (2).

(5) Subtract Δ_J from Y_J. Calculate $F = f(Y_1, \ldots, Y_N)$.
If F > MAX, go to (5), else add Δ_J to Y_J, and go to (6).

(6) Set MAX = F. Set $X_J = Y_J$. Go to (4).

(7) Add 1 to J, if J <= N go to (2).

(8) Divide all Δ_j's by 4. Add 1 to LOOP. If LOOP < END go to (1)

(9) Search complete. Maximum f equals MAX at coordinates X_j.

By reducing the step sizes in (8), the search zooms in on the maximum.

There are countless refinements that can be made to the above algorithm. However, no search algorithm is guaranteed to find the global maximum in a finite number of loops.

To avoid the local maxima and plateau problems, some search procedures begin by calculating f at an array of trial points that are equally spaced along each of the N axes. Hill climbing is then launched at the best trial value.

Another alternative is periodically evaluating f at a randomly chosen point in N-space. If that point is better, the search is restarted there.

Some algorithms attempt to minimize skewed ridge problems by calculating f at all neighboring points in N-space, and then moving to the best of those, simultaneously changing as many variables as required.

The performance of a computer search can be greatly improved by employing physical intuition. Often, some variables are linked by natural phenomena, or have special symmetry properties. The variables $(X_1 + X_2)$ and $(X_1 - X_2)$ might be better search variables than X_1 and X_2.

Also, number the variables with care. Variables with the greatest expected impact on function f should have the smallest indices so that they get searched first. This reduces wasted steps in exploring changes in less important variables.

Many sophisticated search algorithms are described online; you might start looking by googling *function minimization*.

As with any complex tool, you will be more successful using a Search Black Box if you have an understanding of how it works and how it can fail.

4A§20.5 Edge Degradation

Every scientific instrument has imperfections, and most physics experiments involve a variety of instruments, with a variety of imperfections that can combine in intricate ways to bias the final measurement. Such biases, called *systematic errors*, are quite different from the unavoidable *statistical errors* resulting from measuring stochastic variables with a finite number of data points.

We discussed statistical errors in Chapter 9. Let's now discuss systematic errors: how to identify them, and how to properly correct your data.

Let's examine a simple systematic error that often arises due to instrumental imprecision. Assume we are interested in a natural phenomenon with a sharp edge, a rapid change in a small distance. Now imagine trying to quantify that behavior using an instrument that measures distance imprecisely.

To be specific: let the true values Y(x) of some physical entity be constant from x = 1 to x = 5; and let Y drop precipitously to zero for x > 5, as shown by the gray rectangles in Figure 20-5.

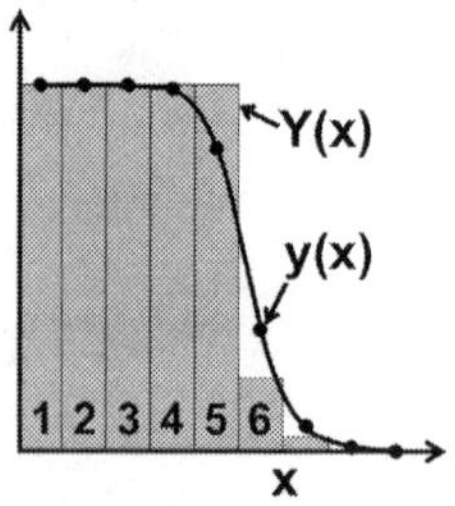

Figure 20-5 Degraded Edge

Because of instrumental uncertainty in our measurement of x, we actually measure y(x), the black dots connected by the black curve. Due to mismeasured x-values, the measured y values decline more gradually than the true Y values.

In this example, we assume: 58% of measured x values fall into the correct bin; 40% are shifted by 1 bin, 20% to a lower bin and 20% to a higher bin; and 2% are shifted by 2 bins, 1% to a lower bin and 1% to a higher bin.

The true Y values and measured y values at each x value are:

x	Y	y
1	100	100
2	100	100
3	100	100
4	100	99
5	100	83
6	20	33
7	4	8
8	1	2
9	0	0

From x = 5 to x = 6, the true Y drops by a factor of 5, while the measured y drops by a factor of only 2.5.

Let me emphasize that this systematic error is an inescapable reality, not a human mistake. Since no instrument is ever perfectly precise, all our measurements are compromised to some degree.

There are at least two ways to try to rectify edge degradation. Both require a precise understanding of our instrument's *point spread function*: the distribution of values that our instrument would measure if the true input values were all the same. This can be determined, for example, by measuring something known to have a perfectly sharp edge.

A logically simple approach to rectifying edge degradation employs search optimization, as described in the prior section. For some assumed true values Z(x), we can calculate z(x), what our instrument would measure due to its point spread function. If z(x) exactly matched y(x), what we actually measured, then the assumed Z(x) would exactly equal the true Y(x). In practice, z(x) and y(x) will differ. We then calculate a χ^2 for how well they match:

$$\chi^2 = \Sigma [\, z(x) - y(x) \,]^2 / [\, \sigma_z^2(x) + \sigma_y^2(x) \,]$$

Here, $\sigma_z^2(x)$ and $\sigma_y^2(x)$ are the variances of z(x) and y(x) respectively. We then search for the Z's that minimize χ^2, and therefore comes closest to the true Y(x).

A more intellectually sophisticated approach to rectifying edge degradation employs Fourier transforms. It can be proven that S(y), the Fourier transform of the measured data y(x), equals the product of S(Y), the Fourier transform of the true values Y(x), multiplied by S(psf), the Fourier transform of the point spread function. We write this:

$$S(y) = S(Y)\, S(psf)$$

The true values Y(x) are therefore equal to the inverse Fourier transform of S(y) / S(psf). We write this:

$$Y = S^{-1} \{ S(y) / S(psf) \}$$

In any real experiment, there will be many measurement imprecision issues, each with an effect like the point spread function discussed above. Monte Carlo methods are the most effective way to calculate and compensate for the impact of these effects on your measured data.

Appendix 1
Trigonometric Identities

For any triangle with sides A, B, and C, each with opposite angles a, b, and c:

$$a + b + c = \pi \text{ radians} = 180 \text{ degrees}$$

Law of Sines: $A / \sin(a) = B / \sin(b) = C / \sin(c)$

Law of Cosines: $A^2 = B^2 + C^2 - 2\,B\,C \cos(a)$

All the following identities are valid for any angles A, B, x, and y:

$$\cos^2A + \sin^2A = 1$$

$$\sin(A + B) = \sin A \cos B + \cos A \sin B$$
$$\cos(A + B) = \cos A \cos B - \sin A \sin B$$

$$\sin(2A) = 2 \sin A \cos A$$
$$\cos(2A) = \cos^2A - \sin^2A$$
$$\cos(2A) = 1 - 2\sin^2A = 2 \cos^2A - 1$$

$$\sin^2(A/2) = (1 - \cos A)/2$$
$$\cos^2(A/2) = (1 + \cos A)/2$$

$$\sin A + \sin B = 2 \sin[(A + B) / 2] \cos[(A - B) / 2]$$
$$\cos A + \cos B = 2 \cos[(A + B) / 2] \cos[(A - B) / 2]$$

$$\sin^2A - \sin^2B = \sin(A + B) \sin(A - B)$$
$$\cos^2A - \cos^2B = \sin(A + B) \sin(B - A)$$
$$\cos^2A - \sin^2B = \cos^2B - \cos^2A = \cos(A + B) \cos(B - A)$$

$$\cosh(ix) = \cos x$$
$$\sinh(ix) = i \sin x$$
$$\tanh(ix) = i \tan x$$

$$\cosh^2x - \sinh^2x = 1$$

$$\sinh x = x + x^3 / 3! + x^5 / 5! - \ldots$$
$$\cosh x = 1 + x^2 / 2! + x^4 / 4! - \ldots$$
$$\tanh x = x - x^3 / 3 + 2x^5 / 15 - \ldots$$

$$\sinh(x \pm y) = \sinh x \cosh y \pm \sinh y \cosh x$$
$$\cosh(x \pm y) = \cosh x \cosh y \pm \sinh x \sinh y$$

$$2 \sinh x \sinh y = \cosh(x + y) - \cosh(x - y)$$
$$2 \cosh x \cosh y = \cosh(x + y) + \cosh(x - y)$$
$$2 \sinh x \cosh y = \sinh(x + y) + \sinh(x - y)$$

$$\sinh x + \sinh y = 2 \sinh[(x + y) / 2] \cosh[(x - y) / 2]$$
$$\sinh x - \sinh y = 2 \sinh[(x - y) / 2] \cosh[(x + y) / 2]$$
$$\cosh x + \cosh y = 2 \cosh[(x + y) / 2] \cosh[(x - y) / 2]$$
$$\cosh x - \cosh y = 2 \sinh[(x + y) / 2] \sinh[(x - y) / 2]$$

Appendix 2
Finite & Infinite Series

§X2.1 Infinite Series

Below are infinite series that have finite sums provided that $|x| < 1$:

$$1 / (1 - x) = 1 + x + x^2 + x^3 + \dots$$

$$1 / (1 + x) = 1 - x + x^2 - x^3 + \dots$$

$$\sqrt{(1 + x)} = 1 + x / 2 - x^2 / 8 + x^3 / 16 - \dots$$

$$\sqrt{(1 - x)} = 1 - x / 2 - x^2 / 8 - x^3 / 16 - \dots$$

$$1 / \sqrt{(1 + x)} = 1 - x / 2 + 3x^2 / 8 - 5x^3 / 16 + \dots$$

$$1 / \sqrt{(1 - x^2)} = 1 + x^2 / 2 + 3x^4 / 8 + 5x^6 / 16 + \dots$$

This tricky sum arises in quantum mechanics:

$$\begin{aligned} \Sigma_J \{J\, x^J\} &= 0 + x + 2x^2 + 3x^3 + 4x^4 + \dots \\ &= x\, [1 + x + x^2 + x^3 + \dots] \\ &\quad + x^2\, [1 + x + x^2 + x^3 + \dots] \\ &\quad + x^3\, [1 + x + x^2 + x^3 + \dots] \\ &\quad + x^4\, [1 + x + x^2 + x^3 + \dots] \\ &\quad + \text{infinitely more rows} \dots \end{aligned}$$

$$\Sigma_J \{J\, x^J\} = x / (1 - x)^2$$

§X2.2 Finite Series

For $|x| < 1$:

$$1 + x + x^2 + x^3 + \dots + x^{n-1} = 1 + x + x^2 + x^3 + \dots - x^n \{ 1 + x + x^2 + x^3 + \dots \}$$

$$1 + x + x^2 + x^3 + \dots + x^{n-1} = (1 - x^n) / (1 - x)$$

We can apply this equation to car loans, student loans, and mortgages. If one borrows a principal amount P at an uncompounded annual interest rate i, and makes n monthly payments of M each, the equation for M is:

$$M = P\, (i / 12)\, x^n / (x^n - 1) \text{ with } x = 1 + i / 12$$

§X2.3 Sums of Integer Series

The sums of integers raised to various powers are:

$$1 + 2 + 3 + 4 + \dots + n = n\, (n + 1) / 2$$

$$1^2 + 2^2 + 3^2 + \dots + n^2 = n\, (n + 1)\, (2n + 1) / 6$$

$$1^3 + 2^3 + 3^3 + \dots + n^3 = n^2\, (n + 1)^2 / 4$$

$$1^4 + 2^4 + 3^4 + \dots + n^4 = n\, (n + 1)\, (2n + 1)\, (3n^2 + 3n - 1) / 30$$

Evens: $2^2 + 4^2 + 6^2 + \dots + n^2 = n\, (n + 1)\, (n + 2) / 6$

Odds: $1^2 + 3^2 + 5^2 + \dots + n^2 = n\, (n + 1)\, (n + 2) / 6$

Some infinite sums of integer reciprocals are:

$$1 + 2^{-2} + 3^{-2} + 4^{-2} + \ldots = \pi^2 / 6$$

$$1 + 2^{-4} + 3^{-4} + 4^{-4} + \ldots = \pi^4 / 90$$

$$1 + 3^{-2} + 5^{-2} + 7^{-2} + \ldots = \pi^2 / 8$$

$$1 + 3^{-4} + 5^{-4} + 7^{-4} + \ldots = \pi^4 / 96$$

$$1 - 2^{-2} + 3^{-2} - 4^{-2} + \ldots = \pi^2 / 12$$

$$1 - 1/3 + 1/5 - 1/7 + \ldots = \pi / 4$$

§X2.4 Series for Common Functions

For $-1 < x <= +1$, the natural logarithm is:

$$\ln\{1+x\} = x - x^2 / 2 + x^3 / 3 - x^4 / 4 + x^5 / 5 - \ldots$$

For any x:

$$\exp(x) = e^x = \Sigma_n x^n/n! = 1 + x + x^2 / 2! + x^3 / 3! + \ldots$$

$$\sin(x) = x - x^3 / 3! + x^5 / 5! - \ldots$$

$$\cos(x) = 1 - x^2 / 2! + x^4 / 4! - \ldots$$

$$\tan(x) = x + x^3 / 3! + 2x^5 / 15 - \ldots$$

$$\sinh(x) = x + x^3 / 3! + x^5 / 5! - \ldots$$

$$\cosh(x) = 1 + x^2 / 2! + x^4 / 4! - \ldots$$

$$\tanh(x) = x - x^3 / 3! + 2x^5 / 15 - \ldots$$

Appendix 3
Tables of Gaussian Probability

Recall the normal Gaussian probability distribution from Section 4A§9.5:

$$\text{Prob}(x) = \exp\{ -(x-\mu)^2 / 2\sigma^2 \} / \sqrt{(2\pi\,\sigma^2)}$$

Here, Prob(x) is the probability of value x, μ is the mean or average value of x, and σ^2 is the variance of x.

For $\mu = 0$ and $\sigma = 1$, the following table lists values for Z, Prob(|x|>Z), and the odds of |x| > Z are:

Z	P(|x|>Z)	Odds of |x|>Z
0.5	0.61708	1 in 1.6204
0.7	0.48392	1 in 2.0665
1	0.31731	1 in 3.1515
2	0.04550	1 in 21.978
3	0.00270	1 in 370.40
4	6.33E–5	1 in 15787.
5	5.73E–7	1 in 1.74E+6
6	1.97E–9	1 in 5.07E+8

In the first line, P(|x|>Z) for x = 0.5 means the probability of a measurement x deviating, either positively or negatively, from the mean of a Gaussian distribution by more than 0.5 standard deviations is 61.708%, and the odds of x deviating by more than 0.5 are 1 in 1.6204 = 1 / 0.61708. The last line says |x| exceeds 6 in 1.97 of one billion data sets, and the odds that |x| exceeds 6 are 1 in 507 million.

For $\mu = 0$ and $\sigma = 1$, the following table lists Z values corresponding to various values of Prob(|x|<Z). For example, the top entry on the right side states that 80% of all measured x values are in the interval: $-1.28156 < x < +1.28156$.

P(|x|<Z)	Z	P(|x|<Z)	Z
0.80	1.28156	3 9's	3.29053
0.90	1.64485	4 9's	3.89059
0.95	1.95996	5 9's	4.41717
0.98	2.32635	6 9's	4.89164
0.99	2.57583	7 9's	5.32672
0.995	2.80703	8 9's	5.73073
0.998	3.09023	9 9's	6.10941

On the right side "5 9's" means 0.99999 — five nines following the decimal point. The bottom entry on the right side means that 99.9999999% of all measured x values are in the interval:

$$-6.10941 < x < +6.10941$$

Appendix 4
χ^2 & Degrees of Freedom

Recall the discussions and equations related to χ^2 in chapters 9 and 19. Assume two sets of N measured quantities:

$$n(j) \pm \sigma_n(j), \text{ for } j = 1 \text{ to } N$$

$$m(j) \pm \sigma_m(j), \text{ for } j = 1 \text{ to } N$$

Here, the σ's are the one standard deviation uncertainties in each quantity.

To evaluate whether or not n(j) and m(j) are samples from the same probability distribution, we compute:

$$\chi^2 = \Sigma_j \{ m(j) - n(j) \}^2 / [\sigma_m(j)^2 + \sigma_n(j)^2]$$

If both m(j) and n(j) are the numbers of rare events with purely statistical uncertainties, this equation reduces to:

$$\chi^2 = \Sigma_j \{ m(j) - n(j) \}^2 / [m(j) + n(j)]$$

If the m(j) are numbers of rare events with purely statistical uncertainties, and the n(j) are based on a theory with no statistical uncertainty, this equation reduces to:

$$\chi^2 = \Sigma_j \{ m(j) - n(j) \}^2 / m(j)$$

The *number of degrees of freedom*, #df, equals N the number of quantities being compared minus the number of adjustable parameters. If the m(j) are scaled to have the same sum as the n(j), there is one adjustable parameter. If the m(j) are measured data and the n(j) represent a 4th-order polynomial fit to the m(j), there are 5 adjustable parameters.

The following table provides guidance in interpreting χ^2 for various numbers of degrees of freedom (#df). The table lists X values versus #df vertically and P horizontally, where P is the probability that χ^2 / #df exceeds X. The table assumes χ^2 and #df are calculated as described above. It further assumes a comparison between two sets of quantities that both truly represent the same physical phenomenon. Lastly, it assumes all data fluctuations are entirely random, independent, and free of systematic errors.

#df	10%	5.0%	2.5%	1.0%	0.5%	0.1%
1	2.71	3.84	5.02	6.63	7.88	10.8
2	2.30	3.00	3.69	4.61	5.20	6.91
3	2.08	2.60	3.12	3.77	4.28	5.42
4	1.94	2.37	2.78	3.32	3.72	4.62
5	1.85	2.21	2.57	3.02	3.35	4.10
6	1.77	2.10	2.41	2.80	3.09	3.74
7	1.72	2.01	2.29	2.64	2.90	3.47
8	1.67	1.94	2.19	2.51	2.74	3.27
9	1.63	1.69	2.11	2.41	2.62	3.10
10	1.60	1.83	2.05	2.32	2.52	2.96
12	1.55	1.75	1.94	2.18	2.36	2.74
15	1.49	1.67	1.83	2.04	2.19	2.51
20	1.42	1.57	1.71	1.88	2.00	2.27
30	1.34	1.46	1.57	1.70	1.79	1.99
63	1.23	1.31	1.38	1.46	1.52	1.64
127	1.16	1.21	1.26	1.31	1.35	1.43
511	1.08	1.11	1.13	1.15	1.17	1.20

The last line of this table says: in a series of repeated data sets measuring the same phenomenon, for a fit with 511 degrees of freedom, χ^2 / 511 exceeds 1.08 in 10% of data sets, and exceeds 1.20 in 0.1% of data sets.

Appendix 5
Vector Identities, Operator & Theorems

§X5.1 Vector Identities

These identities are valid for any vectors $\boldsymbol{A}$, $\boldsymbol{B}$, and $\boldsymbol{C}$, and any scalar functions f and g.

$$f(\boldsymbol{A}+\boldsymbol{B}) = f\boldsymbol{A} + f\boldsymbol{B}$$
$$\boldsymbol{A}+\boldsymbol{B} = \boldsymbol{B}+\boldsymbol{A}$$
$$\boldsymbol{A} \bullet \boldsymbol{B} = \boldsymbol{B} \bullet \boldsymbol{A}$$
$$(\boldsymbol{A}+\boldsymbol{B}) \bullet \boldsymbol{C} = \boldsymbol{A} \bullet \boldsymbol{C} + \boldsymbol{B} \bullet \boldsymbol{C}$$
$$\boldsymbol{A} \bullet (\boldsymbol{A} \times \boldsymbol{B}) = 0$$
$$\boldsymbol{A} \bullet (\boldsymbol{B} \times \boldsymbol{C}) = \boldsymbol{B} \bullet (\boldsymbol{C} \times \boldsymbol{A}) = (\boldsymbol{A} \times \boldsymbol{B}) \bullet \boldsymbol{C}$$
$$\boldsymbol{A} \times (\boldsymbol{B} \times \boldsymbol{C}) = (\boldsymbol{A} \bullet \boldsymbol{C})\,\boldsymbol{B} - (\boldsymbol{A} \bullet \boldsymbol{B})\,\boldsymbol{C}$$
$$(\boldsymbol{A} \times \boldsymbol{B}) \times \boldsymbol{C} = (\boldsymbol{A} \bullet \boldsymbol{C})\,\boldsymbol{B} - \boldsymbol{A}\,(\boldsymbol{B} \bullet \boldsymbol{C})$$
$$(\boldsymbol{A} \times \boldsymbol{B}) \bullet (\boldsymbol{C} \times \boldsymbol{D}) = (\boldsymbol{A} \bullet \boldsymbol{C})\,(\boldsymbol{B} \bullet \boldsymbol{D}) - (\boldsymbol{B} \bullet \boldsymbol{C})\,(\boldsymbol{A} \bullet \boldsymbol{D})$$

§X5.2 The Vector Operator

In rectilinear (x, y, z) coordinates, the vector derivative operator $\check{\boldsymbol{D}}$ is:

$$\check{\boldsymbol{D}} = (\partial/\partial x,\ \partial/\partial y,\ \partial/\partial x)$$

The most common operations with $\check{\boldsymbol{D}}$ are:

The *gradient* of scalar function f: $\check{\boldsymbol{D}}\, f = (\partial f/\partial x,\ \partial f/\partial y,\ \partial f/\partial z)$

The *divergence* of vector field $\boldsymbol{A}$: $\check{\boldsymbol{D}} \bullet \boldsymbol{A} = \partial A_x/\partial x + \partial A_y/\partial y + \partial A_z/\partial z$

The *curl* of vector field $\boldsymbol{A}$: $\check{\boldsymbol{D}} \times \boldsymbol{A} = (\ \partial A_z/\partial y - \partial A_y/\partial z,\ \partial A_x/\partial z - \partial A_z/\partial x,\ \partial A_y/\partial x - \partial A_x/\partial y\ )$

The *curl of the curl* of vector field $\boldsymbol{A}$: $\check{\boldsymbol{D}} \times (\check{\boldsymbol{D}} \times \boldsymbol{A}) = \check{\boldsymbol{D}}\,(\check{\boldsymbol{D}} \bullet \boldsymbol{A}) - \check{D}^2 \boldsymbol{A}$

The *Laplacian* $\check{\boldsymbol{D}} \bullet \check{\boldsymbol{D}} = \check{D}^2$ of scalar field f: $\check{D}^2 f = \partial^2 f/\partial x^2 + \partial^2 f/\partial y^2 + \partial^2 f/\partial z^2$

The forms of these operations in non-rectilinear coordinates are given in Section §X5.5.

§X5.3 Vector Operator Theorems

In what follows, V represents a volume within a closed surface S, dV is an infinitesimal element of V, A is any vector field, f is any scalar field, da is an infinitesimal area within S, and $\boldsymbol{n}$ is the vector of unit length that is normal (perpendicular) to da. For a closed surface S, $\boldsymbol{n}$ points outward. We often write ***da*** to represent $\boldsymbol{n}$ da.

The *flux* of vector field $\boldsymbol{A}$ through surface S is given by:

$$\text{flux of } \boldsymbol{A} \text{ through S} = \int_S \boldsymbol{A} \bullet \boldsymbol{n}\, da$$

The *circulation* of a vector field C around a closed path Γ is:

$$\text{circulation around } \Gamma = \int_\Gamma \boldsymbol{C} \bullet \boldsymbol{ds}$$

Here, ***ds*** is the incremental tangent vector at each point along Γ.

Gauss's law equates the (*flux* of ***A*** through a closed surface S) to the (divergence of ***A*** within the volume V that S encloses). By "flux of A" we mean the outward component of ***A*** summed across the defined surface. If ***A*** represents the velocity of gas molecules, the flux of ***A*** through S is the rate at which gas flows outward through that surface.

$$\text{Gauss's law: } \int_S \mathbf{A} \bullet \mathbf{n}\, da = \int_V (\check{\mathbf{D}} \bullet \mathbf{A})\, dV$$

Stokes' theorem equates the *circulation* of ***A*** around a closed loop Γ to the curl of ***A*** normal to the surface that Γ encloses. By "circulation of ***A***" we mean the sum of the tangential component of ***A*** around loop Γ in the counterclockwise direction.

$$\text{Stokes theorem: } \int_\Gamma \mathbf{A} \bullet \mathbf{ds} = \int_S (\check{\mathbf{D}} \times \mathbf{A}) \bullet \mathbf{n}\, da$$

§X5.4 Vector Operator Identities

These operator identities are valid for any vectors ***A***, ***B***, and ***C***, and any scalar functions f and g.

$$\check{\mathbf{D}} \bullet (\check{\mathbf{D}} \times \mathbf{A}) = 0$$

$$\check{\mathbf{D}} \times (\check{\mathbf{D}}\, f) = 0$$

$$\check{\mathbf{D}} \bullet (f\,\mathbf{A}) = f\,\check{\mathbf{D}} \bullet \mathbf{A} + (\check{\mathbf{D}}\, f) \bullet \mathbf{A}$$

$$\check{\mathbf{D}}\,(A \bullet B) = (\mathbf{A} \bullet \check{\mathbf{D}})\,\mathbf{B} + (\mathbf{B} \bullet \check{\mathbf{D}})\,\mathbf{A} + \mathbf{A} \times (\check{\mathbf{D}} \times \mathbf{B}) + \mathbf{B} \times (\check{\mathbf{D}} \times \mathbf{A})$$

$$\check{\mathbf{D}} \bullet (\mathbf{A} \times \mathbf{B}) = (\check{\mathbf{D}} \times \mathbf{A}) \bullet \mathbf{B} - \mathbf{A} \bullet (\check{\mathbf{D}} \times \mathbf{B})$$

$$\check{\mathbf{D}} \times (f\,\mathbf{A}) = f\,\check{\mathbf{D}} \times \mathbf{A} + \check{\mathbf{D}}\, f \times \mathbf{A}$$

$$\check{\mathbf{D}} \times (\mathbf{A} \times \mathbf{B}) = \mathbf{A}(\check{\mathbf{D}} \bullet \mathbf{B}) - \mathbf{B} \bullet (\check{\mathbf{D}} \bullet \mathbf{A}) + (\mathbf{B} \bullet \check{\mathbf{D}})\,\mathbf{A} - (\mathbf{A} \bullet \check{\mathbf{D}})\,\mathbf{B}$$

$$f\,\check{D}^2 g - g\,\check{D}^2 f = \check{\mathbf{D}} \bullet (f\,\check{\mathbf{D}}\, g - g\,\check{\mathbf{D}}\, f)$$

$$\check{D}^2 (f\,g) = f\,\check{D}^2 g + 2\,(\check{\mathbf{D}}\, f) \bullet (\check{\mathbf{D}}\, g) + g\,\check{D}^2 f$$

$$\check{D}^2 (f\,\mathbf{A}) = \mathbf{A}\,\check{D}^2 f + 2\,(\check{\mathbf{D}} f \bullet \check{\mathbf{D}})\,\mathbf{A} + f\,\check{D}^2 \mathbf{A}$$

§X5.5 Vector Operator in Other Coordinates

The vector operator has different forms in non-rectilinear coordinate systems. Below, I show these in the two most common non-rectilinear systems. Due to their greater complexity, it is best to use these operators in rectilinear coordinates whenever possible.

3-D Cylindrical Coordinates

$\check{\mathbf{D}}$ f is the vector (R, Φ, Z), with:

$$R = \partial f/\partial r$$

$$\Phi\, r = \partial f/\partial \phi$$

$$Z = \partial f/\partial z$$

$$r\,\check{\mathbf{D}} \bullet \mathbf{A} = \partial(r\, A_r)/\partial r + \partial A_\phi/\partial \phi + r\, \partial A_z/\partial z$$

$\check{\mathbf{D}} \times \mathbf{A}$ is the vector (R, Φ, Z), with:

$$R\, r = \partial A_z/\partial \phi - r\, \partial A_\phi/\partial z$$

$$\Phi = \partial A_r/\partial z - \partial A_z/\partial r$$

$$Z\, r = \partial(r\, A_\phi)/\partial r - \partial A_r/\partial \phi$$

$r^2 \check{D}^2 f = r\,\partial(r\,\partial f/\partial r)/\partial r + \partial^2 f/\partial ø^2 + \partial^2 f/\partial z^2$

3-D Spherical Coordinates

$\check{\boldsymbol{D}}$ f is the vector (R, Θ, Φ), with:

$R = \partial f/\partial r$

$\Theta\, r = \partial f/\partial\theta$

$\Phi\, r \sin\theta = \partial f/\partial ø$

$r^2 \sin\theta\, \check{\boldsymbol{D}} \bullet \boldsymbol{A} = \sin\theta\, \partial(r^2 A_r)/\partial r + r\,\partial(\sin\theta\, A_\theta)/\partial\theta + r\,\partial A_ø/\partial ø$

$\check{\boldsymbol{D}} \times \boldsymbol{A}$ = is the vector (R, Θ, Φ), with:

$R\, r \sin\theta = \partial (\sin\theta\, A_ø)/\partial\theta - \partial A_\theta/\partial ø$

$\Theta\, r \sin\theta = \partial A_r/\partial ø - \sin\theta\, \partial(r\, A_ø)/\partial r$

$\Phi\, r = \partial(r\, A_\theta)/\partial r - \partial A_r/\partial\theta$

$r^2 \sin\theta\, \check{D}^2 f = \sin\theta\, \partial(r^2\, \partial f/\partial r)/\partial r + \partial(\sin\theta\, \partial f/\partial\theta) / \partial\theta + \partial^2 f/\partial ø^2$

Appendix 6
Common Derivatives

§X6.1 General Rules of Differentiation

For any functions F & G, constants a & b, and variable q:

$$da/dq = 0$$

$$d(a F + b G)/dq = a\, dF/dq + b\, dG/dq$$

$$d(F G)/dq = G\, dF/dq + F\, dG/dq$$

$$d(F G)/dq = d(G F)/dq$$

$$d(F / G)/dq = (1 / G)\, dF/dq - (F / G^2)\, dG/dq$$

$$d(F^{b})/dq = b\, F^{b-1}\, dF/dq$$

§X6.2 Derivatives of Common Functions

$$d\, x^{n} / dt = n\, x^{n-1}\, dx/dt$$

<u>Trig Functions</u>

$$d \sin(x) / dt = + \cos(x)\, dx/dt$$

$$d \cos(x) / dt = - \sin(x)\, dx/dt$$

$$d \tan(x) / dt = \cos^{-2}(x)\, dx/dt$$

<u>Exponentials</u>

$$d\, e^{x} / dt = e^{x}\, dx/dt$$

$$d \ln\{x\} / dt = (1/x)\, dx/dt$$

<u>Hyperbolic Functions</u>

$$d \sinh(x) / dt = \cosh(x)\, dx/dt$$

$$d \cosh(x) / dt = \sinh(x)\, dx/dt$$

$$d \tanh(x) / dt = \cosh^{-2}(x)\, dx/dt$$

<u>Square Roots</u>

For $\mathbf{r = \sqrt{(x^2 + a^2)}}$, with constant a:

$$dr/dx = x / r$$

$$d\, (r)^{-1} /dx = -x / r^3$$

$$d\, (r)^{-2} /dx = -2x / r^4$$

$$d\, (r)^{-3} /dx = -3x / r^5$$

$$d\, (x / r) /dx = a^2 / r^3$$

For $\boldsymbol{r}_{jk} = \boldsymbol{r}_j - \boldsymbol{r}_j$; $r_{jk} = |\boldsymbol{r}_{jk}|$; $\boldsymbol{v}_j = d\boldsymbol{r}_j/dt$:

$$d\, r_{jk}^{-1} /dt = \boldsymbol{r}_{jk} \bullet (\boldsymbol{v}_k - \boldsymbol{v}_j) / \boldsymbol{r}_{jk}^3$$

§X6.3 Differential Reciprocals

It is sometimes easier to calculate dy/dx than dx/dy. If you can calculate either, you automatically have the answer to the other, because as one might guess from the notation:

$$(dx/dy)\,(dy/dx) = 1$$

Here is an example: an equation relating wave number k and wave frequency ω, in a material with natural frequency Ω.

$$k = (1/c) \sqrt{\{\omega^2 - \Omega^2\}}$$

First calculate dk/dω:

$$dk/d\omega = (1/c)\,(1/2)\,[\omega^2 - \Omega^2\,]^{-1/2}\,(2\omega)$$

$$dk/d\omega = [1 - (\Omega / \omega)^2\,]^{-1/2} / c$$

Now calculate the group velocity dω/dk:

$$k^2 c^2 + \Omega^2 = \omega^2$$

$$2\,k\,c^2 = 2\omega\, d\omega/dk$$

$$d\omega/dk = k\,[\,c^2 / \omega\,]$$

$$d\omega/dk = (1/c) \sqrt{\{\omega^2 - \Omega^2\}}\,[\,c^2 / \omega\,]$$

$$d\omega/dk = c\,[1 - (\Omega / \omega)^2]^{+1/2}$$

Hence: (dω/dk) (dk/dω) = 1.

Appendix 7
Common Integrals

Here are some useful integrals that will solve most physics problems.

Powers

$$\int x^b \, dx = x^{(b+1)} / (b+1), \text{ for b not } -1$$

$$\int x^{-1} \, dx = \int (1/x) \, dx = \ln\{x\}$$

Trig Functions (x in radians)

$$\int \sin(x) \, dx = -\cos(x)$$

$$\int \cos(x) \, dx = +\sin(x)$$

$$\int \sin^2(x) \, dx = x/2 - \sin(2x) / 4$$

$$\int \cos^2(x) \, dx = x/2 + \sin(2x) / 4$$

$$\int \sin(x) \cos(x) \, dx = \sin^2(x) / 2$$

Exponentials

$$\int \exp\{x\} \, dx = \exp\{x\}$$

$$\int \ln\{x\} \, dx = x \ln\{x\} - x$$

Hyperbolic Functions

$$\int \sinh(x) \, dx = \cosh(x)$$

$$\int \cosh(x) \, dx = \sinh(x)$$

$$\int \tanh(x) \, dx = \ln\{ \cosh(x) \}$$

Integrals with $r = \sqrt{(x^2 + a^2)}$, with constant a

$$\int (1 / r) \, dx = \ln\{ x + r \}$$

$$\int (1 / r^3) \, dx = x / (a^2 r)$$

$$\int (x / r) \, dx = r$$

$$\int (x / r^3) \, dx = -1 / r$$

Volume Elements are required in integrals in non-rectilinear coordinate systems, as explained in 4A§14.3. Those that physicists commonly use are:

2-D polar: $r \, dr \, d\theta$

2-D rectilinear: $dx \, dy$

3-D rectilinear: $dx \, dy \, dz$

3-D cylindrical: $r \, dr \, d\phi \, dz$

3-D spherical: $r^2 \sin\theta \, dr \, d\theta \, d\phi$

4-D spacetime: $(c \, dt) \, dx \, dy \, dz$

End of ***Feynman Simplified 4A: Math for Physicists***

Feynman Simplified 4B

Everyone's Guide to the Feynman Lectures on Physics

Feynman Simplified gives mere mortals access to the fabled ***Feynman Lectures on Physics***.

Feynman Simplified 4B is an unprecedented catalog and explanation of every key principle and important equation in all of the ***Feynman Lectures***. In addition, we explore the major discoveries of physics in the subsequent half-century.

To fit all this wonderful physics in one book, ***Feynman Simplified 4B*** is concise, with brief explanations and few derivations. For those beginning their exploration of physics, I recommend building your knowledge gradually and systematically, starting with ***Feynman Simplified 1A*** or ***Feynman Simplified Part 1***. This book is for experienced physicists and students who seek quick reminders of Boltzmann's law, or where the minus sign goes in a transformation equation. Everything you need is here, in one convenient source.

Chapter 1 summarizes the most important principles and equations, with references to further discussion in other chapters.

Chapter 15 explores Feynman's best problem-solving tricks.

Finally, an extensive alphabetical index facilitates access to this book's treasures.

In this book, references to in-depth discussions found elsewhere use this format: 1B§14.2 denotes Section §14.2 of ***Feynman Simplified 1B***, while 4B§14.2 refers to the same section number in this book. References to the ***Feynman Lectures*** are denoted V2p12-3, for Volume 2, chapter 12, page 3.

To learn more about the ***Feynman Simplified*** series, to receive updates, and send us your comments, visit:

www.guidetothecosmos.com

Looking for a specific topic? Visit our website for a free downloadable index to the entire ***Feynman Simplified*** series.

If you enjoy this book please do me the great favor of rating it on Amazon.com.

4B Table of Contents

Chapter 1
Primary Principles & Equations

4B§1.1 Symmetry & Conservation

In this section, we explore the symmetries of natural laws, rather than the symmetries of individual objects. For example, a natural law is symmetric in time if the equation representing that law is unchanged by substituting –t for t. Time-symmetry means nature acts according to the same principles if time runs forward or backward.

The absence of a preferred direction in space means nature is symmetric under rotation. Similarly, the absence of a preferred velocity means the speed of light is the same in all reference frames.

1D§49.8 **Noether's theorem** relates the symmetry properties of natural laws to conservation principles. It says:

Each Symmetry Implies a Conservation Law

Below is a list of the most important symmetry properties, each with their corresponding conserved quantity.

Symmetry ←→ Conserved Quantity

Translation in Time ←→ Energy

Translation in Space ←→ Linear Momentum

Rotation ←→ Angular Momentum

Velocity ←→ Constancy of Light Speed

Quantum Phase ←→ Electric Charge

4B§3.11 Special relativity mandates that all conservation laws that are valid **globally** must also be valid **locally**, meaning they must apply to each point in space and moment in time.

4B§1.2 Major Principles

4B§1.4 **Atomic Hypothesis**: everything we see is made of atoms

4B§13.2 **Particle-Wave Duality**: every entity has both particle and wave properties

4B§13.2 **Uncertainty Principle**: limits determination of complementary variables

4B§14.2 **Conservation of Quarks and of Leptons** in all interactions

4B§14.2 **Conservation of Fermions** of each type, except in Weak interactions

4B§8.5 **Entropy Increases** in all macroscopic processes

4B§3.6 **Linear Superposition**: sums of solutions are also solutions

4A§14.4 **Principle of Least Action**: nature minimizes kinetic minus potential energy

4B§1.3 Primary Physical Quantities (Non-Relativistic)

4B§4.1 **Position**:	$\boldsymbol{r} = (x, y, z)$
4B§4.1 **Velocity**:	$\boldsymbol{v} = d\boldsymbol{r} / dt$
4B§4.1 **Acceleration**:	$\boldsymbol{a} = d\boldsymbol{v}/dt$
4B§4.1 **Inertial Mass**:	$m = \boldsymbol{F} \ / \ \boldsymbol{a}$
4B§4.1 **Linear Momentum**:	$\boldsymbol{p} = m \ \boldsymbol{v}$
4B§4.1 **Angular momentum**:	$\boldsymbol{L} = \boldsymbol{r} \times \boldsymbol{p}$
4B§4.1 **Force**:	$\boldsymbol{F} = d\boldsymbol{p}/dt$
4B§4.1 **Kinetic Energy**:	T or Ŧ = $m \ v^2 / 2$
4B§4.2 **Work**:	$W = -\boldsymbol{F} \cdot \boldsymbol{\Delta r}$
4B§4.2 **Power**:	$dW/dt = -\boldsymbol{F} \cdot \boldsymbol{v}$
4B§4.2 **Potential Energy** U:	$dU/dx = -F_x$

4B§1.4 Atoms & Matter

1A§1.6 The **Atomic Hypothesis** — the belief that everything we see is made of atoms — is the most important idea in science, according to Feynman. Atoms are comprised of three types of particles: *electrons*, *protons*, and *neutrons* (for more on the particles, see 4B§14.1). The number of protons in an atom's nucleus is its *element number*, its atomic number Z on the Periodic Table of Elements. For more on atoms, see 4B§13.7.

Many decades after the ***Feynman Lectures***, we now know that only 4.9% of all the energy in the universe is in the form of normal matter, matter comprised of atoms or the particles that make atoms. About 26% of all cosmic energy is in a form called **dark matter**, and about 69% is in a form called **dark energy**. These **dark** entities seem largely inert; they interact only gravitationally.

Only normal matter and its atoms are capable of creating vibrant structures, such as galaxies, stars, planets, trees, and people.

Four primary **States of Matter** or **Phases of Matter** exist, plus some others that are more exotic.

1A§1.9 **Solid Phase**: tightly bound matter that maintains its own shape.

1A§1.7 **Liquid Phase**: less-tightly bound matter, whose shape conforms to its container, and whose volume varies only modestly with pressure.

1A§1.8 **Gas Phase**: loosely-bound, low-density matter that expands to fill any container, and whose volume is strongly dependent on temperature and pressure.

2C§34.3 **Plasma Phase**: matter comprised of free electrons and ionized atoms. Due to its free charges, plasma responds vigorously to external fields, and can independently generate its own electromagnetic fields. Stars and interstellar gas are primarily comprised of plasma, making it the most prevalent form of normal matter.

4B§1.5 Newton's Laws of Motion

Newton's laws of motion and gravity are valid only in **inertial reference frames**, those moving at constant velocity ***v***. As velocities approach c or when gravity is extremely strong, Newton's equations require relativistic modification.

4B§4.2 **First Law**: inertia — absent forces, velocities are constant

4B§4.2 **Second Law**: $\boldsymbol{F} = m\,\boldsymbol{a}$

4B§4.2 **Third Law**: Reaction = – Action

4B§4.3 **Law of Universal Gravity**: $F = G M m / r^2$

4B§1.6 Maxwell's Equations

The **Vector Operator *Ď*** is defined throughout this book as:

$$\check{\boldsymbol{D}} = (\partial/\partial x,\ \partial/\partial y,\ \partial/\partial z)$$

I use ***Ď*** because the standard notation for this vector operator, an inverted Δ, is not supported in all modern communication technologies.

4B§11.2 Maxwell's field equations for electric field ***E***, magnetic field ***B***, charge density ϱ, current density ***j***, and constant ε_0 are:

$$\check{\boldsymbol{D}} \bullet \boldsymbol{E} = \varrho / \varepsilon_0$$

$$\check{\boldsymbol{D}} \times \boldsymbol{E} = -\,\partial \boldsymbol{B}/\partial t$$

$$\check{\boldsymbol{D}} \bullet \boldsymbol{B} = 0$$

$$c^2\, \check{\boldsymbol{D}} \times \boldsymbol{B} = \partial \boldsymbol{E}/\partial t + \boldsymbol{j} / \varepsilon_0$$

Maxwell's equations are valid in the above form only in **inertial reference frames**, those moving at constant velocity ***v***.

4B§1.7 Einstein's Relativity

Special relativity is valid only in **inertial reference frames**, those moving at constant velocity ***v***. General relativity is valid in all reference frames.

4B§9.1 **Principles of Special Relativity**:

- The speed of light, c, is the same in all reference frames.
- Absolute velocity has no physical meaning; only relative velocities are significant.

For velocity v, we define:

$$\beta = v / c$$

$$\gamma = 1 / \sqrt{(1 - \beta^2)}$$

In what follows, E is energy (kinetic plus mass), p is momentum, m_{rel} is relativistic mass, and m is rest mass.

4B§9.6 $E = m_{rel}\, c^2$

4B§9.6 $E^2 = m^2 c^4 + p^2 c^2$

In a frame moving with velocity v relative to our "stationary" frame, we observe time interval t, mass m_{rel}, and length L (along the v-direction) to be different from the corresponding t_0, m, and L_0 in our frame, according to:

4B§9.6 **Time Dilation**: $t = t_0 / \gamma$

4B§9.6 **Length Contraction**: $L = L_0 / \gamma$

4B§9.6 **Mass Increase**: $m_{rel} = m\,\gamma$

4B§9.7 The **Equivalence Principle** of general relativity states:

Uniform gravitational acceleration
is indistinguishable from
constant mechanical acceleration

4B§9.7 **Einstein's Field Equations** of general relativity are:

$$G_{\mu\sigma} = 8\pi\, T_{\mu\sigma}$$

Here, $G_{\mu\sigma}$ is the Einstein tensor that describes the curvature of spacetime, and $T_{\mu\sigma}$ is the stress-energy tensor that describes the density of mass, energy, and stress. John Wheeler said this equation states: mass and energy tell space and time how to curve, while space and time tell mass and energy how to move.

4B§1.8 Quantum Mechanics

In this section, h is Planck's constant, and p is momentum. Quantum mechanics is valid only in **inertial reference frames**, those moving at constant velocity ***v***.

4B§13.1 **Quantization** is the simple notion that many things in nature come in integer quantities. For example, in any physical entity, the number of electrons is always an integer. On a ramp, elevation is **continuous**, but on a staircase, elevation is **quantized**.

4B§13.2 **Particle-Wave Duality**: Every physical entity simultaneously has both particle and wave properties.

4B§13.2 **de Broglie Wavelength:** every particle has a wavelength λ given by:

$$\lambda = h / p$$

4B§13.2 **Heisenberg Uncertainty Equations**:

$$\Delta x \;\; \Delta p_x = \hbar / 2$$

$$\Delta y \;\; \Delta p_y = \hbar / 2$$

$$\Delta z \;\; \Delta p_z = \hbar / 2$$

$$\Delta t \;\; \Delta E = \hbar / 2$$

where = $\hbar = h / 2\pi$

4B§13.6 **Schrödinger's equation** is:

$$i\hbar\, d\psi/dt = -(\hbar^2 / 2m)\, \check{D}^2\, \psi + V\, \psi$$

Here, ψ is the wavefunction of a particle with mass m in energy potential V. This equation is valid for a particle whose velocity is non-relativistic.

Chapter 2
Physical Constants

Math: π: pi = 3.141,592,653,59 …
e: base of natural logarithm – 2.718,284,590,45 …
ø: golden ratio, ø = 1 + 1 / ø = 1.618,033,988,7 …

Light: c: speed of light = 299,792,458 m/s by definition

Gravity: G: gravitational constant = 6.6741×10^{-11} m^3 / kg-sec^2
g: acceleration near Earth's surface = 32.174 feet/sec^2 = 9.806,65 m/s^2
GM-Earth = $3.986\ 004\ 42 \times 10^{14}$ m^3 / s^2
GM-Sun = $1.327\ 124\ 40 \times 10^{20}$ m^3 / s^2

Particles: q_p: proton's charge = $1.602,176,565 \times 10^{-19}$ coulomb
r_p: radius of proton = 0.88×10^{-15} m
m_p: mass of proton = $1.672,621,9 \times 10^{-27}$ kg = 1836.152,673,9 m_e
m_e: mass of electron = $9.109,383,5 \times 10^{-31}$ kg

Electromagnetic: 1 eV (electron-volt) = $1.602,176,565 \times 10^{-19}$ joules
$e^2 = q_p^2 / 4\pi\varepsilon_0$ = 1.439,964,5 eV-nm (nanometer)
$\mu_0 = 4\pi \times 10^{-7}$ newton / $ampere^2$ by definition = $1.256,637,061 \times 10^{-6}$ newton / $ampere^2$
$\varepsilon_0 = 1 / (\mu_0 c^2)$ by definition = $8.854,1878 \times 10^{-12}$ $coulomb^2$ / newton m^2
$1 / 4\pi\varepsilon_0 = 10^{-7} c^2$, by definition = $8.987,55 \times 10^9$ newton m^2 / $coulomb^2$

Gases: N_A: Avogadro's number = 6.022×10^{23} = number of C^{12} atoms in 12 grams
k: Boltzmann's constant = $1.386,49 \times 10^{-23}$ joules per Kelvin
kT = (1 / 40) eV at 63°F (17°C)
R: molar gas constant = 8.314,46 J / mole K
σ: Stephan-Boltzmann constant = $5.670,367 \times 10^{-8}$ W / m^2 K^4

Atoms: Å: Angstrōm, 1Å = 10 nm = 10,000 μm = 10^{-10} m
h: Planck's constant = $6.626,0700 \times 10^{-34}$ joule-sec = $4.135,667,56 \times 10^{-15}$ eV-sec
ħ ("h-bar") = h / 2π = $1.054,5718 \times 10^{-34}$ joule-sec = $6.582,1195 \times 10^{-16}$ eV-sec
ħ c = 197.326,97 MeV-fermi (1 fermi = 10^{-15} m)
α: Fine Structure constant = 1 / 137.035,999,14
a_0 or r_B: Bohr radius = $\hbar^2 / m e^2$ = 0.0529 nm
R_y: Rydberg = 13.61 eV, hydrogen ground state binding energy

Chapter 3
Essential Mathematics

In the original ebook version, Chapter 3 of ***Feynman Simplified 4B*** provides a review of essential mathematics. Since this edition includes the more comprehensive ***Feynman Simplified 4A: Math for Physicists***, some redundant portions of this chapter have been replaced by references to the appropriate sections in ***4A***.

Primary Symbols & Functions of mathematics are presented in 4A§1.1.

Geometry & Trigonometry are reviewed in 4A§1.2, and Trig Identities are listed in Appendix 1.

4B§3.1 Functions

See Sections 4A§1.3 and 4A§1.4 for a more on functions, mappings, fields, and graphs.

In mathematics, functions define relationships among variables, quantities whose values change with location, time, or for other reasons. For example, temperature is a variable that changes with both location and time. We can describe how temperature T varies with location x and time t by using the function f:

$$T = f(x, t)$$

Here, T is a *dependent variable*, and x and t are called *independent variables*, also called its *arguments*. Functions may have one or more independent variables, but they must have exactly one dependent variable. T is a function of both x and t; we are free to choose the values of x and t, and those values uniquely determine the value of T.

Graphs are visual representations that can be extremely helpful in understanding the key properties of functions. Graphs typically plot a function's dependent variable vertically, and the function's independent variable horizontally.

Here, we show the graphs of two very important functions in physics: the *sine* function and the *exponential* function that we will discuss shortly. The upper graph in the image below plots the value of Y that corresponds to each value of X, as defined by the exponential function:

$$Y = A + B\, e^X$$

Here, A and B are constants, X is the independent variable and Y is the dependent variable.

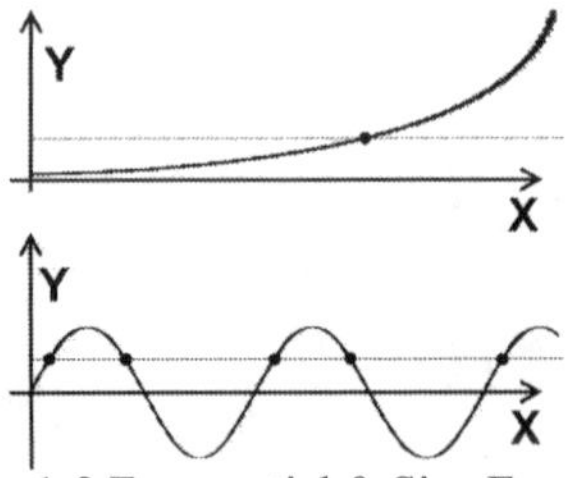

Figure 1-3 Exponential & Sine Functions

The lower graph plots the value of Y that corresponds to each value of X, as defined by the sine function:

$$Y = A \sin(X)$$

In the upper graph, there is only one black dot along the dotted line. In fact, for any Y value, there is *only one* value of X for which $Y = A + B\, e^X$. This means exponentials have the special property of having an *inverse function*. If a function f(x) has an inverse function g, we can write:

$$\text{if } y = f(x)$$

$$\text{then } g(y) = g(\, f(x)\,) = x$$

Conversely, in the lower graph, the 5 black dots along the dotted horizontal line indicate 5 values of X for which sin(X) has the same value of Y. This means Y = A sin(X) does not have an inverse function throughout the entire range of X values.

4B§3.2 Trig Functions

See Sections 4A§1.2 and 4A§1.5 for a more in-depth discussion.

The most basic trigonometric functions and identities describe the properties of right triangles.

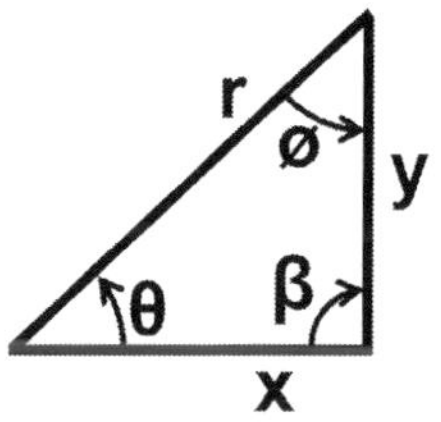

Figure 3-1 Right Triangle

In Figure 3-1, β is the *right angle*, meaning it equals 90 degrees or π / 2 radians. Since the interior angles of all triangles within planar surfaces sum to π radians (180 degrees), this means:

$$\theta + \varnothing = \beta = \pi / 2$$

The three primary trig functions are listed below, with their English names, mathematical notations, and defining equations.

sine: $\sin(\theta) = y / r$

cosine: $\cos(\theta) = x / r$

tangent: $\tan(\theta) = y / x$

4B§3.3 Exponentials & Logarithms

See Section 4A§.7 and Appendix 1 for further discussion and lists of useful equations.

The **Exponential Function** of x is:

$$\exp\{ x \} = e^x = \Sigma_n x^n / n!$$

While e^x is the standard notation, I use exp{x} because popular modern communication technologies are limited to one level each of superscript or subscript and exponents in physics equations can sometimes be quite elaborate.

Logarithms of two types are employed in physics: the traditional base-10 logarithm, and the natural logarithm. The latter is much more common in physics equations (nature doesn't have ten fingers). Logarithms and exponentials are inverse operations. The two logarithms are defined by:

Natural Logarithm: $\ln(\exp\{x\}) = x$

Base-10 Logarithm: $\log(10^x) = x$

The **Hyperbolic Trigonometric Functions** most commonly used in physics are:

hyperbolic sine: $\sinh(x) = (\exp\{+x\} - \exp\{-x\}) / 2$

hyperbolic cosine: $\cosh(x) = (\exp\{+x\} + \exp\{-x\}) / 2$

hyperbolic tangent: $\tanh(x) = \sinh(x) / \cosh(x)$

4B§3.4 Calculus

See Sections 4A, Chapters 11, 12, and 13, for more comprehensive discussions.

Calculus defines two principle operations that are complimentary: differentiation and integration. Derivatives provide the slope of a function, while integrals provide the area under its curve, as illustrated in Figure 3-2.

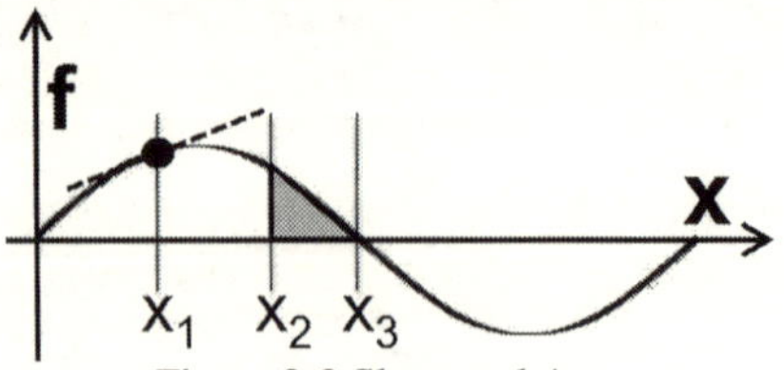

Figure 3-2 Slope and Area

Here, the **derivative** of f(x) at $x = x_1$ is the slope of f(x) at x_1, as indicated by the dashed line. The gray area under the curve f(x) between x_2 and x_3 is the **definite integral** of f(x) between those points.

1A§6.3 **Common Derivatives**

Derivatives describe how rapidly one quantity changes in response to changes in another quantity. As shown above, the derivative of a function f(x) can be visualized as the slope of the curve f(x) when f is plotted versus x. The **general equation** for the derivative of any function f with respect to a variable x is:

$$df/dx = \text{limit as } dx\rightarrow 0 \text{ of } \{ f(x+dx) - f(x) \} / dx$$

The derivatives of common functions are listed in Appendx 6.

1B§22.1 **Partial Derivatives**

Consider a function f of two variables x and y:

$$f(x,y) = x^2 + xy$$

The common derivative with respect to time t is:

$$df/dt = 2x\, dx/dt + y\, dx/dt + x\, dy/dt$$

We need all these terms because both x and y may change over time. But when we want the change in f when y is constant and only x changes, we need the partial derivative with respect to x:

$$\partial f/\partial x = 2x + y$$

When x is constant and y changes, the partial derivative with respect to y is:

$$\partial f/\partial y = x$$

For infinitesimal changes, the change in f for changes in both x and y is:

$$\Delta f = \Delta x\, (\partial f/dx) + \Delta y\, (\partial f/\partial y)$$

Sometimes, Feynman and other authors add subscripts to emphasize the variables that are held constant in each partial derivative, such as:

$$df = dx\, (\partial f/dx)_y + dy\, (\partial f/\partial y)_x$$

1A§6.4 **Common Indefinite Integrals**

The integral of function F(x) can be visualized as the area under the curve F(x) when F is plotted versus x. Unfortunately, there is no general equation for integrals. Note that:

$$\text{if } G = dF/dx, \text{ then } \int G\, dx = F + b$$

This equation highlights two principles of integration. Firstly, integration and differentiation are inverse processes. Secondly, any arbitrary constant b may be added to the result of an indefinite integral. This is because adding any constant to F does not change the value of its derivative G. Generally the integration constant is chosen to match some boundary condition.

The integrals of common functions are listed in Appendx 7.

1A§6.4 **Definite Integrals**

Definite integrals are integrals within specified limits of the independent variable. For example:

$$\int_a^b x^2\, dx = (1/3)\, x^3 |_a^b = (b^3 - a^3) / 3$$

Here, the specified limits are x = a and x = b. To evaluate a definite integral, we first calculate the indefinite integral, then evaluate that at both limits, and lastly subtract the result at x = a from the result at x = b. A definite integral of F(x) corresponds to the area under the curve F(x) between the specified limits. The arbitrary integration constant of the indefinite integral vanishes in the final subtraction.

2A§3.1 The **Path Integral** or **Line Integral** of function f sums the values of f along a specific path Γ from point A to point B through some multidimensional space. The path Γ need not be a straight line. The path integral is written:

$$F = \int_\Gamma f\, ds$$

The value of a path integral depends on every point along path Γ. The path may be arbitrarily complicated; it may be a *closed* loop, in which case A = B, or it may be *open*, where A and B are separated.

The integral from A to B and the integral from B to A have opposite polarities — one must pick the proper direction in each specific situation. If Γ is a closed loop, convention specifies integrating around Γ in a counterclockwise direction, as specified by the right hand rule (see 4A§11.2). Integrating in the clockwise direction erroneously inverts the polarity of ds and F.

2A§4.3 **Multidimensional Integrals** are employed to integrate quantities over a specified area or volume. We evaluate multiple integrals sequentially: we first integrate over one variable, then integrate that result over a second variable, and continue until all integrals are done. The order of integration is arbitrary; regardless of which variables are integrated first or last, one obtains the same ultimate result. Consider this triple integral of function f = x over the volume of a block with one corner at the origin (0, 0, 0) and the opposite corner at (X, Y, Z):

$$F = \int_0^Z \int_0^Y \int_0^X x\, dx\, dy\, dz$$

$$F = \int_0^Z \int_0^Y (x^2 / 2)|_0^X\, dy\, dz$$

$$F = \int_0^Z \int_0^Y (X^2 / 2)\, dy\, dz$$

$$F = \int_0^Z Y\, (X^2 / 2)\, dz$$

$$F = Z\, Y\, X^2 / 2$$

Here are two other volume integrals in spherical (upper) and cylindrical (lower) coordinate systems:

$$F = \int_0^R \int_0^\pi \int_0^{2\pi} f(r,\theta,\varnothing)\, [r^2 \sin\theta]\, d\varnothing\, d\theta\, dr$$

$$F = \int_0^R \int_0^Z \int_0^{2\pi} f(r,\varnothing,z)\, [r]\, d\varnothing\, dz\, dr$$

The factors in [ ]'s are the *volume elements* in these two coordinate systems. Any non-rectilinear coordinate system requires integration with volume elements, as explained in 4A§14.4, which also provides the needed volume elements for the most common coordinate systems.

2B§18.9 **Integration by Parts** for two functions u and v (see 4A§13.4):

$$\int u\, dv = u\, v - \int v\, du$$

2D§36.3 **Variational Calculus** is often used to find the path that minimizes or maximizes some quantity S. The path might be through two or more dimensions of spacetime, or through some more abstract multidimensional space. For example, with variables x and t, find the path x(t) from (x_a,t_a) to (x_b,t_b) that minimizes *action* S.

$$S = \int_a^b \{ \text{Kinetic} - \text{Potential Energy} \}\, dt$$

$$S = \int_a^b \{ (dx/dt)^2\, m / 2 - U(x) \}\, dt$$

As more thoroughly explained in 4A§14.4, the key steps are:

(1) define the true path to be x(t) and an alternate path to be x(t) + w(t)

(2) use integration by parts to eliminate derivatives of w

(3) minimizing S for every possible w(t) yields a solvable equation in x only

4B§3.5 Complex Quantities

1B§12.5 **Complex Quantities** are employed in the analysis of many physical phenomena, including harmonic oscillators, waves, and quantum amplitudes. Complex quantities have the form:

$$z = x + i\,y$$

Here, x and y are real quantities, $i = \sqrt{(-1)}$, and z is complex. Often, we analyze phenomena using complex quantities and then identify the real component of the result as the physical observable. The **real** and **imaginary** components of z are obtained by:

$$x = \mathbf{Re}\,(z)$$

$$y = \mathbf{Im}\,(z)$$

We often graph complex quantities with their real parts along one axis and their imaginary parts along another axis.

Each complex quantity z has a **complex conjugate** z* given by:

$$z^* = x - i\,y$$

The product of a complex quantity and its complex conjugate equals the square of the magnitude of either quantity, as given by:

$$|z|^2 = z\,z^* = (x + i\,y)\,(x - i\,y) = x^2 + y^2$$

1B§12.5 **Euler's Identity** is:

$$\exp\{iy\} = e^{iy} = \cos(y) + i\sin(y)$$

Euler's identity, and the power series of e^x, yield power series for the sine and cosine:

$$e^x = \Sigma_n\, x^n / n!$$

$$e^x = 1 + x + x^2 / 2! + x^3 / 3! + x^4 / 4! + \ldots$$

$$\cos(y) = 1 - y^2 / 2! + y^4 / 4! - \ldots$$

$$\sin(y) = y - y^3 / 3! + y^5 / 5! - \ldots$$

4B§3.6 Linear Systems

Many equations of physics, but not all, are linear. A linear example is:

$$\boldsymbol{F} = m\,\boldsymbol{a}$$

In a linear system, if:

force $\boldsymbol{F}_1$ produces acceleration $\boldsymbol{a}_1$, and

force $\boldsymbol{F}_2$ produces acceleration $\boldsymbol{a}_2$,

then by the principle of **Linear Superposition**:

force $\boldsymbol{F}_1 + \boldsymbol{F}_2$ produces acceleration $\boldsymbol{a}_1 + \boldsymbol{a}_2$

1B§14.3 Any **Linear Differential Equation** with constant coefficients has the following form and solution:

$$f(t) = a_n\, d^nx/dt^n + \ldots + a_1\, dx/dt + a_0\, x$$

$$\text{solution: } x = A \exp\{i\omega t + \phi\}$$

Inserting this solution into the differential equation results in an nth-order polynomial in ω that has n roots. Each root provides one solution to the differential equation. Any linear combination of these solutions is also a valid solution to f(t). Generally, the required solution to any specific problem is determined by boundary conditions, such as starting position and velocity.

4B§3.7 Fourier Analysis

1D§45.4 The **Fourier Series** for f(t), a function that repeats with frequency ω and period T, is:

$$f(t) = \Sigma_n \{ a_n \cos(n\omega t) + b_n \sin(n\omega t) \}$$

The above summation may extend from n = 0 to n = ∞. The coefficients are given by:

$$a_0 = (1/T) \int_T f(t)\, dt$$

$$j > 0: a_j = (2/T) \int_T f(t) \cos(j\,\omega t)\, dt$$

$$b_j = (2/T) \int_T f(t) \sin(j\,\omega t)\, dt$$

1D§45.6 The **Fourier Transform** S(k) is appropriate for a function f(x) that need not repeat, but which has finite extent. S(k) is a function of wave number k given by:

$$S(k) = \int_{-\infty}^{+\infty} f(x) \exp\{ -ik\,x\}\, dx / \sqrt{(2\pi)}$$

4B§3.8 Probability Distributions

Repeated measurements of any real physical quantity have a Gaussian distribution, provided:

(1) All measurements are independent of one another, and

(2) Any systematic errors are constant throughout data collection (rulers don't stretch between measurements).

1D§45.6 For a stochastic quantity x whose Gaussian distribution G(x) has mean μ and standard deviation σ:

$$G(x) = \exp\{ -(x-\mu)^2 / 2\sigma^2 \} / \sqrt{(2\pi\,\sigma^2)}$$

Here, G(x) dx is the probability that a measured value lies in the range x to x + dx. G(x) is called a probability density. The denominator of G(x) is chosen so that:

$$1 = \int_{-\infty}^{+\infty} G(x)\, dx$$

For any continuous distribution f(x), the *mean* μ (the average value of x), and the *variance* σ^2 are given by:

$$F = \int_{-\infty}^{+\infty} f(x)\, dx$$

$$\mu = \int_{-\infty}^{+\infty} x\, f(x)\, dx / F$$

$$\sigma^2 = \int_{-\infty}^{+\infty} (x-\mu)^2 f(x)\, dx / F$$

For any finite set of N measurements, x_j for j=1 to N, the mean μ and variance σ^2 are given by:

$$\mu = \Sigma_j\, x_j / N$$

$$\sigma^2 = \Sigma_j\, (x_j - \mu)^2 / N$$

The **standard deviation** σ equals the square root of the variance, in both continuous and finite data sets.

3C§31.2 **Statistics for Discrete Variables**: for M measurements of a physical quantity E that is restricted to a discrete set of values, E_k for k=1 to K, define m_k to be the number of measured values that equal E_k. The average measured value < E > is given by:

$$< E > = (\Sigma_k\, m_k\, E_k) / M$$

$$< E > = \Sigma_k\, (m_k / M)\, E_k$$

As M→∞, (m_k / M) approaches Prob(k), the probability of obtaining E_k in one measurement: this defines "probability." In any real experiment, M is finite and the (m_k/M)'s fluctuate from the Prob(k)'s, with standard deviations σ_k given by:

$$\sigma_k = \sqrt{\{ M\, p_k\, (1 - p_k)\}} = \sqrt{\{ m_k\, (1 - p_k)\}}$$

where: p_k = Prob(k)

if $p_k << 1$: $\sigma_k = \sqrt{\{ m_k \}}$

The prior equation is a very important result worth remembering. When an experiment detects 100 Higgs bosons, or 100 photons from a galaxy 13 billion light-years away, the one-standard-deviation statistical uncertainty in the value 100 is $\pm\sqrt{100} = \pm 10$.

Appendix 3 provides probability tables for various statistical fluctuations.

4B§3.9 Vectors in 3-D

The **Key Benefit of Vectors** is that any vector equation that is valid in one reference frame and coordinate system is valid without any modification in every other frame and coordinate system.

2D§37.3 **Vectors** are ordered sets of components that transform properly from one coordinate system to another under **Euclidean rotations** (see 4B§3.12). I denote a 3-D vector as ***V*** (in bold italics) and its magnitude as V. In N dimensions, a vector ***V*** has N components, each being the projection of ***V*** along one axis.

A **Unit Vector** has length 1. To specify a unit vector in the direction of velocity ***v***, we sometimes write:

$$\boldsymbol{u} = \boldsymbol{v} / v = \boldsymbol{v} / |\boldsymbol{v}|$$

A **Normal Vector** is a unit vector perpendicular to a surface at a specified point.

1D§49.10 Vectors are either **Polar or Axial**, with polar vectors being more common. The less common axial vectors are also called *pseudovectors*. Both polar and axial vectors transform as above under rotations, but they transform differently under reflection. This is because all axial vectors are defined by cross products (see below). Under reflection in a mirror in the x = 0 plane, all x-coordinate values flip polarity: each x is replaced by –x.

In reflection, vectors transform as:

Polar vector $\boldsymbol{v}$: $(v_x, v_y, v_z) \rightarrow (-v_x, +v_y, +v_z)$

Axial vector $\boldsymbol{\omega}$: $(\omega_x, \omega_y, \omega_z) \rightarrow (+\omega_x, -\omega_y, -\omega_z)$

Examples of polar vectors are:

Position $\boldsymbol{r} = (x, y, z)$

Velocity $\boldsymbol{v} = (v_x, v_y, v_z)$

Linear Momentum $\boldsymbol{p} = (p_x, p_y, p_z)$

Examples of axial vectors are:

Torque $\boldsymbol{\tau} = \boldsymbol{r} \times \boldsymbol{F} = (\tau_x, \tau_y, \tau_z)$

Angular Velocity $\boldsymbol{\omega} = \boldsymbol{r} \times \boldsymbol{v} = (\omega_x, \omega_y, \omega_z)$

Angular Momentum $\boldsymbol{L} = \boldsymbol{r} \times \boldsymbol{p} = (L_x, L_y, L_z)$

1A§6.6 The **Dot Product** is defined for two vectors in any number of dimensions as the sum of the products of corresponding components, such as:

$$\boldsymbol{A} \bullet \boldsymbol{B} = A_x B_x + A_y B_y + A_z B_z$$

The dot product results in a **scalar**, a quantity that has the same value in any coordinate system. The dot product equals the triple product of: the magnitude of vector ***A***; the magnitude of vector ***B***; and the cosine of the angle between those vectors. Graphically, the dot product is the normal projection of either vector onto the other.

1A§6.6 The **Cross Product** of two 3-D vectors is another 3-D vector that is defined as:

$$\boldsymbol{A} \times \boldsymbol{B} = (A_y B_z - A_z B_y,\ \ A_z B_x - A_x B_z,\ \ A_x B_y - A_y B_x)$$

Graphically, ***A*** × ***B*** is a vector whose magnitude equals the area of the parallelogram whose sides are ***A*** and ***B***, and whose direction is normal to that parallelogram, with its polarity set by the right hand rule (see 4A§11.2). The magnitude of ***A*** × ***B*** equals the triple product of: the magnitude of vector ***A***; the magnitude of vector ***B***; and the sine of the angle between those vectors.

2A§2.2 The **Double Cross Product** is often simplified with this identity:

$$\boldsymbol{A} \times (\boldsymbol{B} \times \boldsymbol{C}) = \boldsymbol{B}\,(\boldsymbol{A} \bullet \boldsymbol{C}) - (\boldsymbol{A} \bullet \boldsymbol{B})\,\boldsymbol{C}$$

The parentheses on the left are essential: $\boldsymbol{A} \times (\boldsymbol{B} \times \boldsymbol{C})$ is **not equal** to $(\boldsymbol{A} \times \boldsymbol{B}) \times \boldsymbol{C}$, in general.

4B§3.10 Vector Algebra

The vector operator $\check{\boldsymbol{D}}$ and its square $\check{D}^2$ are provided here in rectilinear coordinates, and in cylindrical and spherical coordinates in Appendix 5.

2A§2.5 We define:

Vector Operator $\check{\boldsymbol{D}} = (\partial/\partial x,\ \partial/\partial y,\ \partial/\partial z)$

Gradient of scalar field T: $\check{\boldsymbol{D}}\,T = (\partial T/\partial x,\ \partial T/\partial y,\ \partial T/\partial z)$

Curl of vector field $\boldsymbol{h}$: $\check{\boldsymbol{D}} \times \boldsymbol{h} = (X, Y, Z)$, with:

$$X = \partial h_z/\partial y - \partial h_y/\partial z$$
$$Y = \partial h_x/\partial z - \partial h_z/\partial x$$
$$Z = \partial h_y/\partial x - \partial h_x/\partial y$$

Divergence of vector field $\boldsymbol{h}$: $\check{\boldsymbol{D}} \bullet \boldsymbol{h} = \partial h_x/\partial x + \partial h_y/\partial y + \partial h_z/\partial z$

2A§2.4 For any scalar field T, and any displacement $\boldsymbol{\Delta r}$, the change in T in $\boldsymbol{\Delta r}$ is:

$$\Delta T = \boldsymbol{\Delta r} \bullet \check{\boldsymbol{D}}T$$

2D§44.4 The **Directional Derivative** of function f along any unit vector $\boldsymbol{u}$ is given by:

$$\boldsymbol{u} \bullet \check{\boldsymbol{D}}\,f = u_x\,\partial f/\partial x + u_y\,\partial f/\partial y + u_z\,\partial f/\partial z$$

The direction derivative of f is the incremental change in f resulting from an incremental motion parallel to $\boldsymbol{u}$. This may be clarified by considering a coordinate system in which the x-axis is parallel to $\boldsymbol{u}$. In this case: $\boldsymbol{u} \bullet \check{\boldsymbol{D}}\,f = (1)\,\partial f/\partial x$.

2A§2.8 The **Second Derivatives** of Fields are:

Laplacian $\check{D}^2$ of scalar field T: $\check{D}^2 T = \partial^2 T/\partial^2 x + \partial^2 T/\partial y^2 + \partial^2 T/\partial z^2$

Laplacian $\check{D}^2$ of vector field $\boldsymbol{h}$: $\check{D}^2 \boldsymbol{h} = \partial^2 \boldsymbol{h}/\partial^2 x + \partial^2 \boldsymbol{h}/\partial y^2 + \partial^2 \boldsymbol{h}/\partial z^2$

Curl of Curl of vector field $\boldsymbol{h}$: $\check{\boldsymbol{D}} \times (\check{\boldsymbol{D}} \times \boldsymbol{h}) = \check{\boldsymbol{D}}\,(\check{\boldsymbol{D}} \bullet \boldsymbol{h}) - \check{D}^2 \boldsymbol{h}$

2A§3.1 The **Path Integral of the Gradient** of a scalar field $\psi(\boldsymbol{r})$, along path Γ from point A to point B, is path-independent and is given by:

$$\int_\Gamma \check{\boldsymbol{D}}\psi \bullet \boldsymbol{ds} = \psi(\boldsymbol{B}) - \psi(\boldsymbol{A})$$

2A§3.2 The **Flux of Vector Field *A*** through surface S is given by:

$$\text{flux of } \boldsymbol{A} \text{ through } S = \int_S \boldsymbol{A} \bullet \boldsymbol{n}\ da$$

Here, $\boldsymbol{n}$ is the unit vector normal to S at each point.

2A§3.3 **Gauss's Theorem** equates the flux of a vector field to its divergence:

$$\int_S \boldsymbol{h} \bullet \boldsymbol{n}\ da = \int_V \check{\boldsymbol{D}} \bullet \boldsymbol{h}\ dV$$

Here, $\boldsymbol{h}$ is any vector field, V is any volume, S is the surface enclosing V, and $\boldsymbol{n}$ is the unit vector normal to S at each point.

2A§3.5 The **Circulation of a Vector Field *C*** around a closed path Γ is:

$$\text{circulation around } \Gamma = \oint_\Gamma \boldsymbol{C} \bullet \boldsymbol{ds}$$

Here, ***ds*** is the incremental tangent vector at each point along Γ.

2A§3.6 **Stokes' Theorem** relates the curl of a vector field to its path integral:

$$\int_S (\boldsymbol{\check{D}} \times \boldsymbol{C}) \bullet \boldsymbol{n}\, da = \oint_\Gamma \boldsymbol{C} \bullet \boldsymbol{ds}$$

Here, ***C*** is any vector field, S is any surface, Γ is the closed loop enclosing S, ***ds*** is the incremental tangent vector at each point along Γ, and ***n*** is the unit vector normal to S at each point.

4B§3.11 Conservation Laws

2A§3.4 **All Conservation is Local**: any quantity that is *globally conserved*, such as electric charge, must also be *locally conserved* at every point in space and every moment in time, as proven by special relativity. Separated events that are simultaneous in one reference frame are not simultaneous in any other frame moving with a different velocity. Thus, if a charge q disappeared in New York, and a charge q reappeared in Cape Town, global charge would not be conserved at every instant in all reference frames. This means charge q must move from New York to Cape Town, and the law of charge conservation, with electric current vector field ***j*** and charge density q, must have the form:

$$\boldsymbol{\check{D}} \bullet \boldsymbol{j} = -\,\partial q/\partial t$$

4B§3.12 Coordinate Transformations

2C§26.2 **Rotational Transformation**: rotating a coordinate system by angle θ about the z-axis transforms 3-vector ***r*** into 3-vector ***r**** according to:

$$\boldsymbol{r} = (x, y, z)$$

$$\boldsymbol{r^*} = (x \cos\theta + y \sin\theta,\ y \cos\theta - x \sin\theta,\ z)$$

Called a Euclidean rotation, this transformation has one plus sign and one minus sign. See 4A§11.2 for the right hand rule's definition of positive θ.

2C§26.2 **Lorentz Transformation:** boosting from a reference frame S to a reference frame S* that is moving at speed v in the +x-direction relative to S is governed by the Lorentz transformation:

$$x^* = \gamma\,(x - \beta\, c\, t)$$

$$c\, t^* = \gamma\,(c\, t - \beta\, x)$$

$$y^* = y$$

$$z^* = z$$

Here, $\beta = v / c$, and $\gamma = 1 / \sqrt{(1 - \beta^2)}$. Note that the Lorentz transformation has $-\beta$ twice.

4B§3.13 Tensors

Tensors are a generalization of vectors. Like vectors, tensors are ordered sets of components that transform properly from one coordinate system to another, as discussed below. And like vectors, each component of a tensor must have the same units — distance, velocity, momentum, etc.

2D§37.1 Like vectors, the **Key Benefit of Tensors** is that any tensor equation that is valid in one reference frame and coordinate system is valid without any modification in every other frame and coordinate system.

2D§37.2 **Tensor Indices and Rank**: a rank N tensor has N indices that all range over the same set of allowed values. A rank N tensor has 4^N components in 4-D, and 3^N components in 3-D.

Tensor indices select desired components, as in:

$$r_\sigma = (ct, x, y, z)$$
$$\text{here: } \sigma = \ t, x, y, z$$
$$\text{or: } \sigma = \ 0, 1, 2, 3$$

You may freely choose to use either the numbers or letters for the values of index σ — physicists use both. There is a critical difference between a *free index* σ that can have any value in the allowed range, and *fixed indices* such as 0, 1, 2, 3, x, y, z, and t. The latter refer to specific components, whereas the former are merely labels that establish relationships. The specific letters we choose for free indices have no significance mathematically or physically: by themselves, x_σ and x_μ mean exactly the same thing, as do $A_{\mu\sigma}$ and $A_{\beta\alpha}$. For example:

$$A_{\mu\sigma} = B_{\sigma\mu}$$

means each component of B equals the component of A on the opposite side of the diagonal. This tensor equation means exactly the same thing for any letters we might substitute for μ and σ.

2D§37.1 **Tensor Addition** simply consists of adding corresponding components, as in:

$$A_\mu = (A_0,\ A_1,\ A_2,\ A_3\ )$$
$$B_\mu = (B_0,\ B_1,\ B_2,\ B_3\ )$$
$$C_\mu = A_\mu + B_\mu$$
$$C_\mu = (A_0{+}B_0,\ A_1{+}B_1,\ A_2{+}B_2,\ A_3{+}B_3\ )$$

2D§37.2 **Tensor Multiplication** consists of multiplying corresponding components and summing the products. Examples are:

$$A_\mu B_\mu = A_0 B_0 + A_1 B_1 + A_2 B_2 + A_3 B_3$$
$$A_{jk\mu} B_\mu = A_{jk0} B_0 + A_{jk1} B_1 + A_{jk2} B_2 + A_{jk3} B_3$$

2D§37.2 The **Einstein Summation Convention** for tensors directs us to sum over all values of any repeated index, such as μ above. This abbreviation obviates the need for the summation symbol Σ.

2D§37.2 **Rotation of Tensors**:

$$\text{Rank 1: } A^*_j = R_{jn} A_n \quad \text{sum n}$$
$$\text{Rank 2: } A^*_{jk} = R_{jn} R_{km} A_{nm} \quad \text{sum n \& m}$$
$$\text{Rank 3: } A^*_{jk\mu} = R_{jn} R_{km} R_{\mu\sigma} A_{nm\sigma} \quad \text{sum n, m \& } \sigma$$

In any tensor equation in which multiple indices are summed, the order of summation makes no difference. The last equation, for example, requires summing over all values of n, all values of m, and all values of σ. The same value of $A^*_{jk\mu}$ is obtained regardless of which sum is done first, which second, and which third.

1C§27.1 **4-Vectors** are rank 1 tensors that are typically common 3-vectors with one added component that is called component number 0. The added component must ensure the 4-vector transforms properly under Lorentz transformations.

In the following, m is an object rest mass, $\beta = v / c$, and $\gamma = 1 / \sqrt{(1 - \beta^2)}$

2C§26.4 The **4-Velocity** rank 1 tensor u_μ is:

$$u_\mu = (\gamma c,\ \gamma v_x,\ \gamma v_y,\ \gamma v_z) = p_\mu / m$$

2C§26.4 The **4-Momentum** rank 1 tensor p_μ is:

$$p_\mu = u_\mu m = (\gamma mc,\ \gamma mv_x,\ \gamma mv_y,\ \gamma mv_z)$$

2D§37.4 The **Lorentz Transformation** 4x4 rank 2 tensor $\Lambda_{\mu\sigma}$ is:

$$\Lambda^{\mu}{}_{\sigma}$$

$$\begin{array}{c|cccc} & \sigma = t & x & y & z \\ \mu = t & \gamma & -\beta\gamma & 0 & 0 \\ x & -\beta\gamma & \gamma & 0 & 0 \\ y & 0 & 0 & 1 & 0 \\ z & 0 & 0 & 0 & 1 \end{array}$$

To transform vector A_μ in reference frame S to the corresponding vector A^*_σ in frame S* that has relative velocity βc in the +x-direction, simply multiply A_μ by the Lorentz tensor.

$$A^*_\sigma = \Lambda_{\mu\sigma} A_\mu \quad \text{sum } \mu$$

2C§26.6 The **d'Alembertian** operator $\Box$ is:

$$\Box = \check{D}_\mu \check{D}_\mu = c^{-2}\, \partial^2/\partial t^2 - \partial^2/\partial x^2 - \partial^2/\partial y^2 - \partial^2/\partial z^2$$

2D§37.5 The **Polarization Tensor** is defined by: $P_j = \Pi_{jk} E_k$

Here, ***P*** is the polarization vector field, ***E*** is the electric vector field, and Π is the 3x3 rank 2 polarization tensor field characterizing the material being polarized.

2D§37.7 The **Inertia Tensor** is defined by: $L_j = \hat{I}_{jk}\, \omega_k$

Here, ***L*** is the angular momentum vector, $\boldsymbol{\omega}$ is the angular velocity vector, and $\hat{I}$ is the 3x3 rank 2 inertia tensor characterizing some solid body.

2D§37.9 The **Stress Tensor** is defined by: $S_{jk} = \Delta F_j / \Delta a_k$

Here, ΔF_j is the j-component of force acting on a small area Δa_k that is perpendicular to the k-axis, and S is the 3x3 rank 2 stress tensor.

2D§37.10 The **Strain Tensor** and **Elasticity Tensor** are related by: $S_{jk} = \Sigma_{nm}\, C_{jknm}\, e_{nm}$

Here, S is the 3x3 rank 2 stress tensor, e is the 3x3 rank 2 strain tensor, and C is the 3x3x3x3 rank 4 elasticity tensor characterizing the material being stressed.

4B§3.14 And More...

2D§42.9 The **Radius of Curvature** R of function y(x) is given by:

$$R = (\,[dy/dx]^2 + 1)^{3/2} / d^2y/dx^2$$

2C§34.2 For any **Function F of a Complex Variable** β:

$$F(\beta) = F(x + iy)$$

$$F(\beta) = U(x,y) + i\, V(x,y)$$

$$\partial U/\partial x = +\, \partial V/\partial y$$

$$\partial V/\partial x = -\, \partial U/\partial y$$

U and V each satisfy the 2-D Laplace equation:

$$\partial^2 U/\partial x^2 + \partial^2 U/\partial y^2 = 0$$

$$\partial^2 V/\partial x^2 + \partial^2 V/\partial y^2 = 0$$

Curves of constant U and curves of constant V always cross orthogonally.

Chapter 4
Basic Newtonian Mechanics

Newton's laws of motion and gravity are valid only in **inertial reference frames**, those moving at constant velocity $\boldsymbol{v}$. Any change in a reference frame's speed v or direction of motion $\boldsymbol{v}$ / v requires an acceleration and makes that frame non-inertial. As velocities approach c or when gravity is extremely strong, Newton's equations require relativistic modification.

4B§4.1 Primary Quantities

In a rectilinear coordinate system with non-relativistic velocities and three mutually orthogonal axes x, y, and z:

1A§6.6	**Position**:	$\boldsymbol{r} = (x, y, z)$, $r = \vert\boldsymbol{r}\vert$
1A§6.6	**Velocity**:	$\boldsymbol{v} = d\boldsymbol{r}/dt$
1A§6.6	**Acceleration**:	$\boldsymbol{a} = d\boldsymbol{v}/dt$
1A§4.3	**Inertial Mass**	$m = \boldsymbol{F} / \boldsymbol{a}$
1A§7.2	**Linear Momentum**:	$\boldsymbol{p} = m\,\boldsymbol{v}$
1D§39.4	**Angular momentum**:	$\boldsymbol{L} = \boldsymbol{r} \times \boldsymbol{p}$
1A§7.2	**Force**:	$\boldsymbol{F} = d\boldsymbol{p}/dt$
1D§39.3	**Torque**:	$\boldsymbol{\tau} = \boldsymbol{r} \times \boldsymbol{F}$
1A§10.1	**Kinetic Energy** (linear):	$T = m\,v^2 / 2$

4B§4.2 Newton's Laws of Motion

1A§7.1 **First Law**: Inertia, absent external forces, an object's velocity remains constant.

1A§7.2 **Second Law**: $\boldsymbol{F} = m\,\boldsymbol{a}$

This establishes the relationship between force and acceleration, with the proportionality constant defining an object's **inertial mass** m.

1A§7.5 **Third Law**: Reaction = – Action

For every action there is an equal but opposite reaction.

1A§10.3 The **Work Energy** expended to move an object a distance $\boldsymbol{\Delta s}$ when that object is subject to force $\boldsymbol{F}$ equals:

$$W = -\boldsymbol{F} \bullet \boldsymbol{\Delta s}$$

1A§10.2 **Power** is the rate of change of energy, released or consumed. The power consumed to move an object at velocity $\boldsymbol{v}$ when that object is subject to force $\boldsymbol{F}$ equals:

$$\text{power} = dW/dt = -\boldsymbol{F} \bullet \boldsymbol{v}$$

1A§10.5 A **Potential Energy** U that varies with position results in a force F given by:

in 1-D: $F = -dU/dx$

in 3-D: $\boldsymbol{F} = -\check{\boldsymbol{D}}\,U$

4B§4.3 Newton's Law of Gravity

See also 4B, Chapter 7.

1A§8.4 According to Newton, all massive objects attract one another with a universal force. For a spherically symmetric body, gravity acts as if all of the body's mass is concentrated at its center (see 1A§10.8). For two spherically symmetric bodies of mass M and m, the attractive force F is given by:

$$F = G\,M\,m\,/\,r^2$$

Here, G is Newton's gravitational constant, and r is the distance between the bodies' centers.

1A§10.9 Within a spherically symmetric shell of any thickness, the force of gravity due to the shell is zero everywhere.

4B§4.4 More Mechanics

1A§7.3 **Hooke's Law** of springs states:

$$F = -\,k\,x$$

Here, F is the force exerted by the spring, k is the spring constant, and x is the spring's displacement from equilibrium. Hooke's law is an idealization that assumes perfect linearity.

1A§9.1 **Friction** is a force that resists an object Θ sliding across a surface S. This force is strongly dependent on the composition and surface properties of both Θ and S. We often roughly characterize friction by:

$$F = \mu\,N$$

Here, μ is the coefficient of friction, N is the force exerted normal to S that pushes Θ against S, and F is the force opposing any motion of Θ across S. In general, μ is larger when Θ is at rest (static friction) than when Θ is moving (sliding friction).

1A§4.3 The **Principle of Virtual Work** focuses on **reversible systems**. Such systems can be driven, with virtually zero work, in either of two opposite directions that we call + and –. Work W is the energy expended (or released) in moving a distance Δx against (or with) a force F:

$$W = F\,\Delta x$$

If W = 0 in both + and – directions, all forces must exactly cancel one another. This precise balance implies the existence of a quantity that remains constant in both + and – motions. For example, the device shown below lifts weights on one side by lowering weights on the other side.

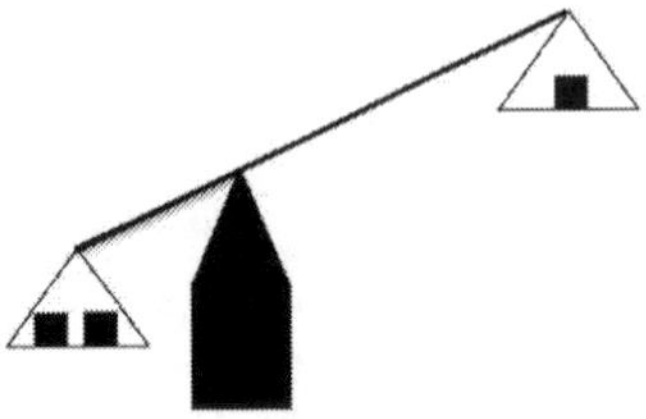

Figure 4-1 Lifting Device

The figure depicts three identical blocks, two on the left and one on the right with the right arm of this device being twice as long as the left arm. Let's assume this machine operates with zero friction, and that all its parts (except the blocks) have zero weight. While unrealistic, such idealizations aid understanding fundamental principles. With these assumptions, the machine is **reversible**: an infinitesimal force can raise either side.

The conserved quantity here is mgh: weight multiplied by height.

1A§9.9 **Pseudo Forces** are not real forces, but rather the perceived consequences of inappropriately applying Newtonian mechanics in non-inertial reference frames.

The most familiar pseudo force is the **Centrifugal Force** experienced in a turning car. As a car turns left, passengers perceive being pushed to the right. In fact, the passengers are continuing in the original direction of motion until pushed left by their seats, seatbelts, and car sidewalls.

2D§38.5 The **Coriolis Force** is a pseudo force. Children tossing a ball to one another on opposite sides of a merry-go-round perceive the ball veering off the straight line between them. In fact, when viewed from a stationary location above, the ball travels in a straight line as the children spin around. The ball arrives where the child was, while the child moves away. In a reference frame rotating at frequency $\boldsymbol{\omega}$, the apparent Coriolis force $\boldsymbol{F}$ on an object with mass m and velocity $\boldsymbol{v}$ is:

$$\boldsymbol{F} = -2m\, \boldsymbol{\omega} \times \boldsymbol{v}$$

Chapter 5
Mechanics of Angular Motion

The mechanics of angular motion discussed in this chapter is valid only in **inertial reference frames**, those moving at constant velocity ***v***.

4B§5.1 Basics of Rotation

Feynman defines rotation as turning a coordinate system by an angle ø around an axis Θ. For continuous rotations, the rate of rotation dø/dt is most commonly quantified as ω radians per second, or alternatively as f full cycles per second (f = ω / 2π). Vector ***ω*** has magnitude ω, and points along Θ, with its polarity defined by the right hand rule (see 4A§11.2).

We typically align one of our coordinate axes, often the z-axis, along the rotation axis Θ, thus ensuring the origin of our coordinate system remains stationary during rotation. We then measure distances from Θ with x-y coordinates.

For a rotation of a coordinate system about the z-axis by angle ø, the (x,y) coordinates of an object P become (x*,y*), as given by:

$$x^* = + x \cos(ø) + y \sin(ø)$$

$$y^* = - x \sin(ø) + y \cos(ø)$$

The following correspondence of linear and angular quantities is explored in ***Feynman Simplified 1D***, Chapter 39.

Variable	Linear Motion	Rotational Motion
Position	***r***	ø
Velocity	***v*** = d***r***/dt	ω = dø/dt
Acceleration	***a*** = d***v***/dt	α = dω/dt
Force / Torque	***F***= m ***a***	***τ*** = ***r*** x ***F***
Momentum	***p*** = d (m ***r***)/dt	***L*** = ***r*** x ***p***
Inertia	m	Î = m r²
Kinetic Energy	m v² / 2	Î ω² / 2

4B§5.2 Angular Momentum

An object's linear momentum ***p*** is a polar vector, whose magnitude is unaffected by our choice of coordinates. In contrast, an object's angular momentum ***L*** is an axial vector defined by a cross product, whose magntitude and direction vary depending on our arbitrary choice of a reference axis Θ. It is often convenient to align Θ with an actual axis of rotation, but we are free to choose any Θ.

The relationships of some key quantities are illustrated below.

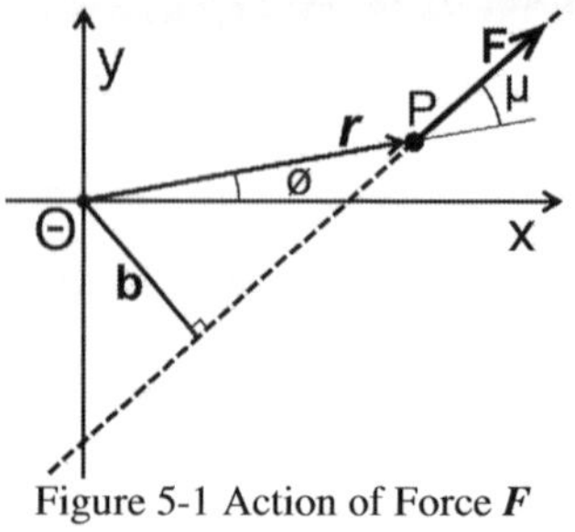

Figure 5-1 Action of Force ***F***

In Figure 5-1, *μ* is the angle between force ***F*** and ***r***, the vector from Θ to P, the position at which ***F*** acts. The dashed line extends ***F*** in both directions, and its closest approach to the axis Θ is at distance b, the **lever arm** of force ***F*** about Θ.

1D§39.4 The **Angular Momentum** $\boldsymbol{L}$ about Θ of a single-point object P with linear momentum $\boldsymbol{p}$ is:

$$\boldsymbol{L} = \boldsymbol{r} \times \boldsymbol{p}$$

Here, $\boldsymbol{r}$ is the vector from Θ to P.

1D§39.3 The **Torque** $\boldsymbol{\tau}$ about Θ due to force $\boldsymbol{F}$ is given by each of these equivalent equations:

$$\boldsymbol{\tau} = \boldsymbol{r} \times \boldsymbol{F}$$

$$\boldsymbol{\tau} = d\boldsymbol{L}/dt$$

$$\tau = b\, F$$

Here, $\boldsymbol{r}$ is the vector from axis Θ to the point at which force $\boldsymbol{F}$ acts, and $d\boldsymbol{L}/dt$ is the rate at which torque $\boldsymbol{\tau}$ changes angular momentum $\boldsymbol{L}$. The last equation says the magnitude of torque τ equals the magnitude of force F multiplied by its **lever arm** b.

1D§39.3 The **Work** W expended by a torque rotating through an angle $\Delta\phi$ is:

$$W = \tau\, \Delta\phi$$

W is positive if torque τ opposes an increase in angle ϕ.

4B§5.3 Moments of Inertia

1D§40.1 The position $\boldsymbol{R}$ of the **Center of Mass** (CM) of a collection of objects is:

$$\boldsymbol{R} = \Sigma_j\, m_j\, \boldsymbol{r}_j$$

Here, each object, denoted by the subscript j, has mass m_j and is at position $\boldsymbol{r}_j$. A body's CM lies within the envelope containing all its parts. The CM of a two-part object lies along the line connecting the CM of each individual part. The CM of a symmetric body lies along its line of symmetry.

1D§39.6 The **Moment of Inertia** Î about axis Θ of a collection of objects is:

$$\hat{I} = \Sigma_j\, m_j\, r_j^2$$

Here, each object, denoted by the subscript j, has mass m_j and is at a distance r_j from axis Θ.

1D§39.7 The **Angular Momentum** $\boldsymbol{L}$ about Θ of a single-point object P is:

$$L = r \times p = m\, r \times v$$

$$L = m\, r \times (\omega \times r)$$

$$\boldsymbol{L} = m\, r^2\, \boldsymbol{\omega}$$

$$\boldsymbol{L} = \hat{I}\, \boldsymbol{\omega}$$

Here, object P has mass m and linear velocity $\boldsymbol{v}$, and its moment of inertia Î and angular velocity $\boldsymbol{\omega}$ are defined about Θ.

1D§40.2 The **Pappus' Centroid Theorem** says any planar surface S rotated about an external axis Θ, sweeps through a volume V equal to AD, where A is the area of S and D is the distance that its *centroid* moves. The centroid is the point whose coordinates equal the average coordinates of all points in S. This preassumes S and Θ lie within one common plane.

1D§40.4 The **Parallel-Axis Theorem** relates moments of inertia about two different parallel axes:

$$\hat{I} = M\, R_{CM}^2 + \hat{I}_{CM}$$

A body's moment of inertia Î about an axis Θ equals the body's moment of inertia about its own CM, $\hat{I}_{CM}$, plus MR_{CM}^2, its mass M multiplied by the square of R_{CM}, the distance of its CM to axis Θ. It is essential that the axis about which $\hat{I}_{CM}$ is calculated is parallel to the axis Θ.

1D§40.4 **Summing Moments about One Axis**: when multiple objects rotate about the same axis Θ, the moment of inertia of all objects combined equals the sum of the moments of each object separately. The equation is:

$$\hat{I}_{all} = \Sigma_j \hat{I}_j$$

1D§40.5 The moment of inertia of a uniform **Sphere** of mass M and radius R about any of its diameters is:

$$\hat{I} = 2 M R^2 / 5$$

1D§40.5 The moment of inertia of a uniform **Cylinder** of mass M, radius R, and length L about its central axis is:

$$\hat{I} = M R^2 / 2$$

1D§40.5 The moment of inertia of a uniform **Cylinder** of mass M, radius R, and length L, about axis Θ that passes through the cylinder's center orthogonally to the cylinder's central axis, is:

$$\hat{I} = M (3R^2 + L^2) / 12$$

1D§40.5 The moment of inertia of a uniform cuboid **Block** of mass M, with side lengths a, b, and c, is:

$$\hat{I} = M (a^2 + b^2) / 12$$

Here, Î is about axis Θ that is parallel to side c and passes through the block's center.

1D§40.5 The moment of inertia of a uniform **Ring** of outer radius R, inner radius r, and thickness Z is:

about its central axis: $\hat{I} = M (R^2 + r^2) / 2$

about any diameter: $\hat{I} = M (R^2 + r^2) / 4$

1D§39.7 The **Rotational Kinetic Energy** of an object spinning with frequency ω about axis Θ is:

$$T = \omega^2 \hat{I} / 2$$

Here, Î is the object's moment of inertia about Θ.

4B§5.4 Precession

1D§41.2 As shown below, gravity exerts a force ***F*** pulling downward on a spinning top. We assume ***F*** acts at the top's center of mass, CM.

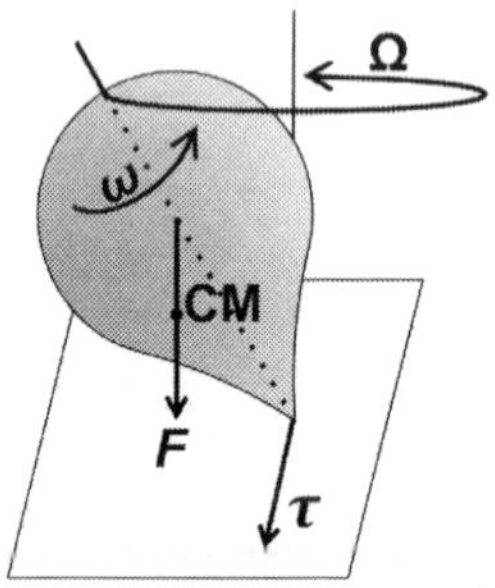

Figure 5-2 Spin & Precession

Force ***F*** creates a horizontal torque τ that turns the top's axis of rotation. As the top spins rapidly about its own axis with angular velocity ω, that axis slowly **precesses** at frequency ***Ω*** about the vertical axis, the direction of gravity's pull. The equations are:

$$\boldsymbol{\tau} = \boldsymbol{R}_{CM} \times \boldsymbol{F}$$

$$\boldsymbol{\tau} = \boldsymbol{\Omega} \times \boldsymbol{L}$$

$$\boldsymbol{\tau} = \boldsymbol{\Omega} \times \boldsymbol{\omega} \hat{I}$$

Here, the top's moment of inertia and angular momentum about its own axis are Î and ***L***, respectively. Also, $\boldsymbol{R}_{CM}$ is the position of its center of mass. The direction of precession is horizontal, perpendicular to the applied force.

Chapter 6
Waves & Oscillators

4B§6.1 Harmonic Oscillation

Harmonic oscillators move in repeating cycles. This section addresses oscillations characterized by single-frequency sinusoidal motion.

1B§12.2 **Free Harmonic Oscillation** is the repetitive motion of a system without any dissipative or external forces. In many cases, such a system oscillates at a specific frequency ω determined by its intrinsic properties; ω is called the **natural frequency**. For a mass m on a spring, with spring constant k, the governing equations are:

$$m\, d^2x/dt^2 = F_{res}$$

$$F_{res} = -\,k\, x$$

$$\omega = \sqrt{(k / m)}$$

$$x = C \cos(\omega t) + B \sin(\omega t)$$

$$x = A \cos(\omega t + \phi)$$

$$x = A \cos(\omega\, [t - t_0]\,)$$

Here, x(t) is the displacement of the mass from its equilibrium position, F_{res} is the spring's restoring force, and the constants A, B, C, ø, and t_0 are chosen to match initial conditions. The three equations for x(t) are all equivalent.

1B§12.4 **Undamped Forced Oscillation** is the repetitive motion of a system that is driven by an oscillating external force, but without damping. For a mass m on a spring, with spring constant k, the governing equations are:

$$m\, d^2x/dt^2 = -\,k\, x + F_{drv}$$

$$\omega = \sqrt{(k / m)}$$

$$F_{drv}(t) = f \cos(\beta t)$$

$$x = D \cos(\beta t)$$

$$D = f / (k - m\, \beta^2) = f / [m\, (\omega^2 - \beta^2)]$$

Here, f is the amplitude and β is the frequency of the driving force F_{drv}, x(t) is the displacement of the mass from its equilibrium position, D is the amplitude of that displacement, and ω is the system's natural frequency. We have arbitrarily chosen the phase of the driving force so that the force is maximum at t = 0.

1B§13.2 **Damped Forced Oscillation** is the repetitive motion of a system that is driven by an external force, with a damping force that resists the system's motion, such as friction. For a mass m on a spring, with spring constant k, the governing equations are:

$$m\, d^2x/dt^2 = -\,k\, x + F_{drv} + F_{fric}$$

$$\omega = \sqrt{(k / m)}$$

$$F_{drv}(t) = f \cos(\beta t)$$

$$F_{fric} = -\,\mu\, m\, dx/dt$$

$$D = (f / m) / [\omega^2 - \beta^2 + i\mu\beta])$$

$$\text{let } D = \varrho \exp\{-i\theta\}$$

$$\varrho = (f / m) / \sqrt{[\,(\omega^2 - \beta^2)^2 + (\mu\beta)^2\,]}$$

$$\sin\theta = \mu\, \beta\, m\, \varrho / f$$

$$x = \varrho \cos(\beta t - \theta)$$

Above, f is the amplitude and β is the frequency of the driving force F_{drv}, F_{fric} is the frictional damping force that opposes motion, x(t) is the displacement from equilibrium, and μ is the coefficient of friction per unit mass. The complex constant D is expressed as a real magnitude ϱ and a real phase angle θ that is always positive. ϱ is the amplitude of the displacement oscillation. θ is the **lag angle**, the difference between the phase angles of the driving force and the displacement. A system's response (displacement) occurs after (lags) the driving force. We have arbitrarily chosen the phase of the driving force so that the force is maximum at t = 0.

1B§13.5 **Energy in Damped Forced Oscillation** is described by the average over a full cycle of various forms of energy: kinetic, potential, total energy, and also power dissipation.

kinetic: $< T > = m\, \varrho^2\, \beta^2 / 4$

potential: $< U > = m\, \omega^2\, \varrho^2 / 4$

total: $< E > = < T > + < U >$

total: $< E > = m\, \varrho^2\, (\omega^2 + \beta^2) / 4$

$< \text{Power} > = \mu\, m\, \varrho^2\, \beta^2 / 2$

$\varrho = (f / m) / \sqrt{[\, (\omega^2 - \beta^2)^2 + (\mu\beta)^2\,]}$

Here, < > means average over a full oscillation cycle, m is the oscillator mass, ω is the system's natural frequency, β and f are the driving force frequency and magnitude, μ is the damping force per unit mass, and ϱ is the amplitude of the oscillation.

1B§13.5 **Resonance** curves describe how a system's oscillation amplitude varies with driving frequency β. We define $Q(\beta)$ as (the average oscillation energy) divided by (the average power dissipation per radian). Using the above results, $Q(\beta)$ is given by:

$Q(\beta) = < E > / (< \text{Power} > / \beta)$

$Q(\beta) = \beta\, \{m\varrho^2\, (\omega^2 + \beta^2) /4\} / (\mu m\, \varrho^2\, \beta^2 /2)$

$Q(\beta) = (\omega^2 + \beta^2) / 2\, \mu\, \beta$

Quality Factor $Q = \omega / \mu$ is the value of $Q(\beta)$ at $\beta = \omega$, the resonance peak.

4B§6.2 Oscillators & Transients

This section addresses the transient behavior of oscillators whose driving forces suddenly cease. The same logic can be applied to driving forces that begin suddenly, or that are impulsive (short duration).

Let's assume the driving force ceases at t=0. We employ these parameters, defined in 4B§6.1, for stable forced harmonic oscillators with damping:

restoring force: $F_{res} = - k\, x$

natural frequency: $\omega = \sqrt{(k / m)}$

driving force: $F_{drv}(t) = f \cos(\beta t)$

friction: $F_{fric} = - \mu\, m\, dx/dt$

displacement: $x(t) = \varrho \cos(\beta t - \theta)$

$\varrho = (f/m) / \sqrt{[\, (\omega^2 - \beta^2)^2 + (\mu\beta)^2\,]}$

$\sin\theta = \mu\, \beta\, m\, \varrho / f$

Here, k is the restoring force's proportionality constant, m is the oscillating mass, ϱ is the displacement amplitude, and θ is the phase angle by which the displacement lags the driving force.

1B§14.1 The differential equation for an unforced oscillator with damping is:

$$m\, d^2\boldsymbol{X}/dt^2 + k\, \boldsymbol{X} + \mu\, m\, d\boldsymbol{X}/dt = 0$$

Here, we allow $\boldsymbol{X}$(t) to be a complex number; the actual motion is the real part of $\boldsymbol{X}$(t).

The most general solution of this equation is:

$$X(t) = A \exp\{ia\} \exp\{-t\mu/2) \exp\{+it\sigma\}$$
$$+ B \exp\{ib\} \exp\{-t\mu/2) \exp\{-it\sigma\}$$
$$\text{with } \sigma^2 = \omega^2 - \mu^2/4$$

For now we allow σ to be complex. The real constants A, B, a, and b are chosen to match initial conditions, including the displacement magnitude and rate of change at t=0. All solutions share the factor $\exp\{-t\mu/2\}$ that drives exponential decay, and all fall into one of three categories illustrated in Figure 6-1.

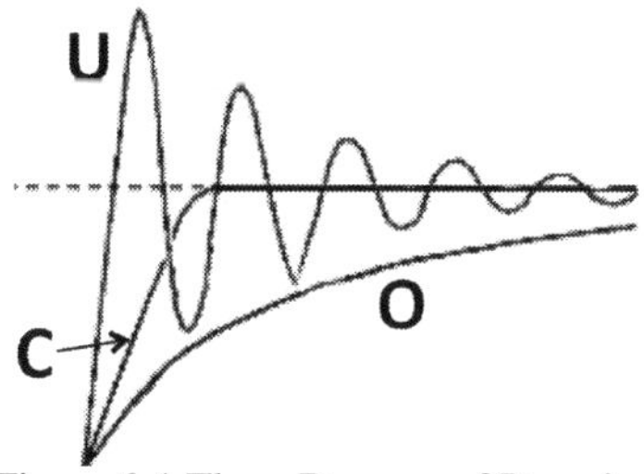

Figure 6-1 Three Degrees of Damping

In this figure, the three decay curves are labeled:

U for Underdamped

O for Overdamped

C for Critically damped

1B§14.2 When $\sigma^2 = \omega^2 - \mu^2/4 > 0$, **Decay is Underdamped**, yielding:

$$x(t) = 2A \exp\{-t\mu/2\} \cos(a + \sigma t)$$

Here, a real solution requires B = A and b = –a, with A and a allowing us to match initial conditions.

1B§14.2 When $\mu = 2\omega$ and $\sigma = 0$, **Decay is Critically Damped**, yielding:

$$x(t) = A \exp\{-\omega t\}$$

Here, the displacement declines exponentially at the fastest possible rate without oscillating.

1B§14.2 When $\Omega^2 = \mu^2 - 4\omega^2 > 0$, **Decay is Overdamped**, yielding:

$$x(t) = A \exp\{-t(\mu + \Omega)/2\}$$
$$+ B \exp\{-t(\mu - \Omega)/2\}$$

Here, A and B allow us to match initial conditions. The displacement declines exponentially without oscillating, but at a slower rate than when critically damped.

4B§6.3 Wave Basics

Waves are motion. They move through space and change with time. A single-frequency sinusoidal wave is shown below.

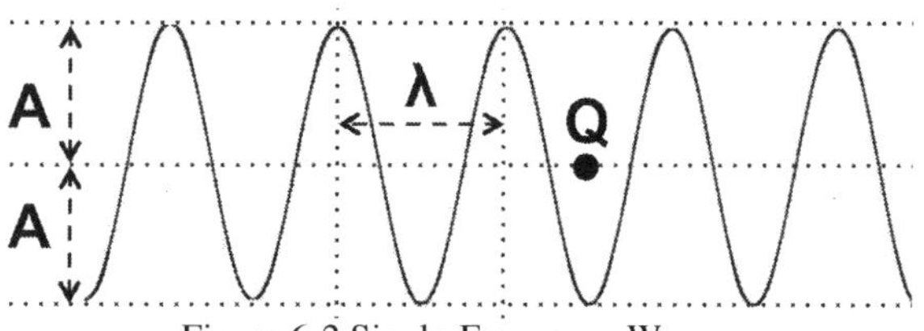

Figure 6-2 Single-Frequency Wave

As the above wave moves, its entire shape moves in unison as if it were a solid object. If we focus on a specific location, such as point Q, we will see the wave height going up and down as time passes, oscillating between +A and –A. We define a full cycle as one oscillation through the entire repeating pattern; for example, from one peak to the next peak. If the wave is described by $\sin\phi$, ϕ increases by 2π radians in one full cycle. Waves are characterized by:

λ: **wavelength**, spatial extent of one full cycle

A: **amplitude**, how much a wave rises and falls from its mean

f: **frequency**, number of full cycles / second

$\omega = 2\pi f$: **angular frequency**, number of radians / second

$v = \lambda f$: **velocity**, distance moved / second

$k = 2\pi / \lambda$: **wave number**, number of radians per unit distance

1D§43.5 **Phase Velocity** v_{ph} is the velocity of a single-frequency wave. Such waves cannot describe a real particle or an information-bearing signal, because they can never begin or end, and are 100% predictable. (Information is by definition not predictable). v_{ph} is given by:

$$v_{ph} = \lambda f = \omega / k$$

1D§43.5 **Group Velocity** v_g is the velocity of a multi-frequency wave packet (see 4B§13.2) that may represent a particle or an information-bearing signal. v_g is given by:

$$v_g = d\omega/dk$$

4B§6.4 Sound Waves

1D§42.4 The **Intensity I** of **Sound Waves** is measured in decibels (db), as defined by:

$$\mathbf{I} = 20 \log_{10}(\Delta P / P_{ref}) \text{ decibels}$$

Here, $P_{ref} = 2\times10^{-10}$ bar, and ΔP is the pressure change from equilibrium. 1 bar is defined to be 10^5 pascal = 10^5 newton/m^2 = 10^5 kg/m-sec^2. At sea level, the average atmospheric pressure is 1.01325 bars.

1D§40.6 The **Speed of Sound Waves** c_s in a medium is given by:

$$c_s = \sqrt{(dP/d\varrho)_0}$$

Here, P is pressure, ϱ is mass density, and the subscript "0" indicates the derivative evaluated at equilibrium pressure and density. This equation applies to small pressure deviations from equilibrium. Large pressure changes, such as those from shock waves, change the equilibrium conditions, generally increasing temperature and pressure, reducing mass density, and increasing c_s.

4B§6.5 Waves in Matter

A **Bow Wave** created by a high-speed object moving through stationary matter is shown below.

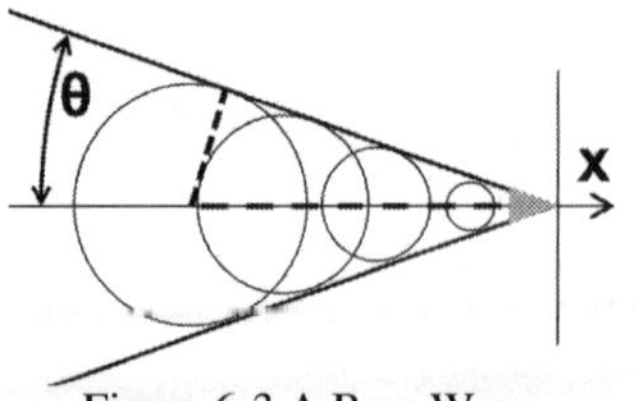

Figure 6-3 A Bow Wave

The high-speed object (small gray triangle) disturbs the matter through which it passes. These disturbances propagate outward as waves. We define x=0 at the object's tip at time t=0. The figure shows four representative waves that emanated from the object's tip at times t= –1, –2, –3, and –4. All such waves accumulate (**interfere constructively**) on the surface of a cone whose half-angle is θ.

1D§46.1 For an object with velocity v, in a medium whose characteristic wave velocity u is less than v, the bow wave cone half-angle θ is given by:

$$\text{for } v > u, \sin\theta = u / v$$

When the medium is air, u is the speed of sound. The bow wave from a super-sonic jet is called a **sonic boom**.

In a transparent medium of refractive index n, u = c / n, where c is the speed of light. The bow wave from a very high velocity particle is **Cherenkov radiation**, light emitted by the medium's accelerated electrons.

In water, waves can have a wide range of wavelengths. We consider two limiting cases: long waves and ripples. For each, we assume wavelengths much smaller than the water's extent in all directions. The group velocities are:

1D§46.7 For **Long Waves**: $v_g = (1/2)\sqrt{(g / k)}$

1D§46.8 For **Ripples**: $v_g = (3/2)\sqrt{(k\,\tau / \varrho)}$

Here, g is the acceleration of gravity, τ is water's surface tension, and ϱ is water's density

4B§6.6 Waves Without Sources

Here, we examine waves so remote in space and time from their originating sources that their behavior is entirely determined by the waves' inherent properties.

1D§42.1 The **1-D Wave Equation** without sources is:

$$\partial^2\psi/\partial x^2 - \partial^2\psi/\partial t^2 / v^2 = 0$$

The most general solution of the one-dimensional wave equation is a linear combination of any two arbitrary functions f and g of the form:

$$\psi(x,t) = f(x - vt) + g(x + vt)$$

Here, f represents a wave moving toward +x, and g represents a wave moving –x, both at velocity v. Note that, for any time increment Δt, f(x –vt) is unchanged if x increases by Δx = v Δt — which is the meaning of a wave moving with velocity v.

2B§20.6 Feynman often represents a 1-D wave as:

$$\psi = \cos(k\,x)\cos(k\,vt)$$

While this form appears different, it does conform to the general solution above, as this trigonometric identity proves:

$$\cos(A + B) + \cos(A - B) = 2\cos A\cos B$$

Hence:

$$2\cos(k\,x)\cos(k\,vt) = \cos(k\,x + k\,vt) + \cos(k\,x - k\,vt)$$

2B§20.1 **Plane Waves** are the simplest 1-D wave equation solutions. A single-frequency plane wave has the form:

$$\phi(x,y,z,t) = A\cos(\omega\,t - k\,x)$$

Here, the wave propagates toward +x with velocity v = ω / k. Its amplitude is the same everywhere within each yz-plane perpendicular to the x-axis, hence the name *plane wave*. The wave is unconfined, extending to infinity in all directions, thus allowing any values of ω, λ, and k. These are idealizations that aid understanding but are never fully realized in nature.

2B§20.2 The **Most General Plane Wave** for electromagnetic fields ***E*** and ***B***, for a wave moving along the x-axis, is a superposition of both y and z linear polarizations, as given by:

$$\boldsymbol{E} = (0,\ E_y,\ E_z)$$

$$\boldsymbol{B} = (0,\ B_y,\ B_z)$$

$$E_y = f(x - ct) + g(x + ct)$$

$$E_z = F(x - ct) + G(x + ct)$$

$$c\,B_y = -F(x - ct) + G(x + ct)$$

$$c\,B_z = +f(x - ct) - g(x + ct)$$

Above, f, g, F, and G are arbitrary functions. If the wave travels in only one direction, if for example g = G = 0, the orientations of ***E*** and ***B*** are such that ***E × B*** is parallel to the wave's velocity ***v***.

1D§44.3 **Waves Confined in 1-D** are severely restricted by their boundary conditions. For example, the oscillation of a guitar string must have zero amplitude at its ends that we define to be x = 0 and x = L. Such waves have the form:

$$y(x,t) = \Sigma_n A_n \cos(\omega_n t + \phi_n) \sin(k_n x)$$
$$k_n = n \pi / L$$
$$\omega_n = k_n v = n v \pi / L$$

Here, y(x,t) is the string's lateral displacement from equilibrium, and A_n and ϕ_n are constants chosen to match any specific situation's initial conditions. Clearly, y = 0 at x = 0. Each integral value of n is an **allowed mode** of oscillation, with a corresponding frequency ω_n and a wave number k_n that ensures y = 0 at x = L. The severely restricted set of allowed modes ensures each string plays the desired base note and harmonics. Lastly, the wave velocity v is determined by the string's mass and tension.

Each allowed mode, by itself, represents a **standing wave** — a wave that oscillates in amplitude, but whose shape and position do not change. Such a wave is stationary in the sense that its crests do not propagate along the string length. However, a sum of multiple modes is generally not stationary, since each mode has a different shape and oscillates at a different frequency. Such mixtures result in waveforms that move up and down the string. At each string end, these moving waveforms reverse direction and polarity.

3A§6.2 **Waves Confined in 3-D** are governed in each dimension as in 1-D. In a 3-D volume V, the number of allowed modes dM between wave number k and wave number k+dk is:

$$dM = V k^2 dk / 2\pi^2$$

2B§20.3 The **3-D Wave Equation** without sources is:

$$\partial^2 f/\partial x^2 + \partial^2 f/\partial y^2 + \partial^2 f/\partial z^2 - \partial^2 f/\partial t^2 / v^2 = 0$$

Here, v is the wave velocity of f that may be either a scalar or vector field. For all electromagnetic fields — ***E***, ***B***, ***A***, and ϕ — v = c, the speed of light. Feynman says all 3-D solutions are simply linear superpositions of all possible 1-D solutions: linear sums of plane waves moving in the +x, –x, +y, –y, +z, and –z directions.

2B§20.5 **Spherical Waves** are the special case of 3-D waves that are spherically symmetric. These waves are functions only of time and radial distance r from an origin Θ; they do not vary with either polar or azimuthal angle. The 3-D source-less spherical wave equation, in spherical coordinates, is:

$$\partial^2(r\,\psi)/\partial r^2 - \partial^2(r\,\psi)/\partial t^2 / v^2 = 0$$

This equation says r ψ satisfies the 1-D wave equation. Since the same equations have the same solutions, r ψ has the same form as the function f discussed above. Hence:

$$r\,\psi(r,t) = f(r - vt)$$

$$\psi(r,t) = f(r - vt) / r$$

Spherically symmetric waves are like plane waves, except that they expand in 3-D and their amplitudes decrease inversely with the distance from their origin Θ. We reject the collapsing wave alternative, r ψ = g(r + vt), as not representing any known natural phenomenon.

4B§6.7 Waves With Sources

2B§21.2 The **3-D Sourced Wave Equation** is:

$$\check{D}^2\psi - \partial^2\psi/\partial t^2 / v^2 = - S$$

Here, S is the wave source and v is the wave velocity. For electromagnetic waves, v=c, and S is some combination of charges and currents.

2B§19.6 The 3-D sourced wave equation for the electromagnetic scalar potential ϕ has the above form and its solutions are known (see 4B§11.2):

$$\Box\,\phi = - \check{D}^2\phi + \partial^2\phi/\partial t^2 / c^2 = \varrho / \varepsilon_0$$

The time-dependent solutions, accounting for retarded time at relativistic velocities, are addressed in 4B§11.8. Here, let's consider the simpler static case:

$$\check{D}^2\phi = -\varrho / \varepsilon_0$$

From 4B§11.3, the solution to this equation, for the potential $\phi(\boldsymbol{r})$ from a charge q at position $\boldsymbol{\acute{o}}$, is:

$$\phi(\boldsymbol{r}) = q / 4\pi\, \varepsilon_0 \,|\, \boldsymbol{r} - \boldsymbol{\acute{o}} \,|$$

2A§4.4 By the linearity of Maxwell's equations, we can extend the above to the potential $\phi(\boldsymbol{r})$ from multiple charges q_j at locations $\boldsymbol{\acute{o}}_j$, yielding:

$$\phi(\boldsymbol{r}) = \Sigma_j\, q_j / 4\pi\, \varepsilon_0 \,|\, \boldsymbol{r} - \boldsymbol{\acute{o}}_j \,|$$

2A§4.4 Extending the above to a continuous charge density ϱ within a volume V yields:

$$\phi(\boldsymbol{r}) = \int_V \{\varrho(\boldsymbol{\acute{o}}) / 4\pi\, \varepsilon_0 \,|\, \boldsymbol{r} - \boldsymbol{\acute{o}} \,| \}\, dV$$

We therefore have the most general solution to the Laplace Equation.

2A§6.1 For multiple discrete sources Y_j at locations $\boldsymbol{\acute{o}}_j$, and for a continuous density $Y(\boldsymbol{\acute{o}})$ within a volume V:

Laplace equation: $\check{D}^2 X = -Y$

Discrete: $X(\boldsymbol{r}) = \Sigma_j\, Y_j / 4\pi \,|\, \boldsymbol{r} - \boldsymbol{\acute{o}} \,|$

Continuous: $X(\boldsymbol{r}) = \int_V \{Y(\boldsymbol{\acute{o}}) / 4\pi \,|\, \boldsymbol{r} - \boldsymbol{\acute{o}} \,| \}\, dV$

4B§6.8 Combining Waves

At any fixed point, a **single-frequency** wave has the form:

$$\text{amplitude} = A \cos(\omega t + \phi)$$

$$\text{intensity } \mathbf{I} = A^2$$

Here, intensity **I** is the wave energy per unit area, ω is its frequency, and ø is its phase angle at t = 0.

1C§35.4 **Waves Combining**. At a fixed point, when two single-frequency waves combine, they interfere, and the resultant amplitude A and intensity **I** are given by:

$$A = A_1 \cos(\omega_1 t + \phi_1) + A_2 \cos(\omega_2 t + \phi_2)$$

$$\mathbf{I} = A_1^2 + A_2^2 + 2A_1 A_2 \cos(\Delta\omega\, t + \Delta\phi)$$

$$\Delta\omega = \omega_1 - \omega_2$$

$$\Delta\phi = \phi_1 - \phi_2$$

For two single-frequency waves of equal intensity I_0, their combined intensity oscillates sinusoidally between zero and $4I_0$ at the **beat frequency** Δω.

1C§31.1 **Coherent Waves** have the same frequency and a fixed phase shift relative to one another. Waves with different frequencies or with varying phase shifts are **incoherent**. From the above equation, the intensity **I** of the combination of two coherent waves of equal intensity $\mathbf{I}_0$ is:

$$\mathbf{I} = 2\, \mathbf{I}_0 \{1 + \cos(\Delta\phi)\}$$

$$\Delta\phi = \phi_1 - \phi_2$$

Intensity **I** has a constant value that ranges between zero and $4\mathbf{I}_0$, depending on the waves' phase difference Δø. Only coherent waves with constant phase shift π can totally cancel one another.

Chapter 7
Gravity per Newton & Einstein

Newton's and Kepler's laws of motion are valid only in **inertial reference frames**, those moving at constant velocity v. General Relativity is valid in any reference frame.

4B§7.1 Kepler's Laws

1A§8.1 **Kepler's Three Laws of Planetary Motion** are:

1. Each Planet orbits in an ellipse with its star at one focal point.
2. The radial line to a planet from its star sweeps out equal areas in equal times.
3. A planet's orbital period squared is proportional to its semi-major axis cubed.

Kepler's laws assume each planet is much less massive than the star it orbits, and that the gravitational effects between planets are negligible compared with the gravitational effect of the star.

All of Kepler's laws are consequences of Newton's inverse-squared law of gravity.

The third law, generalized to the motion of any small body around another of much greater mass, is:

$$M^1 T^2 = R^3$$

To help remember the exponents, this is called the **1-2-3 law**. Here, M is the mass of the heavier body divided by the Sun's mass, T is the orbital period measured in Earth years, and R is the small body's semi-major axis measured in AU (Earth's average orbital radius). The Astronomical Unit, 1 AU, equals 149.60 million km or 92.956 million miles.

4B§7.2 Newton's Theory of Gravity

In this chapter, G is Newton's gravitational constant, and F is the attractive force between two spherically symmetric objects of masses M and m whose centers are separated by distance r.

1A§8.4 **Newton's Law of Gravity** is the original **inverse-squared** law:

$$F = G\,M\,m / r^2$$

When the gravitational parameter GM (see below) is known, this equation defines m as the **gravitational mass** of the smaller body. Experiments show that the gravitational mass and the **inertial mass** (see 4B§4.2) are equal to 1 part in 100 billion.

This confirms Einstein's claim that the two masses must be identical, because gravity is a geometric effect rather than a true force (see 4B§7.4).

1A§8.4 **Gravitational Potential Energy** U:

$$U = -G\,M\,m / r$$

1A§4.3 **Near Earth**'s surface:

$$U = m\,g\,h$$

$$g = G\,M / R^2$$

$$g = 32.174 \text{ feet} / s^2$$

$$g = 9.80665 \text{ m} / s^2$$

Here, U is gravitational potential energy, m is the mass of a small body, M is Earth's mass, R is Earth's radius, and h is the height of the small body above any fixed reference elevation, such as sea-level.

1A§6.1 The **Distance S Fallen** in time t near Earth's surface is:

$$S = g\,t^2 / 2$$

1A§8.5 The **Gravitational Parameter GM** is more precisely measured for astronomical bodies than either G or M separately. The measured values for the Earth and the Sun are:

$$\text{Earth: } GM_E = 3.986{,}004{,}42 \times 10^{14}\ m^3/s^2$$

$$\text{Sun: } GM_S = 1.327{,}124{,}4002 \times 10^{20}\ m^3/s^2$$

$$G = 6.6741 \times 10^{-11}\ m^3 / \text{kg-sec}^2$$

4B§7.3 Orbits & Forces in 1/r Potentials

While this section focuses on gravity, everything here applies to any **conservative force** F(r) that obeys an inverse-squared law, and therefore has a corresponding potential U(r) that is proportional to 1/r. In particular, everything here applies to electrostatic fields.

1A§10.4 **Conservative Forces** have energy potentials that depend only on position — no energy is gained or lost by an object traversing any closed loop. All fundamental forces are conservative. However, some forces, such as friction, appear to be non-conservative. Friction converts kinetic energy into heat that can be hard to track, and inadequate energy tracking can lead to the false perception of non-conservation.

1A§8.3 A **Circular Orbit** of radius r requires a **centripetal** (inward) acceleration a given by:

$$a = v^2 / r$$

Here, v is the velocity of the orbiting body, and r is the circular orbit's radius.

3A§3.5 The **Virial Theorem** applies to any body in a stable orbit, subject to a conservative force whose strength is proportional to $1/r^2$. The theorem says the body's average kinetic energy Ŧ equals minus one-half its average potential energy U. The proof is particularly simple for a circular orbit, using the above result:

$$U = -k / r$$

$$F = k / r^2$$

$$F = m\,a = m\,v^2 / r$$

$$m\,v^2 / 2 = k / 2r$$

$$Ŧ = -U / 2$$

1A§8.5 The **Moon's Orbital Motion** is the sum of two effects: (1) each minute, the Moon moves forward 61km (38 miles) along its current velocity vector; and (2) drops 4.8m (16 feet) toward Earth. That tiny drop is exactly enough to pull the Moon away from a straight-line trajectory and back onto its orbit around Earth.

1A§8.6 **Earth's Tides** are primarily due to slight differences in the Moon's gravitational attraction on different parts of our planet. Let's define Earth's mass and radius to be m and r, respectively, and define the Moon's mass and distance to be M and R, respectively, as shown below (not to scale).

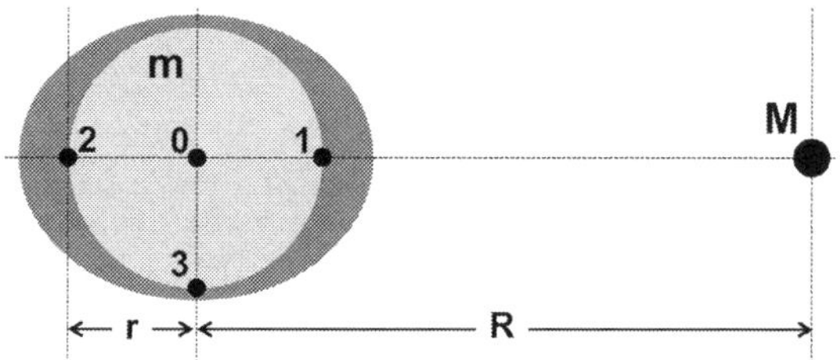

Figure 7-1 Earth's Tides

In Figure 7-1, the forces F_0, F_1, and F_2 at points 0, 1 and 2 are:

$$F_0 = G\, m\, M / R^2$$

$$F_1 = G\, m\, M / (R - r)^2 = F_0\, (1 + 2r / R)$$

$$F_2 = G\, m\, M / (R + r)^2 = F_0\, (1 - 2r / R)$$

Here, we calculate F_1 and F_2 to first order in the small quantity r / R that is nearly 1 / 60. Since $F_1 > F_0 > F_2$, the Moon's gravity pulls more on objects at point 1, and less on objects at point 2, than it does on Earth as a whole. These force differences only modestly deform Earth's solid rock (gray circle above), but they noticeably deform the oceans' free-flowing water (dark gray oval above).

Let F_3 be the force at point 3 in the above figure. Also to first order in r/R, the magnitudes of F_0 and F_3 are equal, but they point in slightly different directions: F_3 has a component pointing toward Earth's center with magnitude:

magnitude of inward component of $F_3 = (r / R)\, F_0$

This analysis shows that the tidal forces are proportional to M / R^3. While the Sun is 27 million times more massive than the Moon, it is also 388 times farther away. The net result is that the Moon's influence is 2.16 times more than the Sun's; hence the Moon and the Sun contribute 68% and 32% of Earth's tides, respectively.

On average, Earth has two high tides each 24 hours and 51 minutes, which is how long the Moon takes to return to the same spot in our sky.

1A§10.7 For an **Infinite Plane** of uniform mass density μ:

$$F = G\, m\, 2\pi \mu$$

Here, F is the gravitational force exerted by the plane on another body of mass m. Note there is no distance dependence (no factor of r). This is because there is nothing to define distance relative to an infinite plane.

1A§10.8 For a **Ball** of mass M that is spherically symmetric, the attractive force F is:

$$F = G\, M\, m / r^2$$

Here, F is the gravitational force exerted by M on another body of mass m, and r is the distance between the centers of M and m. The ball's radius is irrelevant, as is the radial distribution of its mass, provided that distribution is spherically symmetric. The same applies to m. Thus any spherically symmetric body acts as if its mass is contained within a single point. This enormously simplifies computation — rather than needing to calculate the force between each of the 10^{57} particles in the Sun and each of the 10^{51} particles in Earth, with 10^{108} different separations, we need only compute one quantity: $M\, m / r^2$.

4B§7.4 Einstein's Theory of Gravity

In this section, c is the speed of light and G is Newton's gravitational constant. See Chapter 9 for more on Einstein's General Theory of Relativity.

2D§46.2 The **Curvature of Space** can be equal to, greater than, or less than zero. In the table below, P stands for curvature polarity (0, +, –), C / D is the ratio of circumference to diameter of any circle, and $\Sigma\, \theta$ is the sum of the interior angles of any triangle.

P	C / D	Σθ	Example
0	π	π	Plane
+	<π	>π	Sphere
–	>π	<π	Potato Chip

Additionally, parallel lines never cross in a space with zero curvature, but they may cross in spaces with nonzero curvature. The above table demonstrates that the curvature of any space is determinable by measuring circles and triangles entirely within that space. We can determine Earth's curvature with measurements entirely within Earth's surface, and we can measure the curvature of the universe without needing to leave Earth.

2D§46.4 The **4-D Spacetime Metric** outside a spherically symmetric, non-rotating body of mass M was derived by Karl Schwarzchild, and is named in his honor. The metric provides the **proper time dτ** and the **invariant interval ds** between all pairs of events separated by infinitesimal displacements (dt, dr, dθ, dø), as given by:

$$ds^2 = -\,\Omega\, c^2\, dt^2 + dr^2 / \Omega + r^2\, d\theta^2 + r^2 \sin^2\theta\, d\phi^2$$

$$c^2\, d\tau^2 = -\,ds^2$$

$$\text{where } \Omega = 1 - 2\,G\,M / c^2\, r$$

Spacetime curvature slows time by multiplying dt by $\sqrt{\Omega}$, which is less than 1; and it stretches space radially by dividing dr by $\sqrt{\Omega}$.

For Earth

$r = 6371$ km

$G\,M / c^2 = 4.44$ mm

$G\,M / c^2\, r = 6.97\times10^{-10}$

For the Sun

$r = 695{,}500$ km

$G\,M / c^2 = 1.47$ km

$G\,M / c^2\, r = 2.11\times10^{-6}$

2D§46.4 Feynman defines the **excess radius** of a massive body as:

$$\text{excess radius} = G\,M / 3c^2$$

Here, the body must be spherically symmetric, and its mass M must be uniformly distributed. The excess radius of the Earth and the Sun are 1.48mm and 489m, respectively.

2D§46.4 A **Black Hole's Event Horizon** is a sphere of radius r_s, the **Schwarzchild radius**, as given by:

$$r_s = 2\,G\,M / c^2$$

$$\Omega = 0$$

Here, the black hole is assumed to be spherically symmetric, non-rotating, and have mass M. The event horizon is centered on the black hole's **singularity**, an infinitesimal point at which all its mass is located.

2D§46.5 **Time Dilates** (advances at a slower rate) in stronger gravity. In a uniform gravitational field with downward acceleration g, time dilation is given by:

$$\Delta t^* = \Delta t\,(1 - L\,g / c^2\,)$$

Here, we assume two perfect clocks, one of which is elevated at distance L above the other. A time interval that the higher clock measures to be Δt is measured by the lower clock to be Δt^*. Since $\Delta t^* < \Delta t$, the lower clock appears to run slower.

2D§46.6 **Photon Energy and Gravity**. In a uniform gravitational field with downward acceleration g, a photon's frequency changes as it moves in that field, according to:

$$\omega(H) = \omega(0)\,(1 - g\,H / c^2)$$

Here, $\omega(H)$ and $\omega(0)$ are the frequencies at heights H and 0, respectively. This follows from the conservation of energy, and Einstein's equation $E = m\,c^2$. A photon with energy E has an "equivalent mass" m equal to E / c^2. When a mass m moves in a gravitational field, changing its gravitational potential energy by ΔU, its kinetic energy must change by $-\Delta U$. For a photon this means $E = \hbar\,\omega$ changes by $-\Delta U$.

Chapter 8
Statistical Mechanics & Thermodynamics

In this chapter, we define these key quantities:

P: pressure = force per unit area

V: volume

T: temperature in Kelvin

Ŧ: kinetic energy, excluding intramolecular motion

N: number of molecules

N_A: Avogadro's number

n: number of moles = N / N_A

v: molecular velocity

k: Boltzmann's constant

R: gas constant = $k\,N_A$

$< Q >$: the average value of any quantity Q

4B§8.1 Primary Gas Equations

The following equations are valid for a gas at equilibrium, the state in which macroscopic properties, such as temperature and pressure, do not change.

1B§15.2 The **Kinetic Energy** of a gas of N molecules is:

$$Ŧ = N < m\,v^2 / 2 >$$

1B§15.9 The **Ideal Gas Law**, applicable to gases comprised of point-particles that interact only mechanically, without any electromagnetic, chemical, or nuclear reactions, is:

$$P\,V = n\,R\,T = N\,kT$$

1B§15.4 The **Equipartition of Energy** principle says that, at equilibrium, kinetic energy is equally divided among **each degree of freedom**, each available mode of motion. We write this:

$$\text{energy per degree of freedom} = kT / 2$$

$$< m\,v^2 / 2 > = 3\,kT / 2$$

The first equation is the **definition of temperature** in the kinetic theory of gases. The factor of 3 in the latter equation arises because each molecule has 3 independent components of velocity: (v_x, v_y, v_z).

1B§15.2 The **Pressure and Energy** of an **ideal gas** are related by:

$$P = 2\,Ŧ / 3\,V = (2N / 3V) < m\,v^2 / 2 >$$

1B§15.2 The **Specific Heat Ratio** γ is defined by:

$$P\,V = (\gamma - 1)\,Ŧ$$

$\gamma = 5/3$ for a monatomic gas

$\gamma = 4/3$ for a photon "gas", as within plasma

1B§21.4 **Isothermal Processes** occur at a constant temperature, during which:

$$P\ V = \text{constant}$$

1B§21.4 In **Adiabatic Processes**, there is no heat transfer between the gas and its surroundings, during which:

$$P\ V^{\gamma} = \text{constant}$$

4B§8.2 Statistical Mechanics

The following equations are valid for a gas at **equilibrium**, the state in which macroscopic properties, such as temperature and pressure, do not change.

Statistical Mechanics was primarily developed by Ludwig Boltzmann, based on the postulate that matter is comprised of atoms.

1B§17.1 The thermal energy of molecules, atoms, ions, and electrons drives chaotic motion, producing macroscopic consequences, including: **Brownian motion**; **electrical circuit noise**; and **mechanical vibration** in sensitive instruments.

1B§16.2 **Boltzmann's Law** for the relative populations of states of different energy is:

$$N = N_0 \exp\{ -(U - U_0) / kT \}$$

Here, N is the population of a state of energy U, and N_0 is the population of a state of energy U_0.

1B§17.2 The **Random Walk / Drunken Sailor** problem seeks the average distance traveled in a series of N steps of length L, assuming each step is taken in an independent random direction in two dimensions. The answer is:

$$< R > = L \sqrt{N}$$

Here, < > denotes the average over many trials, and R is the distance from the starting point to the end of the last step.

1B§17.3 The **Diffusion Equation** is:

$$< R^2 > = 6\ kT\ t / \mu$$

$$F = -\mu\ v$$

Here, R is the distance moved by a **special particle** through a collection of **background particles** during time t, μ is the **drag coefficient**, v is the average special particle velocity, and F is the drag force that is assumed to be uniform throughout the volume of interest.

1B§18.3 The **Saha Equation** determines the relative densities in a gas of electrons, ions, and neutral particles, according to:

$$n_I\ n_e / n_0 \sim \exp\{ -W / kT\}$$

Here:

n_I is the density of ions

n_e is the density of electrons

n_0 is the density of neutral atoms

W is the ionization energy of one electron

1B§19.1 The **Mean Collision Time** τ is defined to be the average time between collisions of a gas molecule.

1B§19.1 The **Mean Collision Distance** λ, also called the **mean free path**, is given by:

$$\lambda = v\ \tau$$

Here, v is the mean particle velocity, and τ is the mean collision time (see above).

1B§19.1 The **Probability** p of a molecule **Not Colliding** during time t or while traversing distance x is:

$$\text{during time t: } p = \exp\{-t/\tau\}$$

$$\text{in distance x: } p = \exp\{-x/\lambda\}$$

1B§19.1 The **Collision Cross Section** σ, the effective area of a target, is given by:

$$\sigma = 1/\lambda n$$

Here, λ is the mean free path (defined above), and n is the number of targets per unit volume. Typically, targets are identical molecules, atoms, nuclei, or elementary particles. Cross section σ is a combination of the target's physical size and its interaction probability. This equation assumes targets do not shadow one another, which is valid when (σ multiplied by the number of targets per unit area) << 1.

1B§19.2 **Ion Mobility** μ is defined as:

$$\mu = \tau/m$$

Here, τ is the mean collision time (see above), and m is the mass of the ion.

1B§19.2 The **Drift Velocity** v of a gas molecule of mass m subject to force F is:

$$v = F\tau/m$$

Here, τ is the mean collision time (see above). For a molecule with charge q and mobility μ in an electric field E:

$$F = qE$$

$$v = qE\tau/m$$

$$v = qE\mu$$

1B§19.4 The **Diffusion Coefficient** D is given by:

$$D = \mu kT$$

4B§8.3 Black Body Radiation

In this section, ω is the frequency and **I** is the intensity of black body radiation, the radiation emitted by an object with temperature T due solely to its heat energy.

1B§20.2 The **Rayleigh-Jean** law of classical physics for black body radiation is:

$$\mathbf{I}(\omega) = (\omega/\pi c)^2 kT$$

This equation says the intensity of black body radiation increases rapidly as frequency ω increases. Since ω is unlimited, lighting a match should release infinite energy and cremate the entire universe. Called the **ultraviolet catastrophe**, this is an existential failure of classical physics.

1B§20.3 The **Planck Spectrum** of quantum mechanics for black body radiation is:

$$\mathbf{I}(\omega) = \hbar\omega^3 / [\pi^2 c^2 (\exp\{\hbar\omega/kT\} - 1)]$$

Planck derived this equation by assuming, without any physical motivation, that energy is radiated only in integral multiples of ħω. This "solved" the ultraviolet catastrophe by exponentially decreasing black body intensity at high frequency. Einstein later explained that light is comprised of particles, each with energy ħω, and particles must always exist in integral numbers.

1B§20.3 **Harmonic Oscillator Energy** levels are quantized in quantum mechanics, as given by:

$$E_J = (J + 1/2)\,\hbar\,\omega, \text{ for } J = 0, 1, 2, \ldots$$

The lowest energy an oscillator can ever have is the **zero-point energy**: $E_0 = \hbar\,\omega/2$.

1B§20.4 **Total Black Body Radiation**, as given by quantum mechanics, is:

$$\varepsilon = \pi^2 k^4 T^4 / 15 \hbar^3 c^2$$

$$\varepsilon = 4 \sigma T^4$$

$$\text{with } \sigma = \pi^2 k^4 / 60 \hbar^3 c^2$$

Here, ε is the intensity of radiation integrated over all frequencies from the surface of a black body per unit area per unit time. The radiation emanating from a small hole is ε / 4.

4B§8.4 Thermodynamic Laws

1B§21.2 The **Zeroth Law** is:

if A is in equilibrium with B
and B is in equilibrium with C
then A is in equilibrium with C

1B§21.2 The **First Law** restates energy conservation:

$$\Delta Ŧ = \Delta Q + \Delta W$$

Here, Ŧ is kinetic energy, Q is heat energy, and W is work energy.

1B§21.3 The **Second Law** can be stated in many ways, including:

- In any complete **reversible cycle**, entropy does not change.
- In any complete **irreversible cycle**, entropy increases.
- No complete cycle can have the sole effect of converting heat into work.

A reversible cycle is one that neither gains nor loses energy when operating in either the forward or backward directions.

1B§21.6 The **Third Law**, due to Nernst, is:

A system's entropy is minimum, usually zero, at 0 Kelvin.

1B§21.5 The **Efficiency of a Reversible Heat Engine** is given by:

$$\text{efficiency} = W / Q = (T_1 - T_2) / T_1$$

Here, W is the work released when heat energy Q is transferred from a reservoir at temperature T_1 to a reservoir at T_2 with $T_1 > T_2$. Reversible heat engines are idealizations, like frictionless motion. No heat engine can be more efficient than a reversible heat engine, and all reversible heat engines have equal efficiencies.

1B§21.5 **Absolute Temperature** is defined by:

$$T = Q / S$$

$$(\text{temperature}) = (\text{heat}) / (\text{entropy})$$

Feynman shows this definition is equivalent to the kinetic gas theory definition given in 4B§8.1:

$$kT / 2 = \text{average energy / degree of freedom}$$

1B§22.2 The **Specific Heat** C_V per unit volume of a gas is defined by:

$$C_V = (\partial Ŧ/\partial T)_V$$

This says C_V equals the partial derivative of kinetic energy with respect to temperature, while gas volume is held constant.

1B§22.2 Feynman's "Two **Basic Equations of Thermodynamics**" are:

$$dŦ = dQ - P\, dV$$

$$(\partial Ŧ/\partial V)_T = T\, (\partial P/\partial T)_V - P$$

The first equation above says (the change in kinetic energy) equals (the change in heat) minus (pressure multiplied by the change in volume). The second equation says (the partial derivative of kinetic energy with respect to volume, at constant temperature) equals (temperature multiplied by the partial derivative of pressure with respect to temperature, at constant volume) minus pressure.

1B§22.5 The **Clausius-Clapeyron** equation for a liquid/gas mixture of the same substance is:

$$Q / [\, T\, (V_G - V_L)\,] = \partial P_{VAP} / \partial T$$

Here, Q is heat, T is temperature, P_{VAP} is liquid's vapor pressure, and V_L and V_G are the volumes of the substance when it is entirely in the liquid phase and entirely in the gas phase, respectively.

4B§8.5 Entropy

1B§23.6 The **Entropy** S of a system is given by:

$$S = k \ln(\Omega)$$

Here, k is Boltzmann's constant, ln denotes the natural logarithm, and Ω is the number of microscopic rearrangements of the system that are macroscopically indistinguishable — rearrangements that do not change any macroscopic property, such as temperature, pressure, volume, etc. Microscopic rearrangements include swapping atoms of the same type, or reversing all particle velocities.

1B§21.6 The **Nernst Theorem of Entropy** is:

$$S(V,T) = Nk \{ \ln(V) - \ln(T) / (\gamma - 1) \} + C$$

Here, S is the entropy of a gas with specific heat ratio γ, temperature T, and volume V. Also, k is Boltzmann's constant, ln denotes the natural logarithm, and N is the number of molecules. Since only entropy changes have physical significance, the arbitrary constant C is often set to zero.

4B§8.6 Heat Flow

2A§2.7 **Heat Flow** equations include:

$$J = -\varkappa\, (T_1 - T_2)\, A / d$$

$$\boldsymbol{h} = -\varkappa\, \check{\boldsymbol{D}}\, T$$

Here, J is the amount of heat energy flow per unit time, ***h*** is the vector field of heat flow per unit area per unit time, ϰ is the thermal conductivity, A is the cross sectional area, and d is the distance between two parallel isothermal surfaces at temperatures T_1 and T_2. Heat energy always flows from hot to cold. Contrary to the "exhilarating" sensation of jumping into icy water, "cold" is not a physical entity; nothing "flows" from cold to hot.

2A§3.4 The **Heat Diffusion** equation is:

$$dT/dt = (\varkappa / C_v)\, \check{D}^2 T$$

Here, C_v is the specific heat per unit volume, ϰ is the thermal conductivity, and T is absolute temperature.

Chapter 9
Special & General Relativity

Special relativity is valid only in **inertial reference frames**, those moving at constant velocity $\boldsymbol{v}$. General Relativity is valid in any reference frame.

See 4B§7.4 for more on general relativity, Einstein's theory of gravity.

4B§9.1 Principles of Special Relativity

1C§25.1 The **Speed of Light**, c, is the same in all reference frames.

1C§25.1 The **Principle of Relativity**, due to Galileo, says absolute velocity has no physical meaning; only relative velocities are significant.

1C§25.1 The **Michelson-Morley** interferometer experiment of 1887 provided the most definitive validation of these principles at the time Einstein was developing special relativity.

1C§25.5 Einstein showed that the principle of relativity and Maxwell's theory of electromagnetism are compatible only if **light has no medium** and moves at the same speed in all inertial frames.

4B§9.2 Primary Quantities

For **Velocity** $\boldsymbol{v} = (v_x, v_y, v_z)$ and c being the speed of light, define:

$$v = |\boldsymbol{v}|$$
$$\beta = v / c$$
$$\gamma = 1 / \sqrt{(1 - \beta^2)}$$

1C§25.7 **Useful Equations** with β and γ:

$$\text{for any } \beta\text{: } 1 + \gamma^2 \beta^2 = \gamma^2$$
$$\text{for } \beta << 1\text{: } \gamma = 1 + \beta^2 / 2 + 3\beta^4 / 8 + \ldots$$
$$\text{for } \beta \text{ close to 1: } \gamma = 1 / \sqrt{\{ 2 (1 - \beta) \}}$$

4B§9.3 Vectors & Operators in 4-D

Here, the index μ spans the four dimensions of spacetime: $\mu = 0, 1, 2, 3$. Also, $\boldsymbol{v}$ is a velocity 3-vector, $\boldsymbol{F}$ is a force 3-vector, and m is an object's rest mass, the mass measured in its rest frame. Some key 4-vectors and 4-D operators are:

2C§26.2 **4-Position**:	$x_\mu = (ct,\ x,\ y,\ z)$
2C§26.2 **4-Velocity**:	$u_\mu = (\gamma c,\ \gamma v_x,\ \gamma v_y,\ \gamma v_z)$
2C§26.2 **4-Momentum**:	$p_\mu = m\, u_\mu = (E / c,\ p_x,\ p_y,\ p_z)$
2C§27.8 **4-Force**:	$f_\mu = dp_\mu/d\tau = (\gamma \boldsymbol{F} \bullet \boldsymbol{v} / c,\ \gamma F_x,\ \gamma F_y,\ \gamma F_z)$
2C§26.6 **4-Gradient**:	$\check{D}_\mu = (+c^{-1}\, \partial/\partial t,\ -\partial/\partial x,\ -\partial/\partial y,\ -\partial/\partial z)$
2C§26.6 **4-Divergence**:	$\check{D}_\mu A_\mu = +c^{-1}\, \partial A_t/\partial t\ +\ \partial A_x/dx\ +\ \partial A_y/dy\ +\ \partial A_z/dz$
2C§26.6 **d'Alembertian**:	$\square = c^{-2}\, \partial^2/\partial t^2 - \partial^2/\partial x^2 - \partial^2/\partial y^2 - \partial^2/\partial z^2$

2C§26.4 The **Dot Product** of 4-Vectors A_μ and B_μ is:

$$A_\mu B_\mu = +A_t B_t - A_x B_x - A_y B_y - A_z B_z$$

The above equations use **Feynman's sign convention**. The most common modern convention has the opposite signs: minus on the time components, and plus on the spatial components.

1C§25.7 The **Sum of Collinear Velocities** is:

$$w = (u + v) / (1 + u\, v / c^2)$$

Here, w is the result of summing collinear velocities u and v relativistically, with u and v either positive or negative.

1C§25.7 The **Sum of Orthogonal Velocities** is:

$$\text{for } u = (0,\ u_y,\ 0)$$

$$\text{for } v = (v_x,\ 0,\ 0)$$

$$w = (v_x,\ u_y / \gamma,\ 0)$$

$$\gamma = 1 / \sqrt{(1 - v^2 / c^2)}$$

Here, w is the result of summing orthogonal velocities u and v relativistically, with u_y and v_x either positive or negative.

4B§9.4 Key Invariants

Let the displacement of two infinitesimally separated events be dx_μ.

1C§27.2 The **Invariant Interval** ds is given by:

$$ds^2 = -c^2 dt^2 + dx^2 + dy^2 + dz^2$$

Here, ds measures the separation between two nearby events in 4-D spacetime; its value is the same in any reference frame and coordinate system. I use here the most common sign convention: one minus sign on the time component. Feynman typically uses the opposite convention: a plus sign on the time component and a minus sign on each spatial component.

1C§27.2 **Proper Time** $d\tau$ is given by:

$$c^2 d\tau^2 = c^2 dt^2 - dx^2 - dy^2 - dz^2$$

$d\tau$ is the lapsed time measured by a perfect clock that moves between two nearby events in 4-D spacetime; its value is the same in any reference frame and coordinate system. This is the most common sign convention and also the one used by Feynman.

4B§9.5 Lorentz Transformation

For **two reference frames**:

frame q that we call "stationary"

frame Q that we see moving toward +x at constant velocity v

the coordinates in q are: (ct, x, y, z)

the coordinates in Q are: (cT, X, Y, Z)

1C§25.6 The **Lorentz transformation** from frame q to frame Q is:

$$X = \gamma\,(x - \beta\, ct)$$

$$cT = \gamma\,(ct - \beta\, x)$$

$$Y = y$$

$$Z = z$$

1C§25.6 The **Lorentz transformation** from frame Q to frame q is:

$$x = \gamma\,(X + \beta\, cT)$$
$$ct = \gamma\,(cT + \beta\, X)$$
$$y = Y$$
$$z = Z$$

4B§9.6 What's Relative

In a frame moving with velocity v relative to our "stationary" frame, we observe time t, mass m_{rel}, and length L (along the v-direction) to be different than the corresponding t_0, m, and L_0 in our frame, according to:

1C§26.1 **Time Dilation**: $t = t_0 / \gamma$

We observe time running slower in a moving frame. Events of fixed duration T, such as one Earth-day, seem to take more time, γT, in a frame moving relative to Earth.

1C§26.2 **Length Contraction**: $L = L_0 / \gamma$

A length that is L_0 in a moving frame, appears to us have a shorter length L.

1C§26.3 **Relativistic Mass**: $m_{rel} = m\,\gamma$

We observe objects having a greater mass m_{rel} in a moving frame than their rest mass m.

1C§26.4 **Einstein's Famous Equation** is: $E = m_{rel}\, c^2 = \gamma\, m\, c^2$

Here, m_{rel} is the relativistic mass, and m is the rest mass.

1C§26.4 **Energy, Rest Mass, and Momentum** are related by:

$$E^2 = m^2 c^4 + p^2 c^2$$

4B§9.7 Illustrative Examples

Let's employ the Lorentz transformation to solve some simple problems. In relativity, even seemingly simple problems can be tricky. Define a stationary reference frame with coordinates (ct, x, y, z) that contains a clock and a ruler. Also define a second reference frame moving with velocity v in the +x-direction with coordinates (cT, X, Y, Z). Let both frames have the same origin, so that t = x = y = z = 0 is the same event as T = X = Y = Z = 0. This is illustrated below, where the axes are slightly displaced only for clarity.

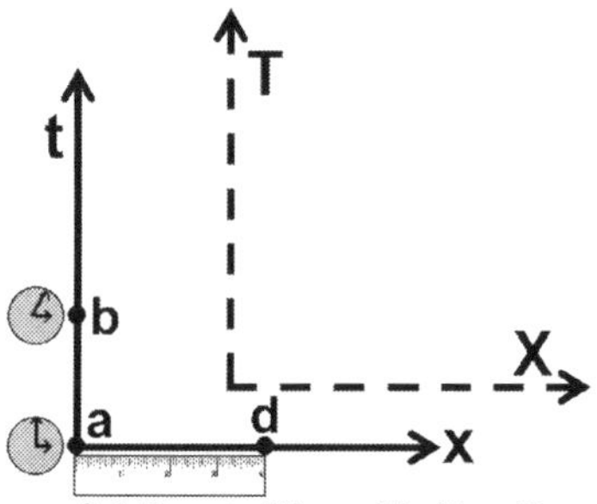

Figure 9-1 Events Viewed in Two Frames

Figure 9-1 shows the clock in the xt frame at two different times: t = 0 and $t = t_0$. The ruler is shown at t = 0, with its ends at x = 0 and $x = L_0$. In the stationary frame, the key events have coordinates:

$$a = (0,\ 0,\ 0,\ 0)$$
$$b = (ct_0,\ 0,\ 0,\ 0)$$
$$d = (0,\ L_0,\ 0,\ 0)$$

The corresponding Lorentz transformed coordinates in the XT frame are:

$$A = (0, \quad 0, 0, 0)$$

$$B = (\gamma c t_0, \quad -\gamma v t_0, 0, 0)$$

$$D = (-\gamma v L_0 / c, \quad \gamma L_0, 0, 0)$$

In the XT frame, the ruler's length seems to be:

$$D_X - A_X = \gamma L_0$$

This is *longer* than L_0, its length in the xt frame. What happened to length contraction? The problem here is that we made an improper comparison. In the xt frame, events a and d are simultaneous, but in the XT frame, events A and D have different time coordinates. It makes no sense to calculate the length of a moving ruler by subtracting its end locations at two different times. To properly compare ruler lengths in two different frames, both ruler end positions must be measured *at the same time* t in the xt frame, and *at the same time* T in the XT frame.

The left end of the ruler is at A at T = 0. Let's find where the right end is at T = 0. Since D_t is negative, let's wait a time interval:

$$\Delta T = -D_t / c > 0$$

and then measure the location of the right end. During time ΔT, an observer in XT sees the ruler move a distance ΔX, given by:

$$\Delta X = \Delta T \, (-v)$$

$$\Delta X = (-D_t / c)\,(-v)$$

$$\Delta X = (+\gamma v L_0 / c^2)\,(-v)$$

$$\Delta X = -\gamma v^2 L_0 / c^2$$

dX is negative, because in the XT frame, the ruler is moving toward –X at velocity v. The new X position is:

$$D_X + \Delta X = \gamma L_0 - \gamma v^2 L_0 / c^2$$

$$D_X + \Delta X = \gamma L_0 (1 - v^2 / c^2)$$

$$D_X + \Delta X = \gamma L_0 / \gamma^2 = L_0 / \gamma$$

When both ends of the ruler are measured at the same time in the XT frame, its length is indeed *contracted*, just as special relativity claims.

Let's now check the clock readings. In the XT frame, the time interval seems to be:

$$(B_t - A_t) / c = \gamma t_0$$

This is *longer* than t_0, the time interval in the xt frame. What happened to time dilation? Again, this is an improper comparison. In the XT frame, events A and B have different X-coordinates. To properly compare the passage of time, we must compare clock readings *at the same location*. Even if two clocks count time at exactly the same rate, we would observe a difference in their readings if one is close and the other is far away, due to the finite speed of light. Compared to your wristwatch, a clock on the moon seems 1.3 seconds late, even if both keep exactly the same time. It makes no sense to measure your heart rate by subtracting your wristwatch reading at one beat from the lunar clock reading at the next beat.

We must compare two clocks at the same location, in this case X = Y = Z =0. We have the clock at A with those coordinates, so we need another clock that passes through X = Y = Z = 0 in the XT frame. Let's find an event g in the xt frame, with the same clock time t_0 as event b, and with some unknown x-coordinate called λ, that transforms to event G with X = Y = Z = 0.

$$g = (c t_0, \quad \lambda, 0, 0)$$

$$G = (\gamma c t_0 - \gamma v \lambda / c, \quad \gamma \lambda - \gamma v t_0, 0, 0)$$

for X = 0, we require $\lambda = v t_0$

$$G = (\gamma c t_0 - \gamma t_0 v^2 / c, \quad 0, \quad 0, \quad 0)$$

$$G = (c t_0 / \gamma, \quad 0, \quad 0, \quad 0)$$

We can now properly compare the time difference t_0 / γ between events G and A in the XT frame with the time difference t_0 between events g and a in the xt frame. The XT observer sees the moving clock running *slower* by the factor γ, just as special relativity claims.

When in doubt, the safest approach is using the Lorentz transformations to properly compare events in different reference frames.

4B§9.8 Gravity Waves

Einstein's theory of general relativity says spacetime is curved by energy and mass. Furthermore, that curvature is dynamic, continuously changing as energy and mass move. These changes can be rapid and substantial, particularly with massive **binaries**, two celestial bodies orbiting one another in close proximity. The most interesting cases are binaries formed by neutron stars or black holes. Binaries curve spacetime in repetitive cycles governed by a wave equation, creating ripples of spacetime curvature that propagate outward in the binary's orbital plane at the speed of light. These ripples, sketched below, are called **Gravity Waves**.

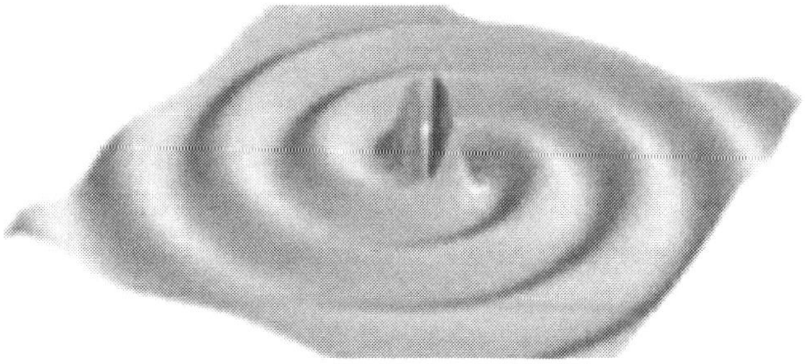

Figure 9-2 Expanding Gravity Wave

Russell Hulse and Joseph Taylor achieved the first indirect detection of gravity waves by precisely measuring the orbit of a binary comprised of two neutron stars. These neutron stars have masses of 1.440 and 1.389 solar masses. Their minimum separation is 1/200th of the Earth-Sun distance, and their orbital period, nearly 8 hours, is known to 13 decimal digits. For this binary, general relativity predicts gravity waves carry away enough energy to shorten the orbital period by 75.818 microseconds per year. After 35 years of observation, the measured rate of orbital decay is (99.7±0.2)% of the rate predicted by general relativity — a remarkable triumph of both theory and experiment.

Unlike other waves, gravity waves have a quadrupole action — simultaneously oscillating in both transverse directions. At one instant, a gravity wave propagating in the z-direction stretches space in the x-direction while squeezing space in the y-direction. Later, it squeezes along x and stretches along y. As stretching and squeezing sinusoidally alternate, there are intermediate moments of zero distortion, as shown below for a wave coming out of the page.

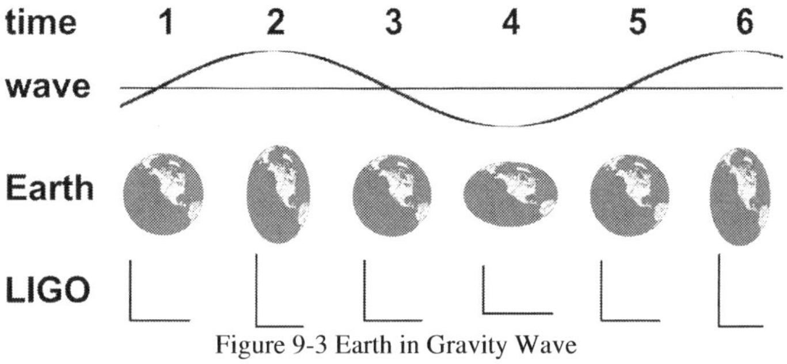

Figure 9-3 Earth in Gravity Wave

The most advanced gravity wave detector is LIGO, the Laser Interferometer Gravitational-Wave Observatory that now has identical interferometers at Hanford, Washington and Livingston, Louisiana, and hopefully more to come. The lowest row in the above image shows the alternating stretching and compression of LIGO's two arms that form an "L". As in the Michelson-Morley interferometer, LIGO measures the travel time difference of light along the two arms. As one arm stretches and the other shrinks, light travel times change.

LIGO's sensitivity is incredible: it can detect length changes of less than one part in 10^{21}.

In February 2016, LIGO announced its first gravity wave detection — a short burst from the merger of two black holes that were 1.3 billion light-years away, and had masses of 29 and 36 times our Sun's mass. It is estimated that 3 solar masses were converted into gravity wave energy in this merger. In an instant, this black hole merger generated more energy than all the stars in our observable universe. Hanford's interferometer signal is shown below.

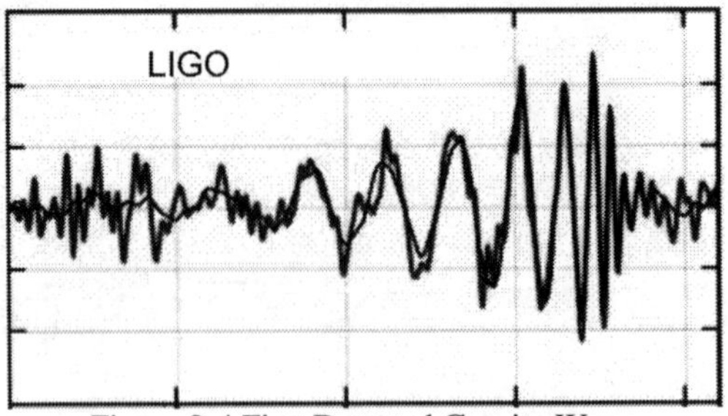

Figure 9-4 First Detected Gravity Wave

In this figure, the horizontal axis is time, with tic marks every 0.05 seconds, and the vertical axis is fractional length change, with tic marks every 5×10^{-22}.

Livingston detected the same wave 0.007 seconds earlier than Hanford, confirming the detection and providing a rough indication of the direction of the wave's origin.

A color image comparing the signal from Livingston and Hanford is on the back cover of this book. There, the Livingston signal is delayed by 7 ms to match Hanford.

This opens a new field of science: gravity wave astronomy.

Chapter 10
Physics of Light

4B§10.1 Basic Parameters of Light

Recall that h is Planck's constant and ħ ("h-bar") = h / 2π.

1C§31.1 **All forms of light** are characterized by:

Velocity is always c = 299,792,458 m/s

Frequency f in cycles / sec (Hertz)

Frequency ω in radians / sec = 2π f

Wavelength λ = c / f = h / p

Wave number k = w / c = 2π / λ

Energy E = h f = ħ ω = p c

Momentum p = E / c = h / λ = ħ k

1C§31.2 Although **all photons are intrinsically identical**, different names are commonly assigned to light with observed frequencies in different ranges, as listed below.

f (Hz)	λ (m)	Name
$> 1\times10^{19}$	$< 3\times10^{-11}$	gamma rays
3×10^{18}	1×10^{-10}	x-rays
3×10^{15}	1×10^{-7}	ultraviolet
7.5×10^{14}	4×10^{-7}	visible – blue
4.3×10^{14}	7×10^{-7}	visible – red
$< 3\times10^{11}$	$> 1\times10^{-3}$	radio waves

4B§10.2 Interference & Diffraction

At any fixed point, a **single-frequency** wave has the form:

$$\text{amplitude} = A \cos(\omega t + \phi)$$

$$\text{intensity } \mathbf{I} = A^2$$

Here, intensity **I** is the wave energy per unit area, ω is its frequency, and ø is its phase angle at t = 0.

1C§35.4 When two single-frequency **Waves Combine**, they interfere, and the resultant amplitude A and intensity **I** at a fixed point, are given by:

$$A = A_1 \cos(\omega_1 t + \phi_1) + A_2 \cos(\omega_2 t + \phi_2)$$

$$\mathbf{I} = A_1^2 + A_2^2 + 2A_1 A_2 \cos(\Delta\omega\, t + \Delta\phi)$$

$$\Delta\omega = \omega_1 - \omega_2$$

$$\Delta\phi = \phi_1 - \phi_2$$

For two single-frequency waves of equal intensity $\mathbf{I}_0$, their combined intensity oscillates between zero and $4\mathbf{I}_0$ at the **beat frequency** Δω. In tightly controlled conditions, interference effects can have dramatic consequences. An example is steering a **phased array** radio beam from Pasadena to either Honolulu or Alberta. But when two wave paths differ by billions of wavelengths, not unrealistic for visible light, even tiny fractional changes can shift their relative phases by many radians and completely alter their interference. The 4mm-pupil diameter of a human eye corresponds to 10,000 wavelengths of blue light. A viewing angle change of 0.01° shifts phases by 10 radians, obliterating interferences effects.

Frequency variations also limit interference in the macro-world. For visible light, a frequency difference between two sources of even one part per billion nullifies interference in 1 microsecond. In a normal lamp, one atom radiates, then another, and then another. Each radiates for only tens of nanoseconds. No coherence can persist for more than billionths of a second.

In V1p32-5, Feynman says: "Of course, in nature [interference] is always there, but we may not be able to detect it."

Lasers are the exception to this general rule. In 1963, Mandel and Magyar demonstrated interference between two independent ruby lasers, with a $\Delta\omega$ of about 500kHz, a frequency match and stability of one part per billion.

1C§31.3 **Interference** is the most dramatic effect of combining waves. It arises only when **coherent waves** combine — waves that have the **same frequency** and a fixed **phase shift**. Waves with different frequencies or with varying phase shifts are **incoherent**. Phase shift is how much one wave is shifted relative to another wave of the same frequency. Two waves that crest and trough at the same time have zero phase shift.

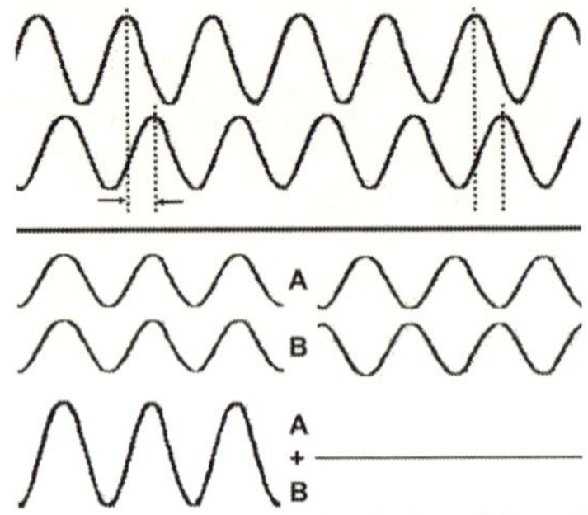

Figure 10-1 Wave Phase Shift & Adding Waves

The two waves shown in the upper portion of Figure 10-1 have a phase shift of 1/4 wavelength = 90° = $\pi/2$ radians. Since waves repeat exactly each cycle, a phase shift of 1.37 wavelengths is indistinguishable from a phase shift of 0.37λ; thus phase shifts are normally described as ranging from 0 to 1 λ, or equivalently from 0 to 2π radians.

1C§31.3 Completely **Constructive Interference** occurs when two waves combine that have the same frequency and zero phase shift, as in the lower left side of Figure 10-1. These combining waves produce a wave whose amplitude equals the sum of the amplitudes of the combining waves.

1C§31.3 Completely **Destructive Interference** occurs when two waves combine that have the same frequency and phase shift of 1/2 wavelength = 180° = π radians, as in the lower right side of Figure 10-1. Two such waves with equal amplitudes **totally cancel one another**.

1C§31.3 **Partial Interference** occurs when two waves combine that have the same frequency and a phase shift that is not an integer multiple of π radians. The result of these combining waves is somewhere between the results of completely constructive and completely destructive interference.

1C§31.6 The **Two Slit Experiment with Light** features a light source S illuminating an opaque barrier with two small slits, resulting in two coherent light waves with zero phase shift exiting the slits, as shown below with the curved arcs representing wave crests.

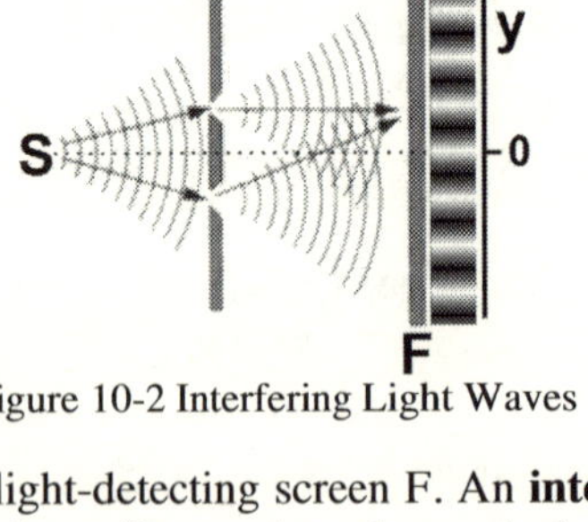

Figure 10-2 Interfering Light Waves

The two coherent waves combine on a light-detecting screen F. An **interference fringe pattern,** alternating black and white bands, is shown on the right of the above figure, along the vertical axis y.

Waves from the upper and lower slits travel the same distance to reach F at y=0. Therefore, their relative phase shift remains zero, as it is at the two-slit barrier. The waves interfere constructively, resulting in high-intensity light at y=0, represented by the central white fringe.

But, as y increases, the light path length from the upper slit decreases, and the path length from the lower slit increases, as shown by the arrows in the figure. At some y value, the difference in travel distance equals λ/2. There, the lower wave arrives one-half cycle after the upper wave, making the interference entirely destructive. The light intensity drops to zero, resulting in a black fringe.

At about twice that value of y, the path length difference grows to λ and the lower wave arrives one full cycle after the upper wave. There, the phase shift is effectively zero, interference is constructive, and a white fringe is produced. This pattern continues up and down the vertical axis, resulting in alternating black and white fringes of nearly equal width.

Interference fringe patterns are hallmark signatures of wave behavior.

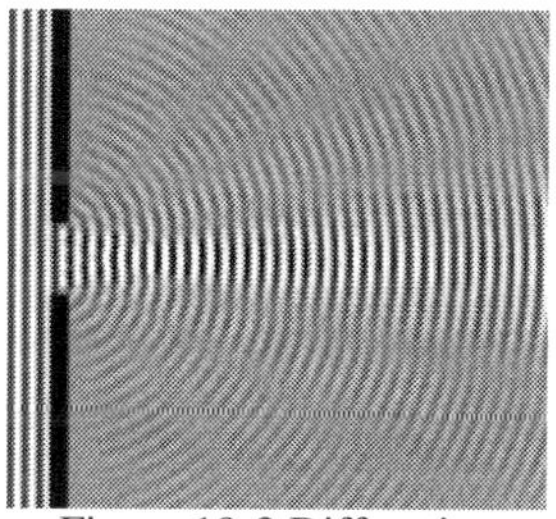

Figure 10-3 Diffraction

1C§31.9 **Diffraction by an Aperture** is shown above. Here a plane wave entering from the left, with white crests and black troughs, hits a bold black barrier with a central aperture. A portion of the wave passes through the aperture, producing a central lobe of high intensity and side lobes of progressively lower intensity, with the lobes separated by null lines of zero intensity. The angle θ of the first null lines on either side of the central lobe is given by:

$$\text{1-D aperture: } \sin\theta = \lambda / W$$

$$\text{2-D aperture: } \sin\theta = 1.22\, \lambda / W$$

Here, λ is the wavelength of incident light, and W is the aperture width.

1C§34.6 **Diffraction by an Aperture**, assuming $\lambda << W$, is best understood as:

(1) no emission comes from the aperture itself;

(2) every atom in the barrier is an emission source; and

(3) the post-barrier field equals minus the field that would be radiated by a barrier that filled the aperture.

1C§31.6 The **Rayleigh Diffraction Limit**, the smallest angle θ_{min} resolvable by an optical system, is given by:

$$\text{in 1-D: } \sin(\theta_{min}) = \lambda / W$$

$$\text{in 2-D: } \sin(\theta_{min}) = 1.22\, \lambda / W$$

Here, W is the system's limiting aperture, and λ is the wavelength of the light employed.

1C§33.2 A **Diffraction Grating** can resolve different wavelengths of light according to:

$$(\lambda - \lambda^*) / \lambda^* = 1 / n\, m$$

Here, λ and λ* are the closest resolvable wavelengths, n is the number diffracting grooves, and m is the **beam order**, the integral number of wavelengths of phase shift between adjacent grooves. This assumes the incident light persists for at least nm wavelengths.

3A§3.4 **X-ray Crystallography** employs x-ray diffraction by the atoms in a crystal, and the resulting constructive interference that occurs at certain angles. These **scattering** angles are given by:

$$\sin\theta = n\, \lambda / 2d$$

Here, n is any integer, λ is the x-ray wavelength, and d is one of the repeating atomic spacings in the 3-D crystal lattice.

4B§10.3 Geometric Optics

Over the centuries, our understanding of light has evolved through several stages.

1C§30.2 The **Principle of Least Distance**, our earliest concept, says light always travels in a straight line. But, this pleasingly simple idea cannot explain refraction.

1C§30.3 Next came **Fermat's Principle of Least Time** that postulates light always takes the fastest route between two points, the path of least time. This concept explains refraction, but fails to explain reflection by elliptical mirrors and other phenomena.

1C§30.4 The ultimate theory is the **Feynman Sum Over Histories**, also called the **Feynman Path Integral Formulation**. Feynman says light travels along **all possible paths** between two points, but only near a path of extremal time (a local maximum or minimum) do many similar paths interfere constructively, establishing the path along which almost all of light's energy travels.

1C§30.4 The **Principle of Reciprocity**, valid in each of the three above concepts of light propagation, says if light can travel from point A to point B, then it can equally well travel from B to A.

1C§30.5 **Any System of Lenses** has two focal points, one on either side of the lens system. Let the material through which light passes, its medium, have refractive index n_1 on side #1 and n_2 on side #2, and let the corresponding focal lengths be f_1 and f_2. These quantities are related by:

$$f_1 / n_1 = f_2 / n_2$$

1C§30.8 **Principal Planes**: any lens system with the same refractive index at opposite ends has two principal planes that are normal to its axis. Such systems act as if everything inside the two planes, including the space between them, can be removed and replaced by a single thin lens.

Let's next consider light from a source S, incident on a medium of refractive index n_2 that has a spherical surface of radius R. Let S reside in a medium with refractive index n_1. (In air, n_1 = 1.0003) Some key quantities are indicated in Figure 10-4.

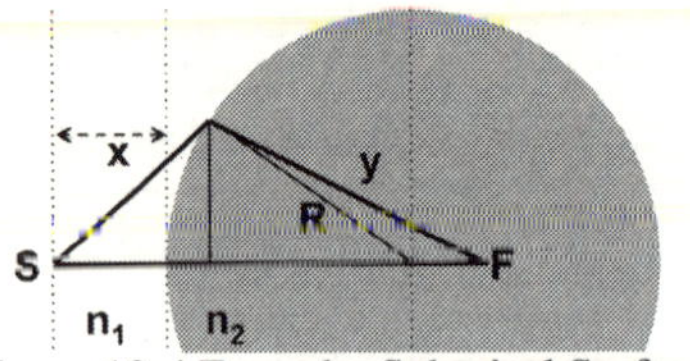

Figure 10-4 Focus by Spherical Surface

1C§30.5 The **Spherical Lens Equation**, applied to the above figure, is:

$$n_1 / x + n_2 / y = (n_2 - n_1) / R$$

Here, x is the source distance outside the sphere (where $n = n_1$), y is the image depth inside the sphere (where $n = n_2$), and R is the radius of the spherical lens. If x = ∞, y becomes the **inside focal length**, and if y = ∞, x becomes the **outside focal length**. These are given by:

$$f_{inside} = n_2 R / (n_2 - n_1)$$

$$f_{outside} = n_1 R / (n_2 - n_1)$$

The following image illustrates the optics of a **thin lens** that has the same refractive index on both sides.

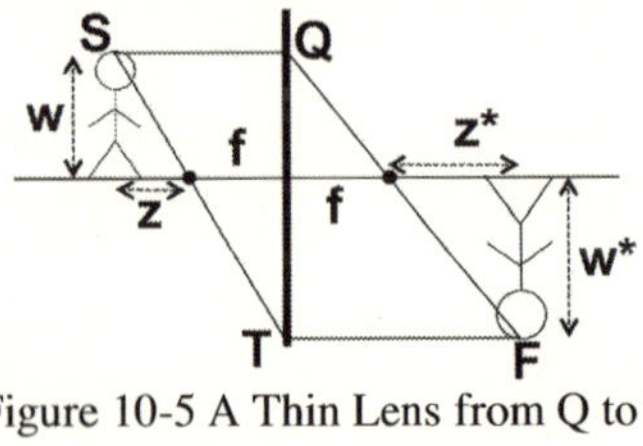

Figure 10-5 A Thin Lens from Q to T

1C§30.6 The **Thin Lens Equation**, with quantities identified in Figure 10-5, is:

$$1 / (f + z) + 1 / (f + z^*) = 1 / f$$

1C§30.7 The **Ray-Tracing Rules** for a thin lens are:

(1) Light rays entering parallel to the lens axis on one side pass through the focal point on the opposite side.

(2) Rays passing through the focal point on one side exit parallel to the lens axis on the opposite side.

1C§30.7 The **Magnification** M of a thin lens, with the quantities identified in Figure 10-5, is:

$$M = w^* / w = f / z = z^* / f$$

4B§10.4 Index of Refraction

In this section, n is the index of refraction, m and q are the electron's mass and charge, N is the number of active electrons per unit volume, and ω is the frequency of incident light.

1C§34.2 **Light's Speed** in refractive matter is **apparently slower** than its speed in vacuum. Matter absorbs and reradiates light, thereby continuously retarding the phase angle of light waves. While light from a single source **always travels at speed c**, combined incident and reradiated waves propagate at **phase velocity** c/n, where n is the refractive index.

1C§34.2 For light traversing **low-density matter** (such as a gas), the refractive index n is given by:

$$n = 1 + q^2 N / \{2\varepsilon_0 m (\Omega^2 - \omega^2)\}$$

This assumes electrons have a single natural harmonic oscillation frequency Ω, with no damping.

2C§32.3 For light traversing **matter of any density**, the **atomic polarizability** α (the polarization of each atom), and the refractive index n, are given by:

$$\alpha(\omega) = (q^2 / m\,\varepsilon_0)\, \Sigma_i\, f_i / (-\omega^2 + i\omega\,\mu_i + \omega_i^2)$$

$$n^2 = 1 + \alpha N / (1 - \alpha N / 3)$$

Here, electrons may have multiple harmonic oscillation frequencies, as denoted by the subscript i. For each oscillatory mode, the frequency is ω_i, the damping coefficient is μ_i, and the amplitude to enter that mode is f_i. This analysis incorporates some features of electron atomic states that are properly described by quantum mechanics.

2C§32.3 The **Clausius-Mossotti** equation (a rearrangement of the above equation) relates the refractive index to atomic polarizability α:

$$3\,(n^2 - 1) / (n^2 + 2) = \alpha N$$

2C§32.5 The **Refractive Index of Mixtures** of different materials, using the **Clausius-Mossotti** equation from above, is given by:

$$3(n^2 - 1) / (n^2 + 2) = \Sigma_k\, \alpha_k\, N_k$$

Here, we sum on the right each component of the mixture, with its own values of α and N. That sum determines the total refractive index n.

2C§32.4 A **Complex Refractive Index** can be separated into its real part n_r and its imaginary part n_i. For low-density matter with a single natural harmonic oscillation frequency Ω, we write:

$$\text{let: } \phi = N\,(q^2 / m\varepsilon_0) / \{ (\Omega^2 - \omega^2)^2 + \omega^2\mu^2 \}$$

$$n = 1 + \phi\,(-\omega^2 - i\omega\mu + \Omega^2)$$

$$n_r = \text{Re}\,\{n\} = 1 + \phi\,(-\omega^2 + \Omega^2)$$

$$n_i = \text{Im}\,\{n\} = + \phi\,(\,i\omega\mu\,)$$

$$E = E_0 \exp\{i\omega\,(t - z / v_{ph})\} \exp\{ -\beta\, z / 2\}$$

$$\text{phase velocity: } v_{ph} = c / n_r$$

$$\text{intensity absp. coeff.: } \beta = 2\, n_i\, \omega / c$$

Above, z is the distance light penetrates the refractive material, v_{ph} is light's phase velocity, ω is its frequency, and E is its electric field strength. n_r reduces phase velocity, corresponding to our normal definition of the refractive index. The wave **intensity** ($\sim E^2$) drops exponentially as $\exp\{-\beta z\}$, where β is the **intensity absorption coefficient**. The damping coefficient μ is positive, hence so are n_i and β.

2C§32.7 **Skin Depth** δ is the absorption coefficient of radiation in a metal, as given by:

$$\delta = \sqrt{(2\varepsilon_0 c^2 / \sigma\, \omega)}$$

$$\text{wave amplitude} \sim \exp\{-z / \delta\}$$

Here, ω is the radiation frequency in radians per second, σ is the metal's conductivity, and z is the distance into the metal. In copper:

$$\delta = 16.7 \text{ cm} / \sqrt{\omega} \text{ (radians/sec)}$$

$$\delta = 6.66 \text{ microns at 10 GHz}$$

2C§34.3 The **Plasma Frequency** ω_p of a gas of free electrons is given by:

$$\omega_p^2 = q^2 n_0 / m\, \varepsilon_0 = 4\pi\, e^2 n_0 / m$$

$$e^2 = q^2 / 4\pi\, \varepsilon_0$$

Here, q and m are the electron charge and mass, and n_0 is number density of electrons at equilibrium. Radiation of frequency ω incident on a free electron gas is reflected if $\omega < \omega_p$, and propagates freely through the gas if $\omega > \omega_p$.

1C§36.3 **Birefringent Materials** have different refractive indices in different directions. These materials are typically comprised of arrays of parallel, long, slender molecules.

When driven by incident light, electrons move easily along the **optic axis**, the lengths of the molecules, but are restricted laterally by the molecules' narrow widths.

Examples are:

- **Polaroid filters** (1C§36.4)
- **optically active** helical molecules that rotate polarization (1C§36.6)
- **anomalous refraction** that produces multiple images (1C§36.9)

4B§10.5 Polarization of Light

A photon moving in the +z-direction has electric field ***E*** and magnetic field ***B*** that oscillate in the xy-plane, orthogonal to the photon's velocity. ***E*** and ***B*** are also orthogonal to one another. Hence, specifying the orientation of ***E*** — the photon's **polarization** — is sufficient to fully specify the photon's electromagnetic field directions.

1C§36.2 **Elliptical Polarization** is the most general polarization state of light. Here the x and y components of light's electric field oscillate at the same frequency, but possibly with different amplitudes and a phase shift. The representation is:

$$\boldsymbol{E} = A\,(\cos\theta \cos[\omega t],\ \sin\theta \cos[\omega t - \phi],\ 0)$$

1C§36.2 **Circular Polarization** is a special case of elliptical polarization. Two convenient basis states are: right-hand-circular (**RHC**) and left-hand-circular (**LHC**) that are represented by:

$$\boldsymbol{E}_{RHC} = (\cos[\omega t],\ +\sin[\omega t],\ 0) / \sqrt{2}$$

$$\boldsymbol{E}_{LHC} = (\cos[\omega t],\ -\sin[\omega t],\ 0) / \sqrt{2}$$

These equations are derived from the elliptical equation above. First, set $\phi = \pi/2$, and use $\cos(\omega t - \pi/2) = \sin(\omega t)$. Then select $\theta = +\pi/4$ for RHC, or $\theta = -\pi/4$ for LHC.

1C§36.1 **Linear Polarization** is the special case of elliptical polarization with $\phi = \pi/2$. Two convenient basis states are x- and y-polarization, corresponding to $\theta = 0$ and $\theta = \pi/2$ respectively, that are represented by:

$$\boldsymbol{E}_x = (\cos[\omega t],\ 0,\ 0)$$

$$\boldsymbol{E}_y = (0,\ \sin[\omega t],\ 0)$$

1C§36.2 **Converting Polarizations** from circular basis states RHC and LHC to x and y linear basis states is as follows:

$$E_x = \{E_{RHC} + E_{LHC}\} / \sqrt{2}$$

$$E_y = \{E_{RHC} - E_{LHC}\} / \sqrt{2}$$

Convert from x-y to RHC-LHC as follows:

$$E_{RHC} = \{E_x + E_y\} / \sqrt{2}$$

$$E_{LHC} = \{E_x - E_y\} / \sqrt{2}$$

4B§10.6 Reflection & Refraction

Consider a slender beam of light incident on a surface S that separates two media whose refractive indices are n_1 and n_2, as shown below.

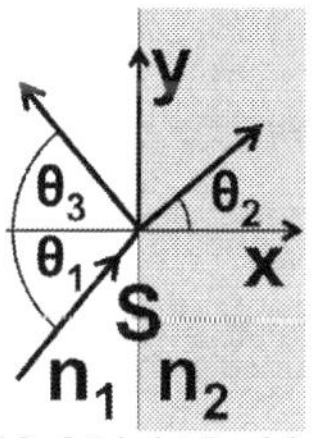

Figure 10-6 Light Incident on S

Here, light is incident at **normal angle** θ_1. "Normal angles" are measured relative to the normal (perpendicular) to the surface at the point of impact. In this case, the normal is the x-axis. The **incident plane** is the plane of the page, the z=0 plane, the plane containing all three light beams shown in the above figure. Let $\mathbf{I_0}$ be the incident light intensity. In general, some light is reflected at angle θ_3, and some is refracted (or transmitted) at angle θ_2.

1C§36.8 **Key Physics** of **Reflection** and **Refraction**. For ease of discussion, consider Figure 10-6 in the case of vacuum ($n_1 = 1$) on the left and a dense medium on the right ($n_2 > 1$). If the atoms of that medium were inert, the incident beam would continue at speed c through that medium on a straight line at angle θ_1. But, what actually happens is that electrons in these atoms, accelerated by the incident light, radiate waves that interfere with the incident light waves, resulting in three effects:

(1) reflected light at angle θ_3 at speed c / n_1

(2) refracted light at angle θ_2 at speed c / n_2

(3) cancellation of incident light in the medium at angle θ_1

1C§36.8 **Separation of Polarization**: the three effects described above occur independently for the two linear polarization states of incident light. Light polarized in the z-direction (its electric field is perpendicular to the page, normal to the incident plane) accelerates electrons in the z-direction only. These electrons radiate secondary waves polarized in the z-direction only. Similarly, light polarized within the incident plane has effects only within the incident plane. This separates the analysis into two more manageable halves.

1C§30.2 **Reflection** is govern by:

reflected angle θ_3 = incident angle θ_1

1C§30.2 **Snell's Law** of **Refraction / Transmission** is:

$$n_1 \sin\theta_1 = n_2 \sin\theta_2$$

1C§31.5 **Total Internal Reflection** occurs when $n_2 < n_1$ and $\theta_1 > \theta_c$, as given by:

$$\sin(\theta_c) = n_2 / n_1$$

When light traversing a dense medium hits the surface S of a less dense medium, 100% of the light is reflected and 0% is transmitted for any incident angle θ_1 equal to or greater than θ_c, the **critical angle**. This phenomenon keeps light within optical fibers.

1C§36.8 The **Fresnel Reflection Formulas**, for incident angle θ_1 and refracted angle θ_2, are:

$$\beta = -\sin(\theta_1 - \theta_2) / \sin(\theta_1 + \theta_2)$$

$$b = -\tan(\theta_1 - \theta_2) / \tan(\theta_1 + \theta_2)$$

For incident light of amplitude 1, β is the reflection amplitude for polarization normal to the incident plane, and b is the reflection amplitude for polarization within the incident plane.

1C§36.7 **Brewster's Angle** θ_B is the angle of incident light that results in reflected light that is entirely polarized normal to the incident plane. θ_B is given by:

$$\tan(\theta_B) = n_2 / n_1$$

At θ_B, the reflected beam at angle θ_3 is orthogonal to the transmitted beam at angle θ_2; hence, transmitted light polarized within the incident plane cannot radiate light at angle θ_3. The reflected beam thus has zero component of polarization within the incident plane.

Due to this effect, sunlight reflected from horizontal surfaces is often primarily polarized horizontally. Polarized sunglasses eliminate most of that reflected glare.

2C§33.6 For light polarized with its electric field ***E*** **normal to the incident plane**:

Reflected intensity $\mathbf{I_S}$ is given by:

$$R_S = \mathbf{I_S} / \mathbf{I_0}$$

$$R_S = \sin^2(\theta_1 - \theta_2) / \sin^2(\theta_1 + \theta_2)$$

Refracted electric field amplitude E_{02} is:

$$E_{02} / E_{01} = 2 \cos\theta_1 \sin\theta_2 / \sin(\theta_1 + \theta_2)$$

2C§33.6 For light polarized with its electric field ***E*** **in the incident plane**:

Reflected intensity $\mathbf{I_{plane}}$ is:

$$R_{plane} = \mathbf{I_{plane}} / \mathbf{I_0}$$

$$R_{plane} = \tan^2(\theta_1 - \theta_2) / \tan^2(\theta_1 + \theta_2)$$

Refracted electric field amplitude E_{02} is:

$$|E_{02}| / |E_{01}| = 2 / (n_2 / n_1 + \cos\theta_2 / \cos\theta_1)$$

2C§33.6 For light at **normal incidence** ($0 = \theta_1 = \theta_2 = \theta_3$):

Reflected intensity $\mathbf{I}(\theta{=}0)$ is:

$$R(\theta{=}0) = \mathbf{I}(\theta{=}0) / \mathbf{I_0}$$

$$R(\theta{=}0) = (n_2 - n_1)^2 / (n_2 + n_1)^2$$

Refracted electric field amplitude E_{02} is:

$$|E_{02}| / |E_{01}| = 2 n_1 / (n_2 + n_1)$$

4B§10.7 Radiation

Consider radiation from the array of electric dipole sources shown below.

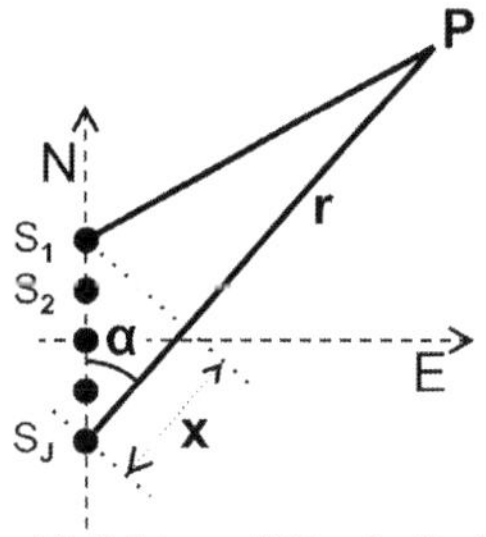

Figure 10-7 Line of Dipole Radiators

In Figure 10-7, we are looking down from above at J dipole radiators, S_1 to S_J, arranged in a straight north-south line. Point P is at angle α relative to due north. P and the dipoles are all at the same elevation.

1C§32.9 An array of **Dipole Radiators**, aligned as in Figure 10-7, produces an electric field $\boldsymbol{E}$ and radiation of intensity **I**, at a remote point P at position $\boldsymbol{r}$, as given by:

$$\boldsymbol{E} = (A / r) \exp(i\omega t) \Sigma_K \{ \exp[i(K-1) u] \}$$

$$\mathbf{I} = \mathbf{I}_1 \sin^2(J u / 2) / \sin^2(u / 2)$$

$$\text{with } \mathbf{I}_1 = (A / r)^2 / 2$$

$$\text{and } u = ø - 2\pi [D / \lambda] \cos\alpha$$

Here, D is the separation between each dipole, and all charges accelerate vertically at frequency ω with the same amplitude A. Also, $\mathbf{I}_1$ is the intensity from one dipole, ø is an adjustable phase shift between successive dipoles, α is the angle of point P relative to the line of dipoles, and Σ_K sums over all dipoles K = 1 to J. Phase angle ø allows us to steer radiation in desired directions, inspiring the name **phased array**.

1C§35.5 The **Radiation from one Electron** being driven sinusoidally at frequency ω, summed over all directions, is given by:

$$\mathbf{I} = \{\varepsilon_0 c E_0^2 / 2\} \sigma(\omega)$$

$$\sigma(\omega) = \sigma_T \{\omega^2 / (\Omega^2 - \omega^2) \}^2$$

$$\sigma_T = 8\pi r_0^2 / 3 = 6.65\times10^{-29} \text{ m}^2$$

Here, **I** is the radiation intensity, E_0 is the amplitude of the driving electric field, Ω is the electron's natural frequency, σ is the electron's **scattering cross section**, and σ_T is the **Thomson scattering cross section**. In particle physics units, σ_T = 0.665 barns, an enormous value. For Ω<<ω, σ(ω) is nearly constant. For Ω>>ω, as it is for visible light in air, σ(ω) is nearly proportional to the fourth power of ω. This is why the **sky is blue** and **sunsets are red**.

4B§10.8 Relativistic Effects

1C§37.8 The **Doppler Effect** relates the frequency of light emitted in one reference frame with the frequency observed in a different frame. Light emitted with frequency Ω in its source frame is observed to have frequency ω by an observer with relative velocity βc, as given by:

$$\omega = \Omega \sqrt{ (1+\beta) / (1-\beta) }$$

If the source and the observer are moving closer, β is positive. If the source and observer are separating, β is negative.

1C§37.9 **Stellar Aberration** is the change in the observed direction of starlight due to Earth's orbital motion around the Sun, as given by:

$$\sin\theta = v / c \sim 0.1 \text{ milliradians}$$

Here, v is Earth's orbital velocity, and θ is the angular shift in the apparent direction of the star. This equation is valid for starlight incident normal to Earth's orbital plane.

Chapter 11
Electromagnetism

The electric force is a trillion, trillion, trillion times stronger than gravity. Unlike gravitational mass that is always positive, electric charge can be positive, negative, or zero, resulting in different forces. A positive charge and a negative charge attract one another, however two positives or two negatives repel one another. Objects with zero charge throughout neither exert nor experience electromagnetic forces. The intense repulsive force of like charges spreads matter out, while the intense attractive force of opposite charges thoroughly mixes positive and negative charges. Electromagnetism determines the characteristics of atoms, giving matter form, strength, and diversity.

Electromagnetism ("EM"), as presented in this book, is valid only in **inertial reference frames**, those moving at constant velocity ***v***.

4B§11.1 Primary Quantities

E: electric vector field, E = | ***E*** |

B: magnetic vector field, B = | ***B*** |

ø: scalar potential

A: vector potential

q and Q: electric charges

J: current, product of charge and velocity

ϱ: charge density per unit volume

j: current density per unit area

Ď: vector operator (∂/∂x, ∂/∂y, ∂/∂z)

ε_0: vacuum permittivity constant

4B§11.2 Primary Equations of EM

1C§34.2 **Linear Superposition** is a fundamental property of electromagnetism. The scalar potential ø due to multiple sources is the sum of the individual ø's from each source. The same is true for each of the three components of the vector fields:

E, the electric field
B, the magnetic field
A, the vector potential

2A§1.7 **Current Density *j***, the product of charge density ϱ and charge velocity ***v***, is given by:

j = ϱ ***v***

2B§14.2 **Charge Conservation** applies at every point in space and moment in time, hence the change in charge within any volume equals the current flow into that volume. This is written:

dϱ/dt = – ***Ď*** • ***j***

2A§1.2 The **Lorentz Force *F*** on a charge q due to fields ***E*** and ***B*** is:

F = q (***E*** + ***v*** × ***B***)

Fields ***E*** and ***B*** are evaluated at the time and location at which they act on charge q.

2A§2.6 **Maxwell's Field Equations** unite electricity and magnetism into a single entity: electromagnetism. Maxwell's equations are valid in all circumstances, but **the following forms are valid only in inertial reference frames free of gravity** (see 4B§11.8 for the fully relativistic equations).

$$\check{\boldsymbol{D}} \bullet \boldsymbol{E} = \varrho / \varepsilon_0$$

$$\check{\boldsymbol{D}} \times \boldsymbol{E} = - \partial \boldsymbol{B}/\partial t$$

$$\check{\boldsymbol{D}} \bullet \boldsymbol{B} = 0$$

$$c^2 \check{\boldsymbol{D}} \times \boldsymbol{B} = \partial \boldsymbol{E}/\partial t + \boldsymbol{j} / \varepsilon_0$$

2A§2.6 **Maxwell's Static Field Equations**, for stationary charges and constant currents, are two pairs of decoupled equations. In this static case only, electricity and magnetism appear unrelated, and are written:

$$\check{\boldsymbol{D}} \bullet \boldsymbol{E} = \varrho / \varepsilon_0$$

$$\check{\boldsymbol{D}} \times \boldsymbol{E} = 0$$

$$\check{\boldsymbol{D}} \bullet \boldsymbol{B} = 0$$

$$\check{\boldsymbol{D}} \times \boldsymbol{B} = \boldsymbol{j} / c^2 \varepsilon_0$$

2A§4.3 **Coulomb's Law** for the force F between stationary charges Q and q that are separated by distance r is:

$$F = Q\,q / 4\pi\varepsilon_0 r^2$$

Force F is attractive if Q and q have opposite polarities ($Q\,q < 0$), and repulsive if they have the same polarity ($Q\,q > 0$).

2A§8.6 and 2B§18.9 **Electromagnetic Field Energy** U within any volume V, and energy density u at any point, are given by:

$$u = \varepsilon_0 c^2 \boldsymbol{B} \bullet \boldsymbol{B} / 2 + \varepsilon_0 \boldsymbol{E} \bullet \boldsymbol{E} / 2$$

$$U = \int_V u \, dV$$

2B§19.6 **Electromagnetic Fields and Potentials** are related by:

$$\boldsymbol{E} = - \check{\boldsymbol{D}}\phi - \partial \boldsymbol{A}/\partial t$$

$$\boldsymbol{B} = \check{\boldsymbol{D}} \times \boldsymbol{A}$$

$$c^2 \check{\boldsymbol{D}} \bullet \boldsymbol{A} = - \partial\phi/\partial t$$

The last equation is called the **Lorentz Gauge**. For **Static Fields** (when all currents are constant and any other charges are stationary), these relationships become (see 2A§4.4 and 2B§15.1):

$$\boldsymbol{E} = - \check{\boldsymbol{D}}\, \phi$$

$$\boldsymbol{B} = \check{\boldsymbol{D}} \times \boldsymbol{A}$$

$$\check{\boldsymbol{D}} \bullet \boldsymbol{A} = 0$$

2B§19.6 The **Wave Equations** that govern the scalar and vector potentials are:

$$\square\, \phi = - \check{D}^2\phi + \partial^2\phi/\partial t^2 / c^2 = \varrho / \varepsilon_0$$

$$\square\, \boldsymbol{A} = - \check{D}^2\boldsymbol{A} + \partial^2\boldsymbol{A}/\partial t^2 / c^2 = \boldsymbol{j} / \varepsilon_0 c^2$$

2B§20.3 The **Wave Equations** for all EM fields in regions without sources are:

$$\square\, \phi = - \check{D}^2\phi + \partial^2\phi/\partial t^2 / c^2 = 0$$

$$\square\, \boldsymbol{A} = - \check{D}^2\boldsymbol{A} + \partial^2\boldsymbol{A}/\partial t^2 / c^2 = 0$$

$$\square\, \boldsymbol{E} = - \check{D}^2\boldsymbol{E} + \partial^2\boldsymbol{E}/\partial t^2 / c^2 = 0$$

$$\square\, \boldsymbol{B} = - \check{D}^2\boldsymbol{B} + \partial^2\boldsymbol{B}/\partial t^2 / c^2 = 0$$

As with harmonic oscillators, we analyze electromagnetic wave phenomena using complex quantities, and ultimately equate actual fields to the real parts of those complex quantities.

2C§33.3 In regions without sources, all EM fields have an exponential dependence of the form:

$$\exp\{ i\omega t - i\boldsymbol{k} \bullet \boldsymbol{r} \}$$

Here, the vector operator $\check{\boldsymbol{D}}$ simplifies to the multiplicative operator $-i\boldsymbol{k}$, as given by:

$$\check{\boldsymbol{D}}\, \phi = -i\boldsymbol{k}\, \phi = \boldsymbol{k}\, \phi \exp\{-i\,\pi / 2\}$$

$$\check{\boldsymbol{D}} \bullet \boldsymbol{E} = -i\boldsymbol{k} \bullet \boldsymbol{E} = \boldsymbol{k} \bullet \boldsymbol{E} \exp\{-i\,\pi / 2\}$$

$$\check{\boldsymbol{D}} \times \boldsymbol{E} = -i\boldsymbol{k} \times \boldsymbol{E} = \boldsymbol{k} \times \boldsymbol{E} \exp\{-i\,\pi / 2\}$$

Note that the –i factor does not make the field imaginary, but merely shifts its phase.

2C§33.3 **Faraday's law** for a single-frequency wave in a region without sources is:

$$\check{\boldsymbol{D}} \times \boldsymbol{E} = -\,\partial \boldsymbol{B}/\partial t$$

$$-i\boldsymbol{k} \times \boldsymbol{E} = -i\omega\, \boldsymbol{B}$$

$$\boldsymbol{B} = \boldsymbol{k} \times \boldsymbol{E} / \omega$$

Thus, ***E*** and ***B*** are always proportional and orthogonal to one another, and both are orthogonal to ***k***, the wave's direction of motion.

2C§28.2 **Poynting's Theorem** for the flow of electromagnetic field energy is:

$$\partial u/\partial t + \check{\boldsymbol{D}} \bullet \boldsymbol{S} + \boldsymbol{E} \bullet \boldsymbol{j} = 0$$

$$u = \varepsilon_0\, c^2\, \boldsymbol{B} \bullet \boldsymbol{B} / 2 + \varepsilon_0\, \boldsymbol{E} \bullet \boldsymbol{E} / 2$$

$$\boldsymbol{S} = \varepsilon_0\, c^2\, \boldsymbol{E} \times \boldsymbol{B}$$

Here, u is the energy density of fields ***E*** and ***B***, ***S*** is the **Poynting vector** of energy flow, and $\boldsymbol{E} \bullet \boldsymbol{J}$ is the change in the energy of matter (assuming only EM forces). The first equation says the EM field energy in any tiny volume dV changes only when fields in dV act on matter ($\boldsymbol{E} \bullet \boldsymbol{J}$) or when EM energy flows into or out of dV ($\check{\boldsymbol{D}} \bullet \boldsymbol{S}$). ***S*** measures energy flow just as ***J*** measures charge flow.

2C§28.5 The **Field Momentum** density ***g*** of ***E*** and ***B*** fields is related to Poynting's vector ***S*** by:

$$\boldsymbol{g} = \boldsymbol{S} / c^2$$

4B§11.3 E Fields from Sources

This section presents equations for calculating electromagnetic fields from known sources: charges and currents.

2C§27.1 The **Most General Field Equations** for ***E*** and ***B*** from a charge q with any motion are:

$$\boldsymbol{E}(\boldsymbol{r},t) = (q / 4\pi\varepsilon_0)\, \{\, \boldsymbol{r} / r^3 \;+\; (r / c)\, d(\boldsymbol{r} / r^3)/dt \;+\; d^2(\boldsymbol{r} / r)/dt^2 / c^2 \,\}$$

$$\mathbf{B}(\mathbf{r},t) = \mathbf{r} \times \mathbf{E} / r\, c$$

Here, the position of charge q is defined to be (0, 0, 0) at **retarded time** $t^* = t - r / c$, and ***r*** is the vector from (0, 0, 0) to the location where ***E*** and ***B*** are evaluated. The derivatives are evaluated at t^*.

2A§4.8 **Gauss's Law** for the electric field E at a distance r from a stationary charge q at (0, 0, 0) is:

$$E(r) = q / 4\pi\varepsilon_0\, r^2$$

E points radially outward if $q>0$ and radially inward if $q < 0$.

2A§4.4 The **Scalar Potential** ø at distance r from a stationary charge q is:

$$\phi(r) = q / 4\pi\varepsilon_0\, r$$

2A§8.1 The **Electric Potential Energy** U shared by two charges Q and q that are separated by distance r is:

$$U = Q\, q / 4\pi\varepsilon_0\, r$$

If the two charges are a proton and an electron, U becomes:

$$U = q_p\, q_e / 4\pi\varepsilon_0\, r = -e^2 / r$$

2A§4.4 The **Scalar Potential** ø at position ***r*** due to **Multiple Discrete Charges** q_j at stationary positions $\boldsymbol{\tilde{o}}_j$ is:

$$ø(\boldsymbol{r}) = \Sigma_j\, q_j / 4\pi\varepsilon_0\, |\boldsymbol{r} - \boldsymbol{\tilde{o}}_j|$$

2A§4.4 The **Scalar Potential** ø at position ***r*** due to a stationary **Continuous Charge Distribution** ϱ is:

$$ø(\boldsymbol{r}) = \int_V \{\varrho(\boldsymbol{\tilde{o}}) / 4\pi\varepsilon_0\, |\boldsymbol{r} - \boldsymbol{\tilde{o}}|\}\, dV$$

Here, the integral is over all points $\boldsymbol{\tilde{o}}$ in the entire volume V enclosing ϱ.

2A§4.3 **Coulomb's Law** for the force on charge q_r at position ***r*** due to a stationary **Discrete or Continuous Charge Distribution** with scalar potential ø is:

$$\boldsymbol{F}(\boldsymbol{r}) = q_r\, ø(\boldsymbol{r})$$

2A§6.2 For an **Electric Dipole** comprised of charges –q and +q separated by distance d, the dipole moment ***μ***, scalar potential ø, and field E, are given by:

$$\mu = q\, d$$

μ points from –q to +q

$$ø(\boldsymbol{r}) = \boldsymbol{\mu} \bullet \boldsymbol{r} / 4\pi\varepsilon_0\, r^3$$

$$ø(\boldsymbol{r}) = -\boldsymbol{\mu} \bullet \boldsymbol{\check{D}}\, \{ 1 / 4\pi\varepsilon_0\, r \}$$

for ***μ*** in +z-direction:

$$E_z = -(\mu / 4\pi\varepsilon_0)\, \{ r^2 - 3z^2 \} / r^5$$

$$E_x = +(\mu / 4\pi\varepsilon_0)\, \{ 3xz / r^5 \}$$

$$E_y = +(\mu / 4\pi\varepsilon_0)\, \{ 3yz / r^5 \}$$

2A§5.4 For an infinitely long **Thin Wire** with charge λ per unit length, the E field at distance r from the wire is:

$$E(r) = \lambda / 2\pi\varepsilon_0\, r$$

E points radially outward if λ>0 and radially inward if λ < 0.

2A§5.5 For an **Infinite Plane** with charge density σ per unit area, the E field is the same at any distance from the plane, as given by:

$$E = \sigma / 2\varepsilon_0$$

E points away from the plane if σ > 0 and toward the plane if σ < 0.

2A§5.6 For a **Ball** of radius R, with uniform charge density per unit volume ϱ, the inside and outside electric fields are:

for r < R: $\boldsymbol{E}(\boldsymbol{r}) = \varrho\, \boldsymbol{r} / 3\varepsilon_0$

for r > R: $\boldsymbol{E}(\boldsymbol{r}) = Q\, \boldsymbol{r} / 4\pi\varepsilon_0\, r^3$

Here, ***r*** is the vector from the ball's center to the point where ***E***(***r***) is evaluated, and $Q = 4\pi\, \varrho\, R^3 / 3$, the ball's total charge.

2A§5.7 **Inside a Conducting Sphere**, the electric field is zero everywhere.

2A§8.1 The electrostatic **Self-Energy of a Ball** that is uniformly charged, and has total charge Q and radius R is:

$$U = (3 / 5)\, Q^2 / 4\, \pi\varepsilon_0\, R$$

2A§8.2 The electrostatic **Self-Energy of a Sphere** that is uniformly charged, with total charge Q and radius R is:

$$U = Q^2 / 8\,\pi\varepsilon_0\,R$$

2C§29.1 An **Electron's Field Energy**, modeled as a uniformly charged sphere of radius R is:

$$U = e^2 / 2R = q_e^{\,2} / 8\,\pi\varepsilon_0\,R$$

2A§5.8 The **Inverse-Squared Law** says, in static conditions, the electric field E(r) from a point charge Q decreases as the square of r, the distance from Q, as given by:

$$E(r) = Q / 4\pi\varepsilon_0\,r^2$$

To test the validity of the inverse-squared law, let's assume the exponent deviates slightly from 2, according to:

$$E(r) = Q / 4\pi\varepsilon_0\,r^{2+\delta}$$

High precision experiments place stringent limits on δ: $\delta = (1.0 \pm 1.2) \times 10^{-16}$.

2A§6.5 The sum of **Two Overlapping Balls**, with uniform charge densities $+\varrho$ and $-\varrho$, and with slightly displaced centers, is equivalent to a single sphere whose surface charge density σ is given by:

$$\sigma = \varrho\,d\,\cos\theta$$

Here, d is the displacement of the balls' centers and d << R, the radius of each ball. Also, θ is the polar angle relative to the displacement axis ($\theta = 0$ at the far end of the positive ball, and $\theta = \pi$ at the far end of the negative ball). The sphere's electric dipole moment μ, and the electric field $\boldsymbol{E}$ within the sphere are:

$$\mu = Q\,d$$

$$Q = \varrho\,4\pi\,R^3 / 3$$

$$E = \varrho\,d / 3\varepsilon_0$$

Here, Q is the total charge of the positive ball, $\boldsymbol{\mu}$ points toward $\theta = 0$ (from the negative charge toward the positive charge), and $\boldsymbol{E}$ is antiparallel to $\boldsymbol{\mu}$.

2C§34.5 The **Electric Field from a Grid** comprised of infinitely long parallel wires can be highly uniform, even unexpectedly close to the grid. Let d be the distance between adjacent identically charged wires, and z be the distance from the grid's plane. For z comparable to or greater than d, the field non-uniformity Δ is:

$$\Delta = A\,\exp\{-2\pi\,z / d\}$$

Here, A is of order 1, and the exponential has the value 0.0019 for z = d. Thus even at z = d, Δ is a small fraction of 1%.

4B§11.4 Field Equations

2A§4.8 **Gauss's Law** is:

$$\int_S \boldsymbol{E} \bullet \boldsymbol{n}\,da = Q / \varepsilon_0$$

Here, Q is the net charge (positive charges minus negative charges) within a volume V enclosed by a surface S, and $\boldsymbol{n}$ is the unit vector normal to S at each point.

2A§3.3 **Gauss's Theorem** is:

$$\int_S \boldsymbol{h} \bullet \boldsymbol{n}\,da = \int_V \check{\boldsymbol{D}} \bullet \boldsymbol{h}\,dV$$

Here, $\boldsymbol{h}$ is any vector field, V is any volume, S is the surface enclosing V, and $\boldsymbol{n}$ is the unit vector normal to S at each point.

2A§3.6 **Stokes' Theorem** is:

$$\int_S (\check{\boldsymbol{D}} \times \boldsymbol{C}) \bullet \boldsymbol{n}\,da = \int_\Gamma \boldsymbol{C} \bullet \boldsymbol{ds}$$

Here, $\boldsymbol{C}$ is any vector field, S is any open (non-closed) surface bounded by curve Γ, $\boldsymbol{ds}$ is the tangent vector to Γ at each point, and $\boldsymbol{n}$ is the unit vector normal to S at each point.

2A§6.1 The **Laplace Equation** and its solution, for multiple discrete sources Y_j at locations $\boldsymbol{ő}_j$, and for a continuous source density $Y(\boldsymbol{ő})$ within a volume V, are:

Laplace equation: $\check{D}^2 X = -Y$

Discrete: $X(\boldsymbol{r}) = \Sigma_j Y_j / 4\pi | \boldsymbol{r} - \boldsymbol{ő}_j |$

Continuous: $X(\boldsymbol{r}) = \int_V \{Y(\boldsymbol{ő}) / 4\pi | \boldsymbol{r} - \boldsymbol{ő} | \} dV$

4B§11.5 Electrical Circuits

2B§22.1 **Electrical Circuits** are closed paths through which current can flow. Such paths may contain any number of **circuit elements** of a wide variety that are connected in series or in parallel. The most common circuit elements are shown below.

Element	Symbol
Wire	
Switch	
Resistor R	
Capacitor C	
Inductor L	
Battery VDC	
Generator VAC	

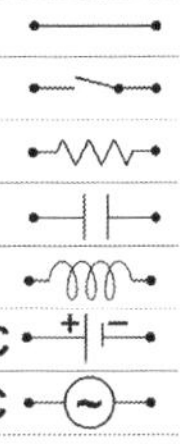

Each circuit element above is a **two-terminal device**: it has exactly two external connections. Current can flow in either terminal and out the other. The graphics in the right column are the device symbols used in circuit diagrams, with terminals represented by the dots at each end of each device. The letters in the middle column are the device symbols used in equations.

In circuit analysis, we assume each basis element is ideal: inductors and capacitors have no resistance, etc. Real world devices are modeled as combinations of ideal circuit elements. For example, a real inductor is modeled as an ideal inductor in series with an ideal resistor and in parallel with an ideal capacitor.

Batteries and generators are **active elements** that power circuits and drive currents. The other elements above are **passive elements** that react to voltages and currents provided by active elements.

2B§22.1 The **Impedance** Z of an ideal two-terminal passive element is the fixed ratio V/J, where V is the voltage difference and J is the current flow between its terminals. When V and J oscillate with frequency ω, the impedances of the ideal passive elements shown above are:

Wire: $Z = 0$

Switch Closed: $Z = 0$

Switch Open: $Z = \infty$

Resistor: $Z = R$

Capacitor: $Z = 1 / i\omega C$

Inductor: $Z = i\omega L$

1B§13.6 **Series & Parallel Impedances**. When two passive circuit elements are combined, their total impedance Z depends on whether they are connected in series (left side of Figure 11-1) or in parallel (right side), as given by:

Series: $Z = Z_1 + Z_2$

Parallel: $1 / Z = 1 / Z_1 + 1 / Z_2$

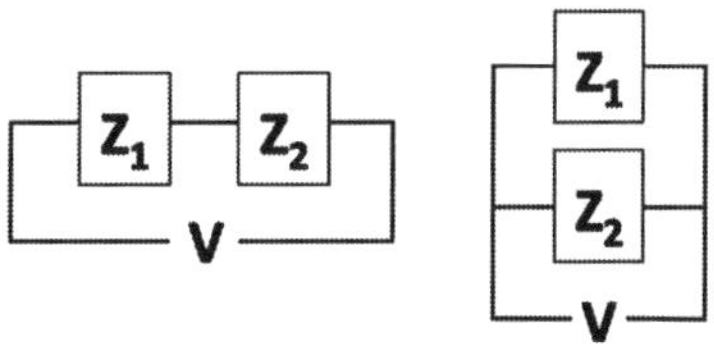

Figure 11-1 Serial & Parallel Impedances

1B§13.3 The **Voltage / Current** relationships for the three primary ideal passive elements are:

Ohm's Law of Resistance: $V = J\,R$

Inductance: $V = L\,dJ/dt = L\,d^2q/dt^2$

Capacitance: $V = Q / C = \int J\,dt / C$

Here, V is voltage, J is current, R is resistance, L is inductance, C is capacitance, and Q and q are charge.

2B§22.1 The **Stored Electrical Energy** U and power dissipated in these elements at voltage V and current J are:

Inductance L: $U = L\,J^2 / 2$

Capacitance C: $U = C\,V^2 / 2$

Resistance R: Power $= V\,J = J^2\,R$

Ideal inductors and capacitors dissipate zero net power in a full cycle of alternating voltage and current. Ideal resistors have no stored electrical energy.

2A§10.3, 2A§8.2, and 2A§7.3 For a **Parallel Plate Capacitor**, whose two identical plates have opposite surface charge densities $+\sigma$ and $-\sigma$, these equations apply:

$$E = \sigma / \varepsilon_0\,\varkappa$$

$$C = Q / C = \varkappa\,\hat{a}\,\varepsilon_0 / d$$

$$Q = \sigma\,\hat{a}$$

$$V = E\,d = Q / C$$

$$V = \sigma\,d / \varepsilon_0\,\varkappa$$

$$U = C\,V^2 / 2$$

$$U = Q^2 / 2C$$

$$U = \sigma^2\,\hat{a}\,d / 2\,\varkappa\,\varepsilon_0$$

Here, E is the electric field between the plates, C is the capacitance, d is the distance between the plates, â is the area of each plate, Q is the total charge on the positive plate, V is the voltage difference between the plates, and U is the stored energy. Lastly, $\varkappa$ is the **dielectric constant** of the material that fills the gap between the plates; $\varkappa = 1.00059$ for air, $\varkappa = 1$ for vacuum, and $\varkappa = 80.4$ for water.

2A§7.3 A **Charged Sphere** of radius R forms one-half of a capacitor — the other half is the constant-zero-potential surface at infinity. The sphere's capacitance C relative to infinity is:

$$C = 4\pi\varepsilon_0\,R$$

2B§22.3 **Kirchhoff's Network Rules** of electrical circuits are:

- The total current flowing into a node equals the total current flowing out of that node.
- The sum of all voltage changes around any closed loop is zero.

In an electric circuit, a node is any connection between two or more circuit elements.

2B§22.8 A **Ladder Network** is a two-terminal device with alternating serial impedances S and parallel impedances P, as shown in Figure 11-2.

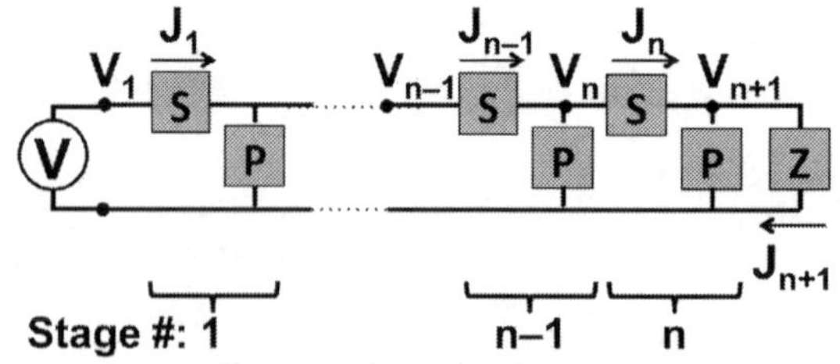

Figure 11-2 Ladder Network

In Figure 11-2, each pair of S and P impedances comprises one **stage**. In the above figure, an external generator drives voltage V across the ladder network's two terminals, the dots to the left of stage #1. We define V_n to be the voltage entering stage #n. Z is the impedance this device would have if it had an infinite number of stages. Terminating a finite ladder with impedance Z, as shown above, ensures a finite ladder of any length has the impedance of an infinite ladder.

2B§22.9 A **Ladder Network**, with serial impedances S and parallel impedances P, with either an infinite number of stages, or terminated as described above, has overall impedance Z between its terminals given by:

$$Z = \{S \pm \sqrt{[S^2 + 4SP]}\ \} / 2$$

2B§22.9 In a **Low-Pass ladder network filter**, these relationships apply:

$V_1 \cos(\omega t)$ is the input voltage

$S = i\omega L$ is the serial impedance

$P = 1 / i\omega C$ is the parallel impedance

$\Omega = \sqrt{(4 / LC)}$ is the **cutoff frequency**

$\beta = V_{n+1} / V_n$ is the **propagation ratio**

$\beta = \{-i\omega + \sqrt{[\Omega^2 - \omega^2]}\ \} / \{+i\omega + \sqrt{[\Omega^2 - \omega^2]}\ \}$

For $\omega > \Omega$, $|\beta|<1$, waves are rapidly attenuated

For $\omega < \Omega$, $|\beta|=1$, waves slowed, but unattenuated

2B§22.9 In a **High-Pass ladder network filter**, these relationships apply:

$V_1 \cos(\omega t)$ is the input voltage

$S = 1 / i\omega C$ is the serial impedance

$P = i\omega L$ is the parallel impedance

$\Omega = \sqrt{(1 / 4\ LC)}$ is the **cutoff frequency**

$\beta = V_{n+1} / V_n$ is the **propagation ratio**

$\beta = \{+i\Omega - \sqrt{[\omega^2 - \Omega^2]}\ \} / \{-i\Omega - \sqrt{[\omega^2 - \Omega^2]}\ \}$

For $\omega < \Omega$, $|\beta|<1$, waves are rapidly attenuated

For $\omega > \Omega$, $|\beta|=1$, waves slowed, but unattenuated

2C§25.1 **Transmission Lines** are governed by:

$$\partial V/\partial x = -L_0\ \partial J/\partial t$$

$$\partial J/\partial x = -C_0\ \partial V/\partial t$$

$$Z = \sqrt{(L_0 / C_0)}$$

$$v = 1 \sqrt{(L_0 C_0)}$$

Here, Z is the transmission line impedance, V is voltage, J is current, and C_0 and L_0 are the capacitance and inductance per unit length. Both V and J satisfy the 1-D wave equation with wave velocity v.

2C§25.1 A **Coax Cable** with vacuum between its conductors has impedance Z given by:

$$Z = \ln(R/r) / 2\pi\varepsilon_0 c$$

Here, R is the radius of the outer conductor, r is the radius of the inner conductor, and ln denotes the natural logarithm.

2C§25.2 In a rectangular **Waveguide**, whose length in the z-direction is much longer than its lateral dimensions X in the x-direction and Y in the y-direction, with X>Y, the wave mode with the lowest frequency is:

$$E = E_y \sin(k_x x) \exp\{i\omega t - ik_z z\}$$

$$k_x = \pi / X$$

$$\Omega = \pi c / X = c\, k_x$$

$$c\, k_z = \pm \sqrt{\{\omega^2 - \Omega^2\}}$$

$$v_{ph} = \omega / k_z = c / \sqrt{\{1 - (\Omega / \omega)^2\}}$$

$$v_{gp} = d\omega/dk_z = c \sqrt{\{1 - (\Omega / \omega)^2\}}$$

$$\lambda_{wg} = 2\pi\, v_{ph} / \omega$$

Here, the electric field E is in the y-direction, Ω is the **cutoff frequency**, v_{ph} is the phase velocity, v_{gp} is the group velocity, and λ_{wg} is the wavelength in the waveguide. For $\omega < \Omega$, waves are rapidly attenuated. Waveguides are typically operated at frequencies with ω slightly greater than Ω, in order to avoid exciting multiple modes.

4B§11.6 E Fields in Dielectrics

Dielectrics are non-electrically-conducting materials (insulators) whose atoms can be polarized by an external electric field. Polarizing means displacing the centroid of each atom's negative charge from the centroid of its positive charge.

2A§10.2 **Dielectric Polarization** is characterized by polarization vector $\boldsymbol{P}$. The simplest dielectrics are homogeneous and isotropic, with $\boldsymbol{P}$ linearly proportional to an external field $\boldsymbol{E}$, for modest field strengths. For such dielectrics, these equations apply:

$$\boldsymbol{P} = \chi\, \varepsilon_0\, \boldsymbol{E}$$

$$\varkappa = 1 + \chi$$

$$\mu = q\, \delta$$

$$P = N \mu$$

$$\boldsymbol{P} = N\, q\, \delta\, (\boldsymbol{E} / E)$$

$$\delta = (\chi\, \varepsilon_0 / N q)\, E$$

Here, $\varkappa$ is the **dielectric constant**, χ is the dielectric's **electric susceptibility**, N is the number of atoms per unit volume, μ is each atom's induced dipole moment, and δ is the separation in each atom between its positive charge +q and negative charge –q.

2A§11.1 A **Non-Polar Dielectric Gas** is comprised of atoms with no intrinsic electric dipole moment. In a static external electric field $\boldsymbol{E}$, positive charge +q and negative charge –q separate by distance δ. The dipole moment μ, **atomic polarizability** α, polarization vector $\boldsymbol{P}$, and dielectric constant $\varkappa$ of a non-polar gas are given by:

$$\delta = q\, E / \Omega^2 m$$

$$\boldsymbol{\mu} = q\, \delta\, (\boldsymbol{E} / E) = \alpha\, \varepsilon_0\, \boldsymbol{E}$$

$$\text{with: } \alpha = q^2 / \varepsilon_0\, \Omega^2 m$$

$$\boldsymbol{P} = N\, \boldsymbol{\mu} = N\, \alpha\, \varepsilon_0\, \boldsymbol{E}$$

$$\varkappa = 1 + N \alpha$$

Here, Ω is the natural oscillation frequency of electrons in these atoms, and N is the number of atoms per unit volume.

2A§11.2 A **Polar Gas** is comprised of atoms with a nonzero intrinsic electric dipole moment μ. In a static external electric field E, the polarization becomes:

$$P = N\, \mu^2\, E / 3kT$$

2A§11.3 **Fields in Dielectrics** are estimated by calculating E_{hole}, the field within a small imaginary hole in the dielectric. For external field E, the fields in two long narrow slots and in a spherical hole are:

$$\text{slot parallel to E:}\quad E_{\mathbf{hole}} = E$$

$$\text{slot orthogonal to E:}\quad E_{\mathbf{hole}} = E + P / \varepsilon_0$$

$$\text{spherical hole:}\quad E_{\mathbf{hole}} = E + P / 3\varepsilon_0$$

2A§11.4 The **Clausius-Mossotti** equation for a dielectric is:

$$\varkappa - 1 = N\,\alpha / (1 - N\,\alpha / 3)$$

Here, $\varkappa$ is the dielectric constant, α is the **atomic polarizability**, and N is the number of atoms per unit volume. This assumes atoms in a dielectric are subject to the same E field that would exist in a tiny spherical hole cut into the material.

2A§11.6 **Ferroelectric** materials exhibit highly nonlinear relationships between an external field E and the material's induced polarization P, as illustrated in Figure 11-3.

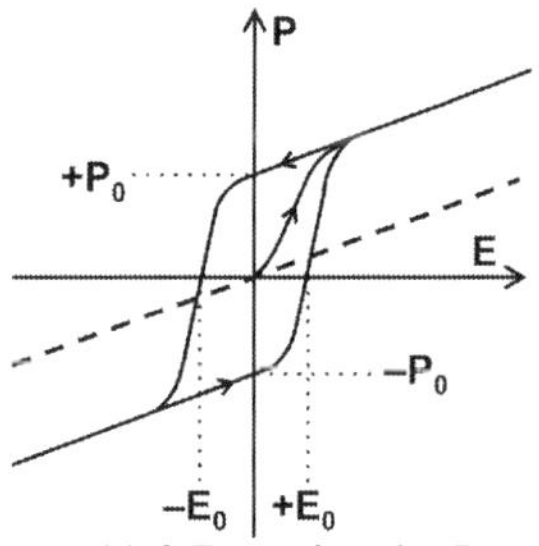

Figure 11-3 Ferroelectric: P vs. E

This figure graphs the polarization P induced by an external field E for a normal dielectric (dashed line) and for a ferroelectric (**hysteresis curve**). For a given E, a ferroelectric's P value depends on its prior history. Once exposed to a strong E field, ferroelectric matter retains a substantial polarization even after the external field is removed. For more on hysteresis behavior see 4B§11.10 after Figure 11-7.

2A§11.6 Since atomic density N varies with temperature, so does $N\alpha$. In some materials, there is a critical temperature T_c at which $N\,\alpha = 3$, making the denominator of the Clausius-Mossotti equation (see above) equal to zero. In such cases, the **Curie-Weiss Law** of **Ferroelectricity** applies. For $T > T_c$:

$$\varkappa - 1 = 9 / \beta\,(T - T_c)$$

Here, α is the **atomic polarizability**, and β is the material's thermal expansion coefficient that may be only a few parts per million per °C. Feynman says that near $T = T_c$, $\varkappa$ can become as large as 100,000.

2D§40.3 **Maxwell's Equations for Dielectrics**, in static conditions with no magnetic fields, are sometimes written:

$$\check{\boldsymbol{D}} \times \boldsymbol{E} = 0$$

$$\check{\boldsymbol{D}} \bullet (\boldsymbol{E} + \boldsymbol{P} / \varepsilon_0) = \varrho_{free} / \varepsilon_0$$

Assuming linearity:

$$\boldsymbol{P} = \chi\,\varepsilon_0\,\boldsymbol{E}$$

$$\varkappa = 1 + \chi$$

$$\check{\boldsymbol{D}} \bullet (\varkappa\,\boldsymbol{E}) = \varrho_{free} / \varepsilon_0$$

Here, $\varrho_{\mathbf{free}}$ is the density of free charges (those not in dielectric matter), $\boldsymbol{P}$ is the polarization vector field, χ is the dielectric's electric susceptibility, and $\varkappa$ is the material's dielectric constant. **Feynman cautions against relying on these equations**, because only in some materials is $\boldsymbol{P}$ linearly proportional to $\boldsymbol{E}$, and then only for modest field strengths.

4B§11.7 B Fields from Sources

2B§14.3 The **Magnetic Force on a Wire** carrying current $\boldsymbol{J}$ is:

$$\boldsymbol{f} = \boldsymbol{J} \times \boldsymbol{B}$$

Here, $\boldsymbol{f}$ is the force per unit length of wire exerted by external magnetic field $\boldsymbol{B}$.

2B§14.4 **Ampere's Law** is:

$$\int_\Gamma \boldsymbol{B} \bullet \boldsymbol{ds} = \text{current J through S} / c^2 \varepsilon_0$$

Here, ***B*** is the magnetic field, S is a surface enclosed by loop Γ, and ***ds*** is the tangent vector at every point along Γ.

2B§14.5 **An Infinitely Long Wire** produces a magnetic field ***B*** and a vector potential ***A*** at position ***r*** outside the wire given by:

$$\boldsymbol{B}(\boldsymbol{r}) = \boldsymbol{J} \times \boldsymbol{r} / 2\pi\varepsilon_0 c^2 r^2$$

$$\boldsymbol{A}(\boldsymbol{r}) = \boldsymbol{J} \ln(w / r) / 2\pi\varepsilon_0 c^2$$

Here, ***J*** is the current flowing in the wire, w is the wire's radius, vector ***r*** extends from the wire's central axis to the point at which the fields are evaluated, and ln denotes the natural logarithm. The magnetic field lines of ***B*** circle the wire.

2B§15.1 For a **Uniform Magnetic Field *B*** pointed toward +z, the vector potential ***A*** at position ***r*** in the xy-plane is:

$$\boldsymbol{A}(\boldsymbol{r}) = \boldsymbol{B} \times \boldsymbol{r} / 2$$

2B§14.5 **Inside a Solenoid**, comprised of wire coiled around a cylinder, the magnetic field B is uniform across the cylinder's cross section, parallel to the cylinder's central axis, and its magnitude is given by:

$$B = n\,J / c^2 \varepsilon_0$$

Here, n is the number of coil windings per unit length, and J is the current flowing in each winding. The polarity of B is determined by the right hand rule (see 4A§11.2).

2B§15.5 **Outside a Solenoid**, whose ***B*** field is in the +z-direction, the vector potential ***A*** is given by:

$$\boldsymbol{A}(x,y) = K\,(-y,\ +x,\ 0) / (\,x^2 + y^2\,)$$

$$\text{with } K = R^2\, n\, J / 2c^2 \varepsilon_0$$

Here, R is the solenoid's radius of , n is the number of coil windings per unit length, and J is the current in each winding.

2B§18.9 A **Solenoid's Inductance** L is given by:

$$L = \pi\, n^2 R^2 Z / \varepsilon_0 c^2$$

Here, Z is the length of the solenoid, R is its radius, and n is the number of windings per unit length.

2B§15.2 For **Current Distribution** $\boldsymbol{j}(\boldsymbol{\sigma})$, the vector potential ***A*** and magnetic field ***B*** are given by:

$$\boldsymbol{A}(\boldsymbol{r}) = \int_V \boldsymbol{j}(\boldsymbol{\sigma})\, dV_\sigma / |\,\boldsymbol{r} - \boldsymbol{\sigma}\,|\, 4\pi\varepsilon_0 c^2$$

$$\boldsymbol{B}(\boldsymbol{r}) = \int_V \boldsymbol{j}(\boldsymbol{\sigma}) \times (\boldsymbol{r} - \boldsymbol{\sigma})\, dV_\sigma / |\,\boldsymbol{r} - \boldsymbol{\sigma}\,|^3\, 4\pi\varepsilon_0 c^2$$

This is the **Biot-Savart Law**. The integral is over all locations $\boldsymbol{\sigma}$ in the volume V enclosing the currents $\boldsymbol{j}(\boldsymbol{\sigma})$.

Sketched below is a **Magnetic Dipole** formed by current J flowing around a gray loop.

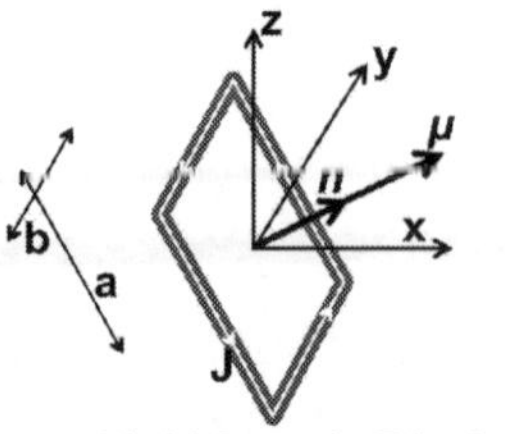

Figure 11-4 Magnetic Dipole

2B§15.7 In any **Magnetic Dipole**, the dipole moment ***μ*** and the vector potential ***A*** at remote locations are given by:

$$\boldsymbol{\mu} = J\, S\, \boldsymbol{n}$$

$$\boldsymbol{A}(\boldsymbol{r}) = \boldsymbol{\mu} \times \boldsymbol{r} / 4\pi\varepsilon_0 c^2 r^3$$

This equation for ***A*** is valid for distances r much greater than the loop dimensions. Here, J is the current flowing through a loop of any shape, S is the area enclosed by the loop, ***r*** is the vector from the loop's center to the point at which fields are evaluated, and ***n*** is the unit normal to S. The polarity of ***n*** is determined by the right hand rule (see 4A§11.2).

2D§38.3 **Dipole Precession** occurs when a magnetic field $\boldsymbol{B}$ exerts a torque $\boldsymbol{\tau}$ on a magnetic dipole moment $\boldsymbol{\mu}$ according to:

$$\boldsymbol{\tau} = \boldsymbol{\mu} \times \boldsymbol{B}$$

This torque seeks to align $\boldsymbol{\mu}$ parallel to $\boldsymbol{B}$. But, since $\boldsymbol{\mu}$ is a gyroscope with angular momentum $\boldsymbol{J}$, $\boldsymbol{\mu}$ maintains its tilt angle relative to $\boldsymbol{B}$ while precessing around $\boldsymbol{B}$ (See 4B§5.4) at frequency ω given by:

$$\omega = \mu \text{ B}$$

3A§4.5 **Particle Spin** is a form of angular momentum that is intrinsic to each elementary particle. While it has some similarities to a spinning top, particle spin is a profoundly quantum mechanical phenomenon that is quantized. All primary fermions — electrons, protons, neutrons, quarks, and leptons — have spin s = 1/2. Most bosons have spin s = 1. Like all forms of momentum, spin is a vector. Any particle's component of spin along any chosen axis, which we call m, must have one of these values:

s = 0: m = 0

s = 1/2: m = +1/2 "up", or –1/2 "down"

s = 1: m = +1, 0, or –1

s = 3/2: m = +3/2, +1/2, –1/2, or –3/2

Massless particles, such as photons, cannot have m=0 along any axis. Any particle's component of spin along any axis can change only by a whole integer, typically ±1. For any spin s, **total spin** S is the length of the 3-D spin vector given by:

$$S = \sqrt{[s\,(s+1)]}$$

$$S = \sqrt{(3/4)} \text{ for spin 1/2 particles}$$

$$S = \sqrt{2} \text{ for spin 1 particles}$$

In all cases, total **spin angular momentum** J and its component j along any axis are given by:

$$J = s\,\hbar$$

$$j = m\,\hbar$$

2D§38.2 For **Atomic and Subatomic Dipoles**, the magnetic dipole moment μ is given by:

$$\mu = g\,(q / 2m)$$

For subatomic particles, q and m are the particle's charge and mass, and g is its **gyromagnetic ratio**.

Each type of subatomic particle has an intrinsic angular momentum called spin (see above), and each type has its own value of g. The g values for the most important particles are:

neutron: $g_n = -3.826{,}085{,}45$

proton: $g_p = 5.585{,}694{,}702$

electron: $g_e = 2.002{,}319{,}304{,}361{,}46$

For the neutron, we use the proton charge for q in the equation for μ. (See 4B§14.4 for more about gyromagnetic ratios.)

The magnetic moments of atoms are typically dominated by their electrons. In the equation for μ, q and m are the electron's charge and mass, and g is called the **Landé factor**. For atomic electrons' orbital angular momentum, g = 1.

In atoms, electrons typically orbit in matched pairs with opposite angular momentum. Additionally, particles of each type typically pair up with opposite spins. Thus, atoms with even numbers of electrons, protons, and neutrons typically have zero net angular momentum and zero net magnetic dipole moment. For other atoms, their total μ is the vector sum of the μ's of any unmatched electron orbital angular momentum plus the μ's of any unmatched particle spins.

2C§34.3 The **Plasma Frequency** ω_p of an electron gas with equilibrium density n_0 is given by:

$$\omega_p^2 = q_e^2\, n_0 / m_e\, \varepsilon_0$$

Radiation of frequency ω incident on plasma is reflected if $\omega < \omega_p$, and propagates freely through the plasma if $\omega > \omega_p$.

2B§18.1 **EMF and Flux**: for any circuit loop Γ enclosing area S, the electromotive force (emf) and the magnetic flux through S are related by:

Definition: $\text{emf} = + \int_\Gamma \boldsymbol{ds} \cdot \boldsymbol{F} / q$

$\text{emf} = + \int_\Gamma \boldsymbol{ds} \cdot (\boldsymbol{E} + \boldsymbol{v} \times \boldsymbol{B})$

Flux Rule: $\text{emf} = - d/dt \{ \int_S \boldsymbol{B} \cdot \boldsymbol{n}\, da \}$

Lenz: emf always opposes flux change

The first equation above defines emf as the integral of the "push" exerted by electromagnetic fields ***E*** and ***B*** driving charges around the closed circuit loop Γ. That "push" is the component of the Lorentz force ***F*** per unit charge that is parallel to the circuit. The flux rule says the total emf around a circuit loop equals minus the rate of change of magnetic flux through the surface S enclosed by the loop Γ. The last line above is **Lenz's law**.

The flux through S may change due to magnetic field changes, or circuit path changes, or any combination thereof.

The flux rule is valid when the *actual atoms* through which current flows *remain the same*. The flux rule *may or may not* work if the circuit changes by redirecting current through different atoms.

When in doubt, go back to Maxwell's equations and the Lorentz force — they are always valid.

Any two current loops have a mutual inductance, even the bizarre loops shown below.

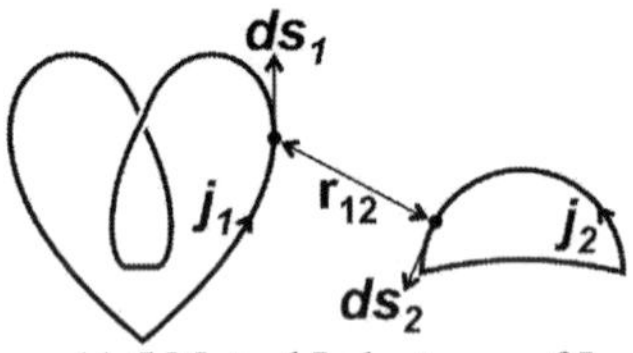

Figure 11-5 Mutual Inductance of Loops

2B§18.7 The **Mutual Inductance** M_{12} of two current loops is given by:

$$M_{12} = \int_{\Gamma 1} \int_{\Gamma 2} \boldsymbol{ds}_2 \cdot \boldsymbol{ds}_1 / r_{12}\, 4\pi\, \varepsilon_0 c^2$$

Here, each integral runs counterclockwise around its loop, $\boldsymbol{ds}_1$ and $\boldsymbol{ds}_2$ are the infinitesimal tangent vectors to loops Γ_1 and Γ_2 respectively, and r_{12} is the distance between the integration points on Γ_1 and Γ_2.

2B§18.9 The **Self-Inductance** L of a current loop is given by:

$$L = \int \varepsilon_0 c^2\, \boldsymbol{B} \cdot \boldsymbol{B}\, dV / J^2$$

Here, ***B*** is the field created when current J flows in the loop, and the integral is over all space.

4B§11.8 Relativistic EM Fields

2C§26.9 **Maxwell's Equations in 4-D** spacetime are:

$$\Box A_\mu = j_\mu / c^2 \varepsilon_0$$

$$\check{D}_\mu j_\mu = 0$$

$\check{D}_\mu A_\mu = 0$, the **Lorentz gauge**

2B§21.3 The **Wave Zone** is all of space that is far from all electromagnetic sources — much farther than the wavelength of any radiation those sources emit. Here these equations apply:

$$\phi(\boldsymbol{r},t) = \int_V \varrho(\boldsymbol{\breve{\sigma}}, t^*)\, dV / 4\pi\varepsilon_0 | \boldsymbol{r} - \boldsymbol{\breve{\sigma}} |$$

$$\boldsymbol{A}(\boldsymbol{r},t) = \int_V \boldsymbol{j}(\boldsymbol{\breve{\sigma}}, t^*)\, dV / 4\pi\varepsilon_0 c^2 | \boldsymbol{r} - \boldsymbol{\breve{\sigma}} |$$

$$\boldsymbol{\check{D}} \cdot \boldsymbol{A} = - \partial\phi/\partial t / c^2$$

$$\boldsymbol{E} = -\boldsymbol{\check{D}}\phi - \partial \boldsymbol{A}/\partial t$$

$$\boldsymbol{B} = \boldsymbol{\check{D}} \times \boldsymbol{A}$$

Above, r = 0 is the center of the sources, **retarded time** $t^* = t - r / c$, and the integrals span all sources at any $\boldsymbol{\delta}$, provided $|\boldsymbol{\delta}| << |\boldsymbol{r}|$.

2C§27.3 For a **Charge q** with **Constant Velocity** v in the +x-direction, the fields are:

$$\boldsymbol{E}(\boldsymbol{R}, t) = \boldsymbol{R}\, \gamma\, q / 4\pi\varepsilon_0 r^3$$

$$\boldsymbol{R} = (x - vt,\ y,\ z)$$

$$r = \sqrt{\{\gamma^2 (x - vt)^2 + y^2 + z^2\}}$$

$$\boldsymbol{B}(\boldsymbol{R}, t) = \boldsymbol{v} \times \boldsymbol{E}(\boldsymbol{R}, t) / c^2$$

$$\gamma = 1 / \sqrt{(1 - v^2 / c^2)}$$

This assumes charge q was at (0,0,0) at time t=0. $\boldsymbol{R}$ is the vector from where charge q is at time t to the location at which the fields are evaluated. The constant velocity requirement ensures that the position of the charge at all future times is precisely predetermined.

2B§21.5 The **Lienard-Wiechert** potentials for a charge q with constant velocity v are:

$$\phi(\boldsymbol{r}, t) = q / 4\pi\varepsilon_0 (r - \boldsymbol{v} \bullet \boldsymbol{r} / c)$$

$$\boldsymbol{A}(\boldsymbol{r}, t) = q\, \boldsymbol{v} / 4\pi\varepsilon_0 c^2 (r - \boldsymbol{v} \bullet \boldsymbol{r} / c)$$

This assumes charge q was at (0, 0, 0) at retarded time $t^* = t - r / c$.

2C§27.11 The **Faraday Tensor** $F_{\mu\sigma}$ generalizes the electromagnetic equations for 4-D relativistic spacetime. $F_{\mu\sigma}$ is written:

$$F_{\mu\sigma} = \begin{array}{c|cccc} & \sigma=t & x & y & z \\ \hline \mu=t & 0 & -E_x/c & -E_y/c & -E_z/c \\ x & E_x/c & 0 & -B_z & +B_y \\ y & E_y/c & +B_z & 0 & -B_x \\ z & E_z/c & -B_y & +B_x & 0 \end{array}$$

2C§27.8 The general **4-Force** f_μ and the **4-D Lorentz Force** are:

$$f_\mu = dp_\mu/d\tau = m\, d^2x_\mu/d\tau^2$$

$$f_\mu = (\gamma\, \boldsymbol{F} \bullet \boldsymbol{v}/c,\ \gamma F_x,\ \gamma F_y,\ \gamma F_z)$$

$$u_\sigma = (\gamma,\ \gamma v_x,\ \gamma v_y,\ \gamma v_z)$$

$$\text{Lorentz force } f_\mu = q\, u_\sigma F_{\mu\sigma}$$

Here, $\boldsymbol{F}$ is the 3-vector force, $F_{\mu\sigma}$ is the **Faraday tensor**. Also, the particle's charge, rest mass, and 4-velocity are q, m, and u_σ.

2B§14.8 The **Relativistic Transformation** equations that transform current 4-vector $(c\,\varrho, \boldsymbol{j})$ from a stationary frame to $(c\,\varrho^*, \boldsymbol{j}^*)$ in a frame moving with velocity v in the +z-direction are:

$$c\,\varrho^* = \gamma\,(c\,\varrho - j_z\, v / c)$$

$$j_x^* = j_x$$

$$j_y^* = j_y$$

$$j_z^* = \gamma\,(j_z - \varrho\, v)$$

$$\gamma = 1 / \sqrt{(1 - v^2 / c^2)}$$

2B§14.7 **Magnetism is a Relativistic Effect**. The magnetic field $\boldsymbol{B}$ at any specific location can always be reduced to zero by a suitable Lorentz transformation. The magnetic force on a moving charge Q in one reference frame corresponds exactly to an electric force in Q's rest frame. One can thus view magnetism as an artifact of not using the simplest reference frame.

2C§27.11 The **Relativistic Transformation** equations that transform fields $\boldsymbol{E}$ and $\boldsymbol{B}$ from a stationary frame to $\boldsymbol{E^*}$ and $\boldsymbol{B^*}$ in a frame moving with velocity $\boldsymbol{v}$ are:

$$E^*_p = E_p$$

$$B^*_p = B_p$$

$$\boldsymbol{E^*}_t = \gamma\,(\boldsymbol{E} + \boldsymbol{v} \times \boldsymbol{B})_t$$

$$\boldsymbol{B^*}_t = \gamma\,(\boldsymbol{B} - \boldsymbol{v} \times \boldsymbol{E} / c^2)_t$$

Here, subscript p denotes a field parallel to $\boldsymbol{v}$, and subscript t denotes the vector sum of fields transverse to $\boldsymbol{v}$. If $\boldsymbol{v}$ is in the +x-direction, two other equivalent versions of the transform equations are:

Version 2

$$E^*_x = E_x$$

$$B^*_x = B_x$$

$$E^*_y = \gamma\,(E_y - v\,B_z)$$

$$E^*_z = \gamma\,(E_z + v\,B_y)$$

$$B^*_y = \gamma\,(B_y + v\,E_z / c^2)$$

$$B^*_z = \gamma\,(B_z - v\,E_y / c^2)$$

Version 3

$$E^*_x = E_x$$

$$B^*_x = B_x$$

$$E^*_y = \gamma\,(\boldsymbol{E} + \boldsymbol{v} \times \boldsymbol{B})_y$$

$$E^*_z = \gamma\,(\boldsymbol{E} + \boldsymbol{v} \times \boldsymbol{B})_z$$

$$B^*_y = \gamma\,\boldsymbol{B} - \boldsymbol{v} \times \boldsymbol{E} / c^2)_y$$

$$B^*_z = \gamma\,\boldsymbol{B} - \boldsymbol{v} \times \boldsymbol{E} / c^2)_z$$

4B§11.9 EM Fields in Matter

2C§33.5 Near a **Sharp Boundary** in the x=0 plane, x-derivatives dominate all other derivatives, and Maxwell's equations show that the following quantities are continuous across the boundary:

$$E_z,\ E_y,\ B_x,\ B_z,\ B_y \text{ and } \{\varepsilon_0\, E_x + P_x\}$$

2C§32.3 The **Index of Refraction** n of low-density matter, such as a gas, is given by:

$$\alpha(\omega) = (q^2 / m\varepsilon_0)\ \Sigma_i\, f_i / (-\omega^2 + i\omega\, \mu_i + \omega_i^2)$$

$$n = 1 + \alpha\, N / 2$$

$$3\,(n^2 - 1) / (n^2 + 2) = \alpha\, N$$

Here, α is the **atomic polarizability**, N is the number of active electrons per unit volume, q and m are the electron's charge and mass, and ω is the frequency of the incident electric field. The summation is over all electron excited states. Excited state i has natural frequency ω_i, damping force $m\omega\mu_i$, and weighting factor f_i. The last equation above is the **Clausius-Mossotti** equation.

2C§32.4 For a **Complex Refractive Index**, we separate n into its real and imaginary parts:

$$n = n_r - i n_i$$

$$E = E_0 \exp\{i\omega(t - z\, n_r / c)\} \exp\{-\beta\, z / 2)\}$$

$$\beta = 2\, n_i\, \omega / c$$

Above, light moves toward +z at velocity c/n_r, with frequency ω, and with exponentially decreasing amplitude. β is the **intensity absorption coefficient**.

2C§32.7 **In Metals**, free electrons in the conduction band dominate refractive effects. The imaginary part of the refractive index can be comparable to or larger than its real part. Free electrons have no excited states, but do experience damping due to electrical resistance. We then have:

$$\sigma = N q^2 \tau / m$$

$$n^2 = 1 + \sigma / \{i\omega\, \varepsilon_0 (i\omega\, \tau + 1)\}$$

Here, τ is the mean collision time and σ is the metal's conductivity. At **"Low Frequencies"** (below 10,000 GHz in copper), where $\omega\tau << 1$ and $\omega << \sigma / \varepsilon_0$, the refractive index approaches:

$$n = (1 - i) \sqrt{(\sigma / 2\omega\, \varepsilon_0)}$$

plasma frequency $\omega_p = \sqrt{(N q^2 / m\, \varepsilon_0)}$

amplitude skin depth $= \sqrt{(2\varepsilon_0 c^2 / \sigma\, \omega)}$

intensity skin depth $= \sqrt{(\varepsilon_0 c^2 / 2\sigma\, \omega)}$

Here, the real and imaginary parts of n have equal magnitude. The amplitude skin depth is the depth into the metal at which the wave amplitude drops by a factor of e. Since intensity is proportional to amplitude squared, the intensity skin depth is half of the amplitude skin depth. The intensity skin depth is 3.33 μm in Cu at 10GHz. Thus high frequency EM fields have only microscopic penetration depths in high-conductivity metals. This is why even very thin coatings of such metals are highly effective in the GHz frequency range.

At **Very High Frequencies**, above plasma frequency ω_p, 1.6×10^{16} Hz in Cu, the index approaches:

$$n^2 = 1 - \omega_p^2 / \omega^2$$

Here, the refractive index is entirely real, and EM waves propagate through the metal without any attenuation — the metal becomes transparent for $\omega > \omega_p$. Since Earth's ionosphere is a free electron gas, communicating with distant spacecraft requires very high frequency signals.

2C§34.4 The **Debye Length** δ applies to colloidal particles, and is defined by:

$$\delta^2 = \varepsilon_0 kT / 2q^2 n_0$$

Here, q is the ion charge, and n_0 is the global ion density. Colloidal particles are electrically charged mid-sized objects — microscopic but much larger than atoms. Colloidal particles of the same type have the same charge, which prevents them from clumping together and precipitating out of solution. In an **electrolyte**, a solution with both positive and negative ions, the appropriate polarity ions attach to colloidal particles, thereby reducing their charge. This **ion cloak**'s density decreases exponentially from the colloidal particle's surface with mean distance δ. At sufficiently high ion density, nearly neutral colloidal particles do clump and precipitate. This is called **salting out a colloid**.

4B§11.10 Magnetic Matter

Feynman sometimes writes the electron charge as q, in which case $q < 0$. Other times he writes $-q_e$, in which case $q_e > 0$. I find that needlessly confusing, and will try to make this less confusing.

When the intended particle is clear, I will denote its charge by q, which may be positive or negative. When there may be uncertainty, I will denote the electron's charge by $-q_p$, where q_p is the proton charge that we all agree is positive.

2D§38.8 The **Bohr Magneton** is defined to be:

$$\mu_B = q_p \hbar / 2m_{electron}$$

2D§38.8 The **Magnetic Energy** U of a particle with spin j in field B is:

$$U = + g\, B\, j\, (q \hbar / 2m)$$

Here, j is the spin component parallel to ***B***, q and m are the particle's charge and mass, and g is the particle's **gyromagnetic ratio** (see 4B§14.4).

2D§39.4 **Magnetic Dipoles Precess** in a magnetic field B according to:

$$\omega = (g\, q / 2m)\, B$$

Here, ω is the precession frequency, q and m are the particle's charge and mass, and g is the particle's **gyromagnetic ratio** (see 4B§14.4). See 4B§5.4 for more on precession.

2D§38.4 **Diamagnetism** is a weak effect that occurs in all materials. This arises when an external magnetic field ***B*** increases each atom's intrinsic magnetic dipole moment, if any, by $\Delta\mu$ as given by:

$$\boldsymbol{\Delta\mu} = -q^2 < r^2 > \boldsymbol{B} / 6m$$

Here, q and m are the electron's charge and mass, while $< r^2 >$ is the average square of the radius of a three-dimensional quantum mechanical electron orbit. The additional dipole moment $\boldsymbol{\Delta\mu}$ is antiparallel to ***B***, consistent with **Lenz's law**, and increases each atom's potential energy by ΔU, as given by:

$$\Delta U = -\boldsymbol{\Delta\mu} \bullet \boldsymbol{B} = +q^2 < r^2 > B^2 / 6m$$

If $\boldsymbol{\Delta\mu}$ is the atom's only magnetic dipole moment, diamagnetism weakly pushes it away from a magnetic field, as nature seeks a lower energy state.

2D§39.5 **Paramagnetic Magnetization** occurs in materials whose atoms have intrinsic nonzero magnetic moments, as described by:

$$< \boldsymbol{\mu} > = \mu_0 \tanh(Z)$$

$$\text{with: } Z = \mu_0\, B / kT$$

$$\boldsymbol{M} = N < \boldsymbol{\mu} >$$

Here, μ_0 is the intrinsic dipole moment of each atom of the paramagnetic material. In a magnetic field B and at absolute temperature T, Boltzmann's law (see 4B§8.2) provides the ratio of dipole moments aligned parallel versus antiparallel to B. This yields an average magnetic moment $< \boldsymbol{\mu} >$ per atom. The magnetization per unit volume ***M*** is the product of N, the number of paramagnetic atoms per unit volume, and $< \boldsymbol{\mu} >$.

Typically, **Z << 1**. Feynman says Z is only 0.02 at room temperature, even in a field of 10,000 gauss. In this case, tanh(Z) approaches Z, and $< \boldsymbol{\mu} >$ approaches:

$$< \boldsymbol{\mu} > = \mu_0^2\, \boldsymbol{B} / kT$$

In this limit, ***M*** and the atomic **magnetic susceptibility** χ become:

$$\chi = N\, \mu_0^2 / 3kT$$

$$\boldsymbol{M} = \chi\, \boldsymbol{B} = N\, \mu_0^2\, \boldsymbol{B} / kT$$

Note that ***M*** is parallel to ***B***. A quantum refinement of this classical analysis replaces μ_0^2 with μ_B^2 (j+1) j, where j is the atomic angular momentum. In the Z << 1 limit, this yields:

$$\boldsymbol{M} = N\, g^2\, \mu_B^2\, j\,(j+1)\, \boldsymbol{B} / 3kT$$

Conversely, **Z >> 1** at very low temperatures, where tanh(Z) approaches 1, and **M saturates** at its asymptotic limit:

$$M = N\, \mu_0$$

At low enough temperatures, **paramagnetism dominates diamagnetism**, and such materials are attracted to magnetic fields, where their potential energy decreases.

2D§40.3 The **Static Ferromagnetic** equations are:

$$\boldsymbol{H} = \boldsymbol{B} - \boldsymbol{M} / \varepsilon_0 c^2$$

$$\check{\boldsymbol{D}} \bullet (\boldsymbol{H} + \boldsymbol{M} / \varepsilon_0 c^2) = 0$$

$$\check{\boldsymbol{D}} \times \boldsymbol{H} = 0$$

2D§40.4 The **Curie Temperature** T_C is defined to be:

$$k\, T_C = \lambda\, \mu^2\, N / \varepsilon_0 c^2$$

Classical physics says λ = 1/3. Actual values are: λ = 900 in nickel, and up to λ = 3000 in other ferromagnetic matter.

2D§41.2 The **Magnetic Field Energy** density u in an iron-core inductor is:

$$u = \varepsilon_0 c^2 \int H\, dB$$

$$\text{with } \boldsymbol{H} = \boldsymbol{B} - \boldsymbol{M} / \varepsilon_0 c^2$$

Since $\boldsymbol{M}$ and $\boldsymbol{B}$ are parallel, they are both parallel to $\boldsymbol{H}$. Therefore $\boldsymbol{H} \bullet \boldsymbol{B} = H\,B$.

2D§41.2 The **Magnetic Permeability** η, for weak fields, is defined as:

$$\boldsymbol{B} = \eta\, \boldsymbol{H}$$

2D§41.1 and 2D§41.9 **Hysteresis** is a general characteristic of ferromagnetic materials, providing a multi-valued relationship between B and H, as illustrated below.

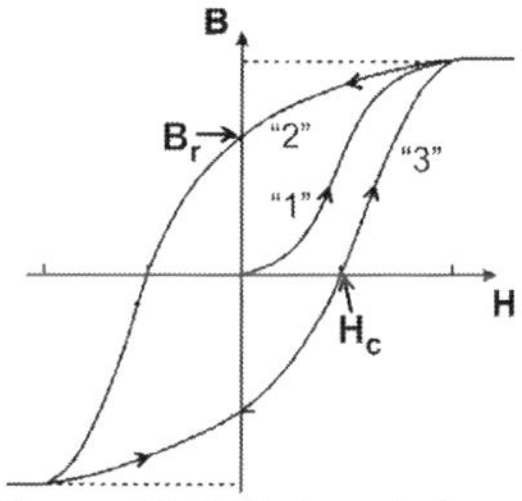

Figure 11-6 Hysteresis Loop

Recall that $\boldsymbol{H} = \boldsymbol{B} - \boldsymbol{M} / \varepsilon_0 c^2$. Un-magnetized "soft iron" starts at B = H = 0. Increasing H drives B up curve "1" to saturation, thereby aligning all atomic magnetic moments to H. Subsequently decreasing H drives B along curve "2". B_r is the **residual magnetic field** B at H = 0. As H becomes more negative, B continues along curve "2" to the opposite saturation limit. As H subsequently increases, B increases along curve "3", eventually becoming zero at $H = H_c$, the **coercive force**.

2C§32.8 **Historical Notation**: Feynman says that in Maxwell's time, physicists did not understand atoms. They did not realize atoms polarize in dielectric materials exposed to external fields. They also did not realize that magnets contain circulating currents due to the atomic states of electrons. Unable to account for polarization charges and currents, and electrons' magnetic effects, they defined vector fields $\boldsymbol{D}$ and $\boldsymbol{H}$ in terms of the observed free charges and currents, ϱ_{free} and $\boldsymbol{j}_{\text{free}}$. Some authors still use these fields. In terms of these defined fields, Maxwell's equations are:

$$\boldsymbol{D} = \varepsilon_0 \boldsymbol{E} + \boldsymbol{P} = \varepsilon\, E$$

$$\mu\, \boldsymbol{H} = \boldsymbol{B}$$

$$\check{\boldsymbol{D}} \bullet \boldsymbol{D} = \varrho_{\text{free}}$$

$$\check{\boldsymbol{D}} \bullet \boldsymbol{B} = 0$$

$$c\, \check{\boldsymbol{D}} \times \boldsymbol{E} = -\, \partial \boldsymbol{B}/\partial t$$

$$c\, \check{\boldsymbol{D}} \times \boldsymbol{H} = j_{\text{free}} + \partial \boldsymbol{D}/\partial t$$

Feynman recommends not using *D* or *H*, saying in V2p32-5: "These relations are only approximately true for some materials and even then only if the fields are not changing rapidly with time. ...We think the right way is to keep the equations in terms of the fundamental quantities as we now understand them." He presents the older equations only for reference.

Chapter 12
Physics of Solids & Liquids

4B§12.1 Linear Stress & Strain

Stress is a force that deforms a solid, creating **strain**, the displacement of the solid's atoms from their natural positions.

In this chapter, we assume stress and displacement are both small enough to be governed by Hooke's law. Each dimension of a solid will change by only a small fraction of its value.

1A§7.3 **Hooke's Law** is:

$$F = \text{constant} \times \Delta L / L$$

A force F that acts to increase a solid's length L is called a **tensile force**. A force F that acts to decrease L is a **compressive force**.

Let's consider two identical blocks, one of which is subject to an elastic deformation, as illustrated in Figure 12-1.

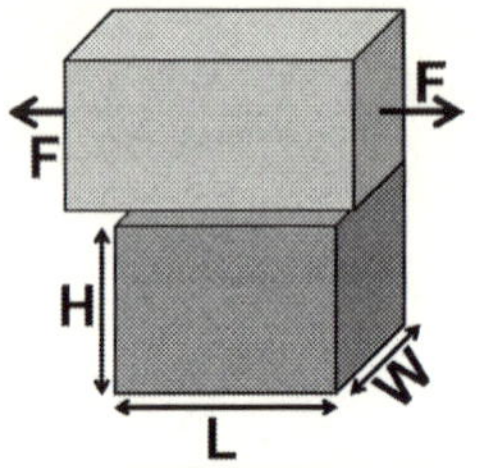

Figure 12-1 Stretching a Block

The lower block has no forces acting upon it, and has height H, width W, and length L.

The upper block is being stretched in the L direction by two equal but opposite forces F that are uniformly applied across the entire surface of each end. No forces are applied along the H and W directions. The deformed block's dimensions are H+ΔH, L+ΔL, and W+ΔW; each Δ may be positive or negative.

2D§42.1 Feynman states three **General Principles of Elasticity** relating the quantities in Figure 12-1.

The **first principle** says the force per unit area (F / H W) equals Y multiplied by (ΔL) / L, the fractional length change in the F direction, as given by:

$$F / H W = Y \Delta L / L$$

The **second principle** says the fractional changes in lateral directions, H and W, equal minus σ multiplied by the fractional length change in the F direction, (ΔL) / L. This minus sign means tension in one direction causes compression in the orthogonal directions, while compression in one direction causes tension in the orthogonal directions, as given by:

$$\Delta H / H = \Delta W / W = -\sigma \Delta L / L$$

The **third principle** is the linear superposition of stress and strain — stresses sum linearly as do strains. This is a direct consequence of the linearity of Hooke's law.

Here, Y is **Young's modulus**, and σ is **Poisson's ratio**. Feynman says σ is always between 0 and 0.5.

2D§42.2 For a **Uniform Pressure** P, squeezing a block on all sides:

$$\Delta L / L = \Delta H / H = \Delta W / W = -(1 - 2\sigma) P / Y$$

$$\Delta V / V = -P / K$$

$$\text{with } K = Y / 3 (1 - 2\sigma)$$

Here, V = L H W is the block's undeformed volume, P is the **volume stress**, ΔV / V is the **volume strain**, and K is the **bulk modulus**.

2D§42.2 If dimensions **H and W are held Constant** (see Figure 12-1), while a tensile force F stretches L, the result is:

$$\Delta L / L = F \beta / Y$$

$$\text{with } \beta = \{1 - 2\sigma^2 / (1 - \sigma)\}$$

$$\text{or } \beta = (1 + \sigma)(1 - 2\sigma) / (1 - \sigma)$$

Feynman says **Poisson's ratio** σ is always between 0 and 0.5, which means β is always between 0 and 1. Hence, more force is required to stretch a block if its sides are held stationary than if they are free to contract.

2D§42.2 **Orthogonal Tension and Compression** by forces of equal magnitude, for example a tensile force F along L combined with a compressive force F along H, results in:

$$\Delta L / L = +(1 + \sigma) P / Y$$

$$\Delta H / H = -(1 + \sigma) P / Y$$

$$\Delta W / W = -\sigma(+P / Y) - \sigma(-P / Y) = 0$$

L increases by the same factor as H decreases, while W is unchanged. This means the solid's volume remains constant.

4B§12.2 Shear & Torsion

2D§42.3 **Torsion** or **shear stress** is a force that acts parallel to an object's surface, causing a twisting deformation. For simplicity, let's assume two equal shear forces act on opposite sides of a solid; this is called **pure shear**. A single shear force F can be separated into a pure shear force of F/2 plus an ordinary linear force of F.

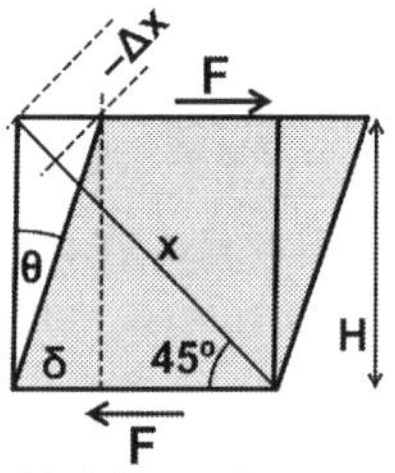

Figure 12-2 Shear Force on Cube

This figure shows a 2-D cross section of a 3-D cube with side lengths H. With no shear forces, the indicated diagonal has length $x=H\sqrt{2}$. Shear forces F acting on the upper and lower faces of the cube squeeze x, changing it by Δx (Δx<0). The opposite diagonal y (not indicated in the figure) suffers a tensile force and increases by Δy. For small shear, the twist angle θ of the deformed cube is given by:

$$\Delta x = -\delta / \sqrt{2}$$

$$\Delta y = +\delta / \sqrt{2}$$

$$\tan\theta = \theta = \delta / H$$

2D§42.3 **Shear Stress** g is the stress per unit area. With two equal shear forces F acting in opposite directions on two opposite faces of a cube of side length H, g is given by:

$$g = F / H^2$$

2D§42.3 For **Shear Stress** g per unit area on each of two opposite faces of the above cube:

$$\text{stretch:} \quad \Delta y / y = \{ +(1 + \sigma) / Y \}\, g$$

$$\text{squeeze:} \quad \Delta x / x = \{ -(1 + \sigma) / Y \}\, g$$

$$\text{twist angle:} \quad \theta = 2\{ (1 + \sigma) / Y \}\, g$$

$$\text{define:} \quad \mu = Y / 2(1 + \sigma)$$

$$\text{shear stress:} \quad g = \mu\,\theta$$

Here, μ is called the **shear modulus** or the **coefficient of rigidity**.

Now consider a solid rod that is twisted about its symmetry axis by shear forces, as shown below.

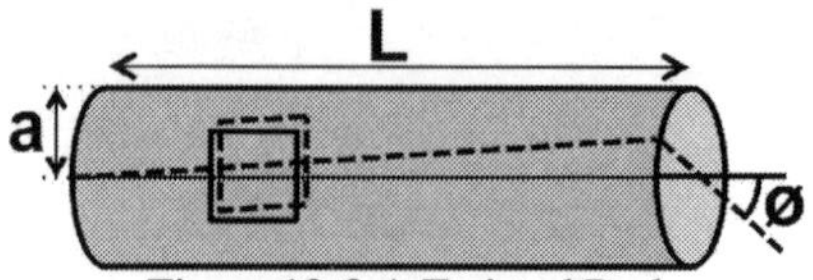

Figure 12-3 A Twisted Rod

2D§42.4 A **Twisted Rod**, as shown above, is governed by:

$$\tau = (\pi\, a^4 \mu / 2L)\, \phi$$

Here, the rod's left end is fixed while its right end is subject to torque τ. The twist angle is ϕ. The rod has length L, radius a, and **shear modulus** μ. Note the fourth power of radius a.

2D§42.5 Time-varying torques produce **Torsion / Transverse Waves** that propagate through a solid with velocity c_{trans} given by:

$$c_{trans} = \sqrt{(\mu / \varrho)}$$

Here, ϱ is the solid's mass density, and μ is its **shear modulus**.

2D§42.6 Time-varying compressive or tensile forces produce **Compression / Longitudinal Waves** that propagate through a solid with velocity c_{long} given by:

$$c_{long} = \sqrt{(Y / \varrho\, \beta)}$$

$$\beta = (1 + \sigma)\,(1 - 2\sigma) / (1 - \sigma)$$

Here, ϱ is the solid's mass density, Y is **Young's modulus**, and σ is **Poisson's ratio**. Longitudinal waves are the same phenomenon as sound, with a similar velocity equation.

4B§12.3 Stressed Beams

Beams resist bending, not because their internal atomic bonds resist bending, but rather because those bonds resist stretching and compression.

Consider bending a large beam of length D into a circular arc of radius R, as shown below.

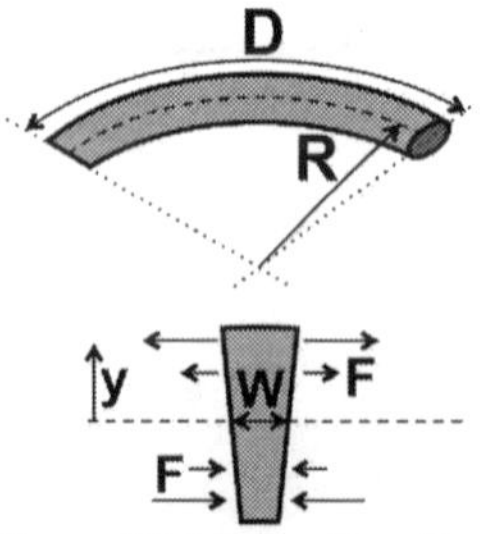

Figure 12-4 Beam Bending

The upper image shows material stretching along the upper side of the beam and compressing along the lower side. In between, there is an undeformed **neutral surface** indicated by the dashed curve. The lower image shows a cross section of the beam. Here, y is the distance above the neutral surface, and W is the undeformed width of this segment. Forces F stretch the beam for $y > 0$ and compress it for $y < 0$, with F being linearly proportional to y. This means the fractional deformation $\Delta W/W$ is also linearly proportional to y.

2D§42.7 The **Bending Moment** M of a beam bent into a circular arc, as shown above, is defined as:

$$M = Y\, \hat{I} / R$$

$$\text{with: } \hat{I} = \int y^2\, da$$

Here, Y is **Young's modulus**, R is the beam's radius of curvature, and y is the distance above the neutral surface. The integral is over the beam's cross section, and $\hat{I}$ is the moment of inertia per unit mass of each thin beam cross section. The product $Y\hat{I}$ is called the **beam stiffness**.

Let's next consider a cantilevered beam of length L supporting a weight w at its right end, as shown below.

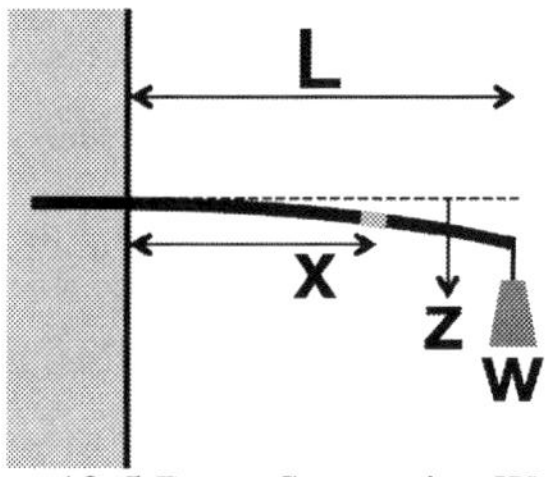

Figure 12-5 Beam Supporting Weight

The beam's left end is embedded in a wall, while a mass m of weight w (w = m g) hangs from the beam's right end. Let z be the vertical axis, with +z pointing downward.

2D§42.7 At its right end, the above **Beam Droops** a distance z, as given by:

$$z(x = L) = w L^3 / 3 Y \hat{I}$$

Here, w is the weight hanging from the beam's right end, L is the beam length, Y is **Young's modulus**, and Î is the moment of inertia per unit mass of each thin beam cross section (see **Bending Moment** above).

2D§42.7 The **Euler Force** F is given by:

$$F = Y \hat{I} (\pi / L)^2$$

Supporting any force G less than its Euler force, a beam remains straight and rigid. But, as G exceeds the Euler force, the **beam suddenly buckles**; it bends in an arc and becomes incapable of fulfilling its intended function. Here, L is the beam length, Y is **Young's modulus**, and Î is the moment of inertia per unit mass of each thin beam cross section (see **Bending Moment** above).

4B§12.4 Elasticity Tensors

Not all solids are homogeneous and isotropic, as assumed above. Tensors are the most convenient tools for analyzing the added complexity of stresses and strains that vary with location or direction.

2D§43.2 The **Strain Vector Field** $\boldsymbol{u}(\boldsymbol{r})$ is a deformed solid's displacement at $\boldsymbol{r}$, as defined by:

$$\boldsymbol{u}(\boldsymbol{r}) = \check{\boldsymbol{r}} - \boldsymbol{r}$$

Here, $\boldsymbol{r}$ is the position of any point P in the undeformed solid, and $\check{\boldsymbol{r}}$ is the position of P after deformation.

2D§43.2 The 3x3 rank 2 **Strain Tensor** is defined by:

$$e_{jk} = (\partial u_j/dk + \partial u_k/dj) / 2$$

Here, j and k are indices that range over the three dimensions of space, and u_j is the j-component of the **strain vector *u***. Each component of e_{jk} may vary with position in the solid.

2D§37.9 The 3x3 rank 2 **Stress Tensor** is defined by:

$$S_{jk} = \boldsymbol{f}_j \bullet \boldsymbol{n}_k$$

Here, $\boldsymbol{f}_j$ is the j-component of force per unit area, and $\boldsymbol{n}_k$ is the unit vector in the k-direction. Each component of S_{jk} may vary with position in the solid.

2D§43.3 The 3x3x3x3 rank 4 **Elasticity Tensor** C is defined by:

$$S_{jk} = \Sigma_{nm} C_{jknm} e_{nm}$$

Here, j, k, n, and m are indices that range over the three dimensions of space, S is the **stress tensor**, and e is the **strain tensor**. Each component of C_{jknm} may vary with position in the solid.

2D§43.3 The **Stored Energy** U of deformed matter is given by:

$$U = \int_V \Sigma_{nm} C_{jknm} e_{jk} e_{nm} dV / 2$$

Here, j, k, n, and m are indices that range over the three dimensions of space, C is the **elasticity tensor**, e is the **strain tensor**, and the integral is throughout the volume V that encloses the deformed matter.

2D§43.3 The **Lamé Elastic Constants** μ and λ of isotropic matter are given by:

$$S_{jk} = 2\mu\, e_{jk} + \lambda\, \Sigma_n e_{nn}\, \delta_{jk}$$

Here, j and k are indices that range over the three dimensions of space, e is the **strain tensor**, and δ_{jk} is the rank 2 unit tensor: $\delta_{jk} = 1$ if j = k, otherwise $\delta_{jk} = 0$.

4B§12.5 Non-Viscous Fluid Dynamics

Fluid dynamics describes gases and liquids.

For gases, zero viscosity can be a good approximation. But, a zero-viscosity fluid, similar to frictionless motion, is an idealization that is instructive but generally unrealistic. Feynman quotes John Von Neumann calling such a fluid "dry water". (Liquid helium is an exception.)

In mechanics, we often examine the motion of individual objects, calculating their positions and velocities. But in fluid dynamics, tracking individual molecules is not very useful. It is more effective to calculate the macroscopic properties of the fluid at different positions and times. Here for example, we define $\boldsymbol{v}(\boldsymbol{r},t)$ not as the velocity of a specific fluid molecule, but rather as the average velocity of fluid molecules that pass position $\boldsymbol{r}$ at time t.

Some **Key Quantities** and **General Principles** of non-viscous fluids are:

ϱ: fluid mass density

P: fluid pressure

f: force per unit volume

$\boldsymbol{v}(\boldsymbol{r},t)$: velocity of fluid at $(\boldsymbol{r},t)$

$\boldsymbol{a}(\boldsymbol{r},t)$: acceleration of fluid at $(\boldsymbol{r},t)$

ø: gravitational potential energy per unit mass

2D§44.3 The **Continuity Equation** expresses the local conservation of mass:

$$\check{\boldsymbol{D}} \bullet (\varrho\, \boldsymbol{v}) = -\, \partial\varrho/\partial t$$

2D§44.3 The **Equation of Motion** of a **non-viscous incompressible** fluid is:

$$\boldsymbol{f} = \varrho\, \boldsymbol{a} = -\check{\boldsymbol{D}}P - \varrho\, \check{\boldsymbol{D}}\text{ø}$$

2D§44.3 **Fluid Acceleration at a Fixed Location** is given by:

$$\boldsymbol{a} = \boldsymbol{v} \bullet \check{\boldsymbol{D}}\boldsymbol{v} + \partial\boldsymbol{v}/\partial t$$

2D§44.2 In **Hydrostatics**, each macroscopic group of fluid molecules has zero average velocity, but each molecule has thermal energy that creates pressure. Generally, the fluid is also subject to gravity.

The net result is that pressure and gravity must balance one another to eliminate net force and maintain static conditions. Hence:

$$F = 0 = \check{\boldsymbol{D}}\, P + \varrho\, \check{\boldsymbol{D}}\, \text{ø}$$

$$P + \varrho\, \text{ø} = \text{some constant } K$$

Here, at each position and moment in time, K has a fixed value, ensuring zero net force F.

2D§44.3 A fluid's **Vorticity** $\boldsymbol{\Omega}$ is defined as:

$$\boldsymbol{\Omega} = \check{\boldsymbol{D}} \times \boldsymbol{v}$$

Vorticity is the circulation of $\boldsymbol{v}$ per unit area perpendicular to $\boldsymbol{\Omega}$. If $\boldsymbol{\Omega} = 0$ everywhere, the fluid is called **irrotational**.

2D§44.3 Combining and rearranging prior equations yields:

$$(\check{\boldsymbol{D}} \times \boldsymbol{v}) \times \boldsymbol{v} + \check{\boldsymbol{D}}\,(\boldsymbol{v} \bullet \boldsymbol{v})/2 + \partial \boldsymbol{v}/\partial t = -\check{\boldsymbol{D}}\, P / \varrho - \check{\boldsymbol{D}}\, \emptyset$$

2D§44.3 The **Field Equations** of a **zero-viscosity, incompressible** fluid are:

$$\check{\boldsymbol{D}} \bullet \boldsymbol{v} = 0$$

$$\check{\boldsymbol{D}} \times (\boldsymbol{\Omega} \times \boldsymbol{v}) + \partial \boldsymbol{\Omega}/\partial t = 0$$

If the fluid is irrotational ($\Omega = 0$), these equations become identical to the equations of electrostatics in empty space:

$$\check{\boldsymbol{D}} \bullet \boldsymbol{v} = 0$$

$$\check{\boldsymbol{D}} \times \boldsymbol{v} = 0$$

2D§44.4 For steady fluid flow, a **streamline** is the locus of points along the path of a fluid molecule. Along each streamline, the fluid velocity vector is always parallel to the streamline. A bundle of adjacent streamlines is called a **stream tube**. In steady flow, fluid never flows across the sides of a stream tube.

2D§44.4 **Bernoulli's Theorem** says, for irrotational flow, ψ is constant along any streamline, where ψ is defined as:

$$\psi = P / \varrho + \emptyset + \boldsymbol{v} \bullet \boldsymbol{v} / 2$$

A consequence of Bernoulli's theorem is:

v ϱ A is constant within any stream tube

Here, $\boldsymbol{v}$ is the fluid velocity, and A is the cross sectional area of the stream tube. If fluid density ϱ is constant, v A is constant within each stream tube. Hence, water flowing past a constriction in a pipe speeds up, thereby decreasing its pressure.

A beautiful example of vorticity is the smoke ring shown in Figure 12-6.

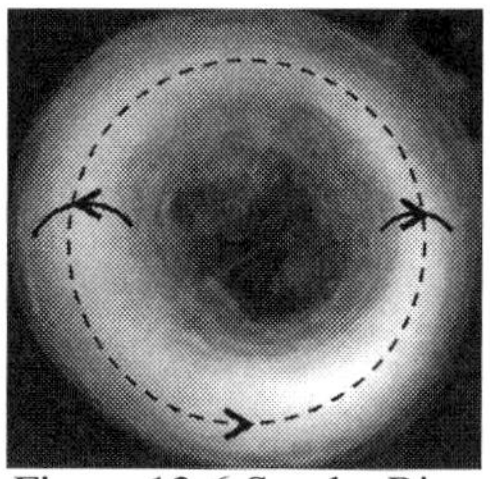

Figure 12-6 Smoke Ring

Here, the two solid arrows are **streamlines**, indicating the flow of smoke particles. The fluid flows around the small diameter of the torus, flowing out of the page along the inside edge of the torus, and back into the page on its outside edge. The single dashed arrow indicates a **vortex line**; vorticity $\boldsymbol{\Omega}$ circles the large diameter of the torus.

4B§12.6 Viscous Fluid Dynamics

2D§44.3 The **Equation of Motion** of a **viscous incompressible** fluid is:

$$\boldsymbol{f} = \varrho\, \boldsymbol{a} = -\check{\boldsymbol{D}}\, P - \varrho\, \check{\boldsymbol{D}}\, \emptyset + \boldsymbol{f}_{visc}$$

Here, $\boldsymbol{f}$ is the force per unit volume, ϱ is the fluid mass density, $\boldsymbol{a}$ is the fluid acceleration, P is pressure, $\emptyset$ is the potential energy per unit mass, and $\boldsymbol{f}_{visc}$ is the sum of the fluid shear forces per unit volume. Water is viscous and nearly incompressible — in extreme ocean depths, where pressures reach 400 atmospheres, water is only 1.8% denser than at the surface.

2D§45.1 In a viscous fluid, the **Sheer Force** per unit area g between two parallel plates, one stationary and the other moving slowly with velocity v_0, is:

$$g = \eta \, v_0 / d$$

Here, η is the fluid's **first coefficient of viscosity**, and d is the plate separation.

2D§45.1 The **Differential Sheer Force** per unit area Δg inside a viscous fluid is given by:

$$\Delta g = \eta \, \Delta v / \Delta y$$

Here, (Δv / Δy) is the differential velocity per unit distance, and η is the fluid's **first coefficient of viscosity**.

2D§45.2 The 3x3 rank 2 **Stress Tensor** S_{jk} of a **viscous compressible** fluid is:

$$S_{jk} = \eta \, (\partial v_j/\partial k + \partial v_k/dj) + \eta^* \, \delta_{jk} \, \check{\boldsymbol{D}} \bullet \boldsymbol{v}$$

Here, j and k are indices that range over the three dimensions of space, $\boldsymbol{v}$ is the fluid velocity, η is the fluid's **first coefficient of viscosity**, η* is the fluid's **second coefficient of viscosity**, and δ_{jk} is the rank 2 unit tensor. For an **incompressible** fluid, η* = 0.

2D§45.2 The **Viscous Shear Force** per unit volume $\boldsymbol{f}_{visc}$ is given by:

$$(\boldsymbol{f}_{visc})_j = \partial S_{jk}/\partial k \quad \text{sum over k}$$

$$\boldsymbol{f}_{visc} = \eta \, \check{D}^2 \boldsymbol{v} + (\eta + \eta^*) \, \check{\boldsymbol{D}} \, (\check{\boldsymbol{D}} \bullet \boldsymbol{v})$$

Here, j and k are indices that range over the three dimensions of space, S_{jk} is the **stress tensor**, $\boldsymbol{v}$ is the fluid velocity, η is the fluid's **first coefficient of viscosity**, and η* is the fluid's **second coefficient of viscosity**. Feynman assumes η and η* are constant throughout time and space.

2D§45.2 The **Equation of Motion** of a **viscous compressible** fluid is:

$$\varrho \, \{ (\boldsymbol{v} \bullet \check{\boldsymbol{D}}) \, \boldsymbol{v} + \partial \boldsymbol{v}/\partial t \} = -\check{\boldsymbol{D}} \, P - \varrho \, \check{\boldsymbol{D}} \, \phi + \eta \, \check{D}^2 \boldsymbol{v} + (\eta + \eta^*) \, \check{\boldsymbol{D}} \, (\check{\boldsymbol{D}} \bullet \boldsymbol{v})$$

$$\partial \boldsymbol{\Omega}/\partial t + \check{\boldsymbol{D}} \times (\boldsymbol{\Omega} \times \boldsymbol{v}) = (\eta / \varrho) \, \check{D}^2 \boldsymbol{\Omega}$$

Feynman says: "It's complicated. But that's the way nature is." Here, ϱ is the fluid mass density, $\boldsymbol{v}$ is the fluid velocity, P is pressure, ø is the potential energy per unit mass, $\boldsymbol{\Omega}$ is the fluid's vorticity, η is the fluid's **first coefficient of viscosity**, and η* is the fluid's **second coefficient of viscosity**.

2D§45.3 The **Force Exerted by an Incompressible Fluid** on a stationary object depends on many variables, including: object size and shape; and fluid density, viscosity, and velocity. Remarkably, four of these quantities combine in the same manner in a wide range of situations. We define **Reynolds Number** Ř as:

$$\check{R} = \varrho \, v \, W / \eta$$

Here, W is the object's **effective area** (see table below), ϱ is the fluid mass density, v is the fluid velocity, and η is the fluid's **first coefficient of viscosity**.

2D§45.3 The **Equation of Motion** of a **viscous incompressible** fluid, with Reynolds number Ř, simplifies to:

$$\partial \boldsymbol{\Omega}/\partial t + \check{\boldsymbol{D}} \times (\boldsymbol{\Omega} \times \boldsymbol{v}) = \check{D}^2 \boldsymbol{\Omega} / \check{R}$$

Here, $\boldsymbol{\Omega}$ is the fluid's vorticity, and $\boldsymbol{v}$ is the fluid velocity.

2D§45.4 The **Drag Coefficient** C_d is normally defined as:

$$C_d = F_d \, (2 / \varrho \, v^2 \, W)$$

Here, F_d is the drag force, ϱ is the fluid mass density, v is the fluid velocity, and W is the **effective area** (see table below). Feynman uses a different definition that is not standard and seems less effective.

2D§45.4 **Drag Coefficient Table**. Drag coefficients vary enormously with object geometry and Reynolds number. Drag coefficients for some common geometric shapes are listed below. T. O. Jesse provided these graphics and drag coefficients at a mid-range Reynolds number.

<u>Drag Coefficients</u>

0.47 Sphere	○
0.42 Half sphere	◐
0.50 Cone	◁
1.06 Cubo	□
0.80 Diamond	◇
0.82 Cylinder, long	▭
1.15 Cylinder, short	□
0.04 Airfoil	⌒
0.09 Half airfoil	～

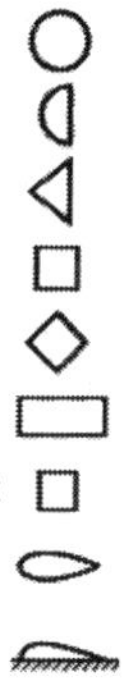

The **effective area** is the object's largest cross sectional area, except in the case of airfoils, where it is the nominal wing area.

2D§45.3 For a **Compressible Fluid**, one additional parameter is required to describe its flow. This is the **Mach number**, v/c_s, the fluid velocity divided by the speed of sound in the fluid. Feynman says two flow problems have the same solution if both have the same Mach number and Reynolds number.

Chapter 13
Quantum Mechanics

"If quantum mechanics hasn't profoundly shocked you,
you haven't understood it yet"
— Niels Bohr

"I think that I can safely say that
no one understands quantum mechanics"
— Richard Feynman

Quantum mechanics correctly describes nature's fundamental processes, the nuts-and-bolts of reality at its core. This often-strange theory is one of the most extensively tested and precisely confirmed creations of the human mind. The everyday world we perceive is a hazy, superficial, diluted version of the tempestuous reality of the quantum micro-world.

The **two major principles** of quantum mechanics are:

• Quantization

• Particle-Wave Duality

Throughout this chapter, h is Planck's constant, the number that sets the scale at which quantum effects become dominant, and ħ ("h-bar") equals h / 2π.

Quantum mechanics is valid only in **inertial reference frames**, those moving at constant velocity ***v***. Since quantum mechanics was developed to describe atoms, where particle velocities are << c, this chapter employs non-relativistic equations.

4B§13.1 Quantization

Quantization is the simple notion that many things in nature come in integer quantities — the number of electrons is always an integer, as are the number of people. The image below illustrates the difference between something that is **continuous** and something that is **quantized**.

Figure 13-1 Staircase and Ramp

On a ramp, elevation is continuous: every value of elevation is possible. On a staircase, elevation is quantized: only a few discrete values are possible. On a staircase, elevation changes abruptly and substantially. The micro-world of atoms and particles is replete with significant staircases — the steps are large and they dramatically impact natural processes in this realm. We, however, live on a much larger scale, billions of times larger. As sizes scale up from the microscopic toward the macroscopic, the steps in nature's staircases appear ever smaller and ever more numerous, as depicted below.

Figure 13-2 Staircase Viewed On Ever-Larger Scales

Eventually, the steps become too small to be individually discernible, and we observe smooth ramps instead of staircases. Nature does not have one set of laws for the atomic scale and a different set of laws for the human scale. **Nature's laws are universal; staircases exist always and everywhere**. But at our scale, these steps are so small that they almost never make a distinguishable difference.

4B§13.2 Particles & Waves

3A§1.7 The **de Broglie Wavelength**. Einstein said light is both a wave and a particle. Louis de Broglie extended this idea, saying every particle is also a wave whose wavelength λ is given by:

$$\lambda = h / p$$

Here, h is Planck's constant, and p is the particle's momentum. This remarkable combining of particles and waves, now called **Particle-Wave Duality**, is the essential foundation of quantum mechanics, and the reason that key quantities are quantized.

A normal wave spreads throughout space and time; conversely, particles seem localized. How can a particle be comprised of waves?

3A§2.1 **Particles are Wave Packets** — waves that are partially localized — as illustrated in Figure 13-3.

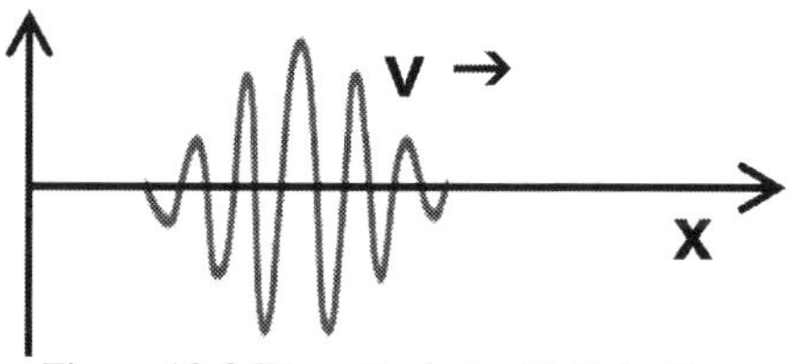

Figure 13-3 Wave Packet with Velocity v

The above wave packet is the superposition of many waves with slightly different frequencies that are clustered around a central frequency f_0. Inevitably, as mathematically proven with Fourier Transforms (see 4B§3.7), the greater the frequency range Δf, the smaller the wave packet's width Δx; the narrower the frequency range, the broader the packet's width.

The result is: the product of Δf and Δx has an irreducible minimum value.

3A§2.1 The **Heisenberg Uncertainty Principle** is a direct consequence of particles being wave packets. It comprises four equations:

$$\Delta x \;\; \Delta p_x = \hbar / 2 = h / 4\pi$$
$$\Delta y \;\; \Delta p_y = \hbar / 2$$
$$\Delta z \;\; \Delta p_z = \hbar / 2$$
$$\Delta t \;\; \Delta E \; = \hbar / 2$$

The uncertainty products given above are the very best case. They are achieved only when all distributions are Gaussian, and each Δ is defined to contain half the distribution. In less favorable circumstances, the products of the uncertainties are larger.

3A§2.2 **Phase Space is** an abstract space whose dimensions are distance and momentum. One can think of the uncertainty principle as saying that each particle-wave requires its own patch of phase space — that it cannot be squeezed into anything smaller than some minimum size. This is sketched below for two dimensions: x and p_x.

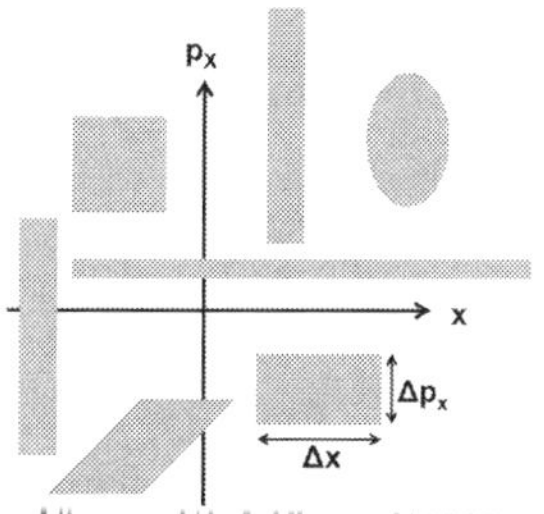

Figure 13-4 Phase Space

The area of phase space that a particle-wave occupies can have various shapes, as shown above. But, the uncertainty principle says that area can never be less than ħ / 2.

3A§9.5 **Barrier Penetration** is another consequence of every entity having a wavelength and being governed by a wave equation. When a quantum particle-wave hits a classically impenetrable barrier, the wave amplitude does not immediately drop to zero. Instead, the magnitude of the wave decreases exponentially as the wave penetrates deeper into the classically forbidden region that is shown in gray below.

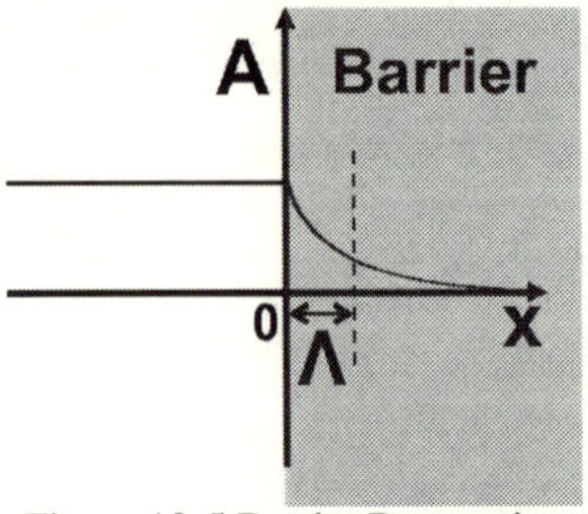

Figure 13-5 Barrier Penetration

In the above figure, a particle-wave-packet with wave amplitude A enters from the left. At x = 0, it encounters an "impenetrable" barrier: the particle-wave-packet does not have enough energy to scale the barrier. Classically, the particle would bounce off the barrier's surface and never penetrate the x > 0 region.

In quantum mechanics, things are different: the energy potential, the particle's kinetic energy, and the particle-wave amplitude are:

For x < 0:

energy potential = 0

kinetic energy = T

amplitude = A

For x > 0:

energy potential = V

kinetic energy = T – V = – ε

for T < V, let $\Lambda = \hbar / \sqrt{2m\varepsilon}$

amplitude = A exp{ – x / Λ}

Here, m is the particle's mass. If the particle-wave's kinetic energy T is less than the barrier height V, ε is positive, Λ is real, and the wave penetrates modestly into the classically forbidden region. As T approaches V, ε decreases and the penetration depth increases.

4B§13.3 States & Basis States

3A§4.1 A **Quantum State** defines an entity's variable properties, such as position, linear momentum, energy, and angular momentum. It need not define intrinsic properties such as a particle's charge and mass. In general, different types of particles can be put into a given state, and can be moved from one state to another.

It is beneficial to think of **a quantum state as a vector** that defines a location in the abstract space of all possible properties. An example is shown below, where each dot is one possible quantum state in the abstract 2-D space of energy along the vertical axis and the z-component of angular momentum along the horizontal axis.

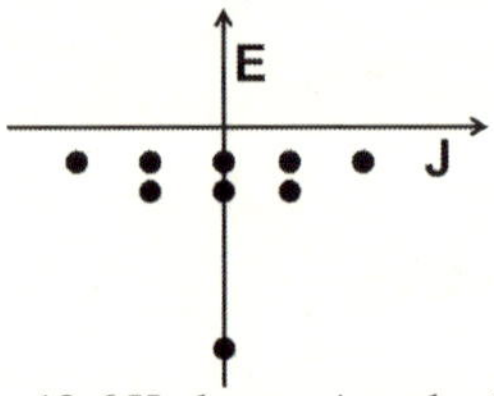

Figure 13-6 Hydrogen Angular States

Each dot in Figure 13-6 represents an allowed quantum state of an electron in a hydrogen atom that has energy E and angular momentum $J = m\hbar$ along the z-axis. See 4B§13.7 for an explanation of these states.

3A§4.1 The **Dirac Bra-Ket** convention denotes quantum states and amplitudes as:

a bra: $< B |$

a ket: $| A >$

a bra-ket: $< B | A >$

Reading from right to left, $< B | A >$ is the **amplitude** that state A results in, or goes to, state B. Quantum amplitudes, generally complex quantities, are used to calculate the probabilities of various outcomes, as discussed in 4B§13.4. Often, $< B |$ and $| A >$ are called **state vectors**.

Pursuing our analogy of a quantum state as a vector in property space, $<B|A>$ is analogous to $\boldsymbol{B} \bullet \boldsymbol{A}$, the dot product of 3-D vectors A and B: $< B | A >$ measures the degree to which A and B have the same properties. We can also consider $< B | A >$ to be the **projection** of state $< B |$ onto state $| A >$, just as $\boldsymbol{B} \bullet \boldsymbol{A}$ is the projection of vector $\boldsymbol{B}$ onto vector $\boldsymbol{A}$.

One **key difference** between 3-D vectors and quantum state vectors is that the former commute but the latter do not. Thus:

$\boldsymbol{B} \bullet \boldsymbol{A} = \boldsymbol{A} \bullet \boldsymbol{B}$ for any two vectors in 3-D

however, $< B | A >$ does not equal $< A | B >$ in general

See **Principle #7** in 4B§13.4 for more on how $< B | A >$ realtes to $< A | B >$.

3A§7.5 **Quantum Basis States** are logically similar to unit vectors along the axes of a Cartesian coordinate system. Any 3-vector $\boldsymbol{A}$ can be represented by the sum of multiples of the unit vectors along each axis. It is sometimes easier to work with the components of $\boldsymbol{A}$ than with $\boldsymbol{A}$ itself. Similarly, it is often convenient to find the "components" of quantum state $|A>$ along various "axes", analyze each component separately, and then recombine the results. To be fully effective, we seek a **complete set** of **orthonormal basis states** (the quantum equivalent of axes). The choice of basis states is arbitrary — any complete set of orthonormal basis states can be employed to describe any phenomenon. Some choices may simplify the mathematics, but any choice will yield the same final results, eventually.

3A§7.5 Let's first consider a **Discrete Basis Set** with a finite number of basis states that we denote $| J >$, J=1 to N.

A basis set is **orthonormal** if:

$< J | J > = 1$ for each basis state $|J>$

$< J | K > = 0$ whenever K is not J

A basis set is **complete** if every possible state $|A>$ equals some linear combination of basis states, as in:

$$| A > = \Sigma_J < J | A > | J >$$

Here, the amplitude $< J | A >$ is analogous to $\boldsymbol{J} \bullet \boldsymbol{A}$: $< J | A>$ is the "component" of "vector" $| A >$ along the $| J >$ "axis".

If $| J >$, J=1 to N, form a complete orthonormal set of quantum basis states, then for any states $| A >$ and $| B >$:

$$| A |^2 = \Sigma_J | < A | J > |^2 = \Sigma_J < A | J > < A | J >^*$$

$$< A | B > = \Sigma_J < A | J > < J | B >$$

These equations are exactly analogous to the corresponding equations of vector algebra. Recall from 4B§3.5 that, for any complex quantity z, its complex conjugate z*, and the magnitude of either are given by:

$$z = x + iy$$

$$z^* = x - iy$$

$$| z |^2 = z z^* = x^2 + y^2$$

3B§22.3 Let's next consider a **Continuous Basis Set** with an infinite number of basis states. Continuous basis sets are required to analyze many phenomena involving continuous physical quantities, such as position and momentum.

We define the **wave function** $\psi(x)$ to be the amplitude that a particle-wave is at position x, with x being a continuous variable spanning every real number from $-\infty$ to $+\infty$. $\psi(x)$ is the continuous analog of $< x \mid \psi >$ which might also occur when using a discrete basis if x were restricted to a finite set of values. With continuous basis sets, we replace summations over finite sets with integrals over continuous variables. For example, compare the discrete (D) and the corresponding continuous (C) equations:

$$\text{D: } < B \mid A > = \Sigma_x < B \mid x > < x \mid A >$$

$$\text{D: } < B \mid A > = \Sigma_x < x \mid B >^* < x \mid A >$$

$$\text{C: } < ø \mid \psi > = \int_{-\infty}^{+\infty} < ø \mid x > < x \mid \psi > dx$$

$$\text{C: } < ø \mid \psi > = \int_{-\infty}^{+\infty} ø^*(x)\, \psi(x)\, dx$$

The orthonormal requirement for a continuous basis set is defined as:

$$\text{if } X = x, < X \mid x > = 1, \text{ otherwise } < X \mid x > = 0$$

$$< X \mid \psi > = \int_{-\infty}^{+\infty} < X \mid x > < x \mid \psi > dx$$

$$\psi(X) = \int_{-\infty}^{+\infty} \delta(X-x)\, \psi(x)\, dx \text{ (see next)}$$

2B§22.5 The **Dirac Delta Function** $\delta(u)$ satisfies the above requirement, and is defined as:

$$\delta(u) = \text{limit as } \varepsilon \to 0 \text{ of } \exp\{-u^2 / \varepsilon^2\} / \varepsilon \sqrt{\pi}$$

$$\delta(u) \text{ is zero for all u except } u = 0$$

$$\text{and: } \int_{-\infty}^{+\infty} \delta(u)\, du = 1$$

Clearly, $\delta(u)$ is not an ordinary, continuous function; it is infinitely narrow and infinitely large at $u = 0$. It is a mathematical tool that is properly used only within integrals.

4B§13.4 Quantum Probability

Quantum mechanics forces us to abandon all hope of **exactly predicting the future**. Instead, we must resign ourselves to the goal of **predicting the probabilities** of various possible future outcomes.

3A§4.9 Feynman lists **Seven Principles of Probabilistic Quantum Mechanics**. For all these, ø(A) and Prob(A) are defined to be the **amplitude** and **probability** of **event A**, respectively.

Note that we use the word "amplitude" in two different ways. In 4B§6.3, we defined amplitude as the amount a wave goes up and down. In quantum mechanics, we still use "amplitude" in that sense, but we also use it in a probabilistic sense as described below, where an amplitude might be written:

$$ø(A) = \exp\{ (E\,t - \boldsymbol{r} \cdot \boldsymbol{p}) / i\hbar\}$$

Principle #1: If event A can occur in only one way:

$$\text{Prob}(A) = | ø(A) |^2 = ø(A)\, ø^*(A)$$

Here, $| z |^2 = z\, z^*$ is the square of the magnitude of the complex quantity z.

Principle #2: If event A can occur in N different ways, with amplitudes $ø_1(A)$ to $ø_N(A)$, and the amplitudes are **coherent**, the amplitudes are summed before squaring, as in:

$$\text{Prob}(A) = | ø_1(A) + \ldots + ø_N(A) |^2$$

Particle-wave amplitudes are *coherent* if they have the same frequency and a definite phase relationship, and are *incoherent* otherwise. Coherent waves start with a definite phase relationship, and up through event A, suffer no interaction sufficient to disrupt those phase relationships. (See 4B§10.2) Absent a sufficient disturbance, it is impossible to distinguish which one of the N paths the particle-wave took to reach event A. If it is impossible to distinguish which path was taken, quantum mechanics says the particle-wave took all possible paths simultaneously.

Principle #3: If event A can occur in N different ways, with amplitudes $ø_1(A)$ to $ø_N(A)$, and the amplitudes are **incoherent**, the amplitudes are squared before summing, as in:

$$\text{Prob}(A) = |\, ø_1(A)\,|^2 + \ldots + |\, ø_N(A)\,|^2$$

Principle #4: The amplitude for a sequence of events equals the product of the amplitudes of each separate event, as in:

$$ø(1 \text{ then } 2 \text{ then } 3) = ø(1)\, ø(2)\, ø(3)$$

Principle #5: If ø(1) is the amplitude of particle 1 going from S_1 to F_1, and ø(2) is the amplitude of particle 2 going from S_2 to F_2, then the amplitude ø(1 and 2) for both events to occur is the product of the individual amplitudes:

$$ø(1 \text{ and } 2) = ø(1)\, ø(2)$$

Principle #6: When two identical particles can enter, exit, or be in the same state, their amplitudes interfere. If the particles are bosons governed by **Bose-Einstein Statistics**, their amplitudes add. If the particles are fermions governed by **Fermi-Dirac Statistics**, their amplitudes subtract. There is no third alternative. For two identical particles 1 and 2, and any two states A and B, the combined amplitude ø is:

$$\text{Fermions: } ø = \langle 1 | A \rangle \langle 2 | B \rangle - \langle 1 | B \rangle \langle 2 | A \rangle$$

$$\text{Bosons: } ø = \langle 1 | A \rangle \langle 2 | B \rangle + \langle 1 | B \rangle \langle 2 | A \rangle$$

If A and B are the same state:

$$\text{Fermions: } ø = 0$$

$$\text{Bosons: } ø = 2 \langle 1 | A \rangle \langle 2 | A \rangle$$

Particles with different spins are not in the same state.

Principle #7: The amplitude for A to go to B equals the complex conjugate of the amplitude for B to go to A. We write this:

$$\langle B | A \rangle = \langle A | B \rangle^*$$

This assumes the total probability of being somewhere is always 100%. For particles that decay, this principle applies only if one properly accounts for all the decay products.

3A§5.4 **Bosons are Gregarious**, as **Principle #6** shows. Let P be the probability of boson B entering a state Q if Q does not already contain any bosons identical to B. Then, the probability of boson B entering a state Q if Q already contains N bosons identical to B is:

$$\text{Prob(adding B to N prior B's)} = (N+1)\, P$$

Hence, the probability that N Identical Bosons are in state Q is larger by a factor of N factorial (N!) than the probability of N non-identical bosons being in that state.

3C§4.7 **Fermions are Antisocial**, as **Principle #6** shows. Two identical fermions cannot occupy the same quantum state. If two electrons share a common momentum and position state, such as the 1s ground state of helium, they must have opposite spins: one spin up and the other spin down, leaving no room for a third electron. Called the **Pauli Exclusion Principle**, this is the basis of the electron shell structure reflected in the Periodic Table of Elements (see 4B§13.7).

3C§32.1 The **Probability Current** $\hat{\boldsymbol{J}}$ and the **Conservation of Probability** for a particle of mass m are written:

$$\hat{\boldsymbol{J}} = \{ (\psi^* \boldsymbol{P}\, \psi) + (\psi^* \boldsymbol{P}\, \psi)^* \} / 2m$$

$$\partial(\text{Probability})/\partial t + \check{\boldsymbol{D}} \cdot \hat{\boldsymbol{J}} = 0$$

4B§13.5 Operators & Matrices

3A§12.5 **Operators** transform one state vector into another. In quantum mechanics, every action that changes particle states is represented by an operator. For example:

$$<\chi \mid A \mid \psi> = \Sigma_{JK} <\chi \mid K><K \mid A \mid J><J \mid \psi>$$

The left side of this equation denotes the product of state χ with the state resulting from A operating on state ψ. The right side expands this in terms of basis states. Knowing $<K \mid A \mid J>$ for all basis states J and K completely determines the actions of operator A on any quantum state.

Operators are often represented as N×N matrices, where N is the number of basis states. Matrix component A_{KJ} equals $<K \mid A \mid J>$.

Quantum mechanics precludes the possibility of multiple operators acting on any state simultaneously. If multiple operators act on a state sequentially, we express that using the product of those operators. Neither operators nor matrix products are commutative; their order is critical. The first operator to act holds the right-most position, with the second immediately on its left, and continuing on to the last operator to act in the left-most position. For two operators, A followed by B, we write:

$$<\chi \mid B\,A \mid \psi> = <\chi \mid B\,(\Sigma_{JK} \mid K><K \mid A \mid J><J \mid \psi>)$$

$$<\chi \mid B\,A \mid \psi> = \Sigma_{JKL} <\chi \mid L><L \mid B \mid K><K \mid A \mid J><J \mid \psi>$$

3A§10.7 The **Hamiltonian Operator** of any system is its energy operator H that determines how the system evolves over time. For a system with **wavefunction** $\psi(x,t)$, the **Hamiltonian Equation** is:

$$i\hbar\, d \mid \Psi(x,t)> / dt = H \mid \Psi(x,t)>$$

$$i\hbar\, d <k \mid \Psi> / dt = \Sigma_n <k \mid H \mid n><n \mid \Psi>$$

$$i\hbar\, d\Psi/dt = \Sigma_{kn} \mid k><k \mid H \mid n><n \mid \Psi>$$

$$i\hbar\, d\Psi/dt = H\,\psi$$

Here, k and n denote any basis state, and the sums are over all basis states. If $\psi(x,t)$ has the usual time dependence, the last equation reduces to its most recognized form:

$$\psi(x,t) = \exp\{E\,t / i\hbar\}\, ø(x)$$

$$E\,\psi = H\,\psi$$

3B§14.4 **Eigenvalues and Eigenvectors**. Any non-singular square matrix, including most Hamiltonians, can be **diagonalized**, meaning transformed to a representation in a new set of basis states in which all off-diagonal matrix components are zero. In that basis, diagonal component H_{nn} is called the **eigenvalue** E_n, and the corresponding basis state is called the **eigenvector** V_n. For each n:

$$H\,V_n = E_n\,V_n$$

Since all off-diagonal components of the Hamiltonian are now zero, a particle in eigenvector V_n remains there forever and never transitions to any other state, absent external interaction. Hence **each eigenvector V_n is a stationary state of definite energy E_n**.

3B§13.1 **Two-State Systems**, the simplest non-trivial systems, have exactly two basis states. For example, a singly-ionized hydrogen molecule is comprised of two protons sharing a single electron. Two obvious basis states are:

$\mid 1>$ electron around proton #1

$\mid 2>$ electron around proton #2

By symmetry, both basis states must have the same energy — kinetic plus potential — that we call E_0. Let's assume the electron is bound to that proton, and therefore $E_0 < 0$ — the electron's kinetic energy is less than its potential in the attractive electric field of a proton. Classically, this is the end of the story, an electron starting in either basis state would remain there indefinitely, absent external interaction.

However, due to the quantum phenomena of barrier penetration (see 4B§13.2), the electron has a nonzero amplitude to switch partners. Again by symmetry, the amplitude for the electron to go from proton #1 to proton #2 must equal the amplitude to go from #2 to #1. We call that amplitude –E*.

The 2x2 Hamiltonian has these components:

$$H_{11} = H_{22} = E_0$$

$$H_{12} = H_{21} = -E^*$$

Solving the Hamiltonian equation yields the two stationary states of definite energy — the system's eigenvectors — that are:

$$|+> = \{ |1> - |2> \} / \sqrt{2} \text{ with } E_+ = E_0 + E^*$$

$$|-> = \{ |1> + |2> \} / \sqrt{2} \text{ with } E_- = E_0 - E^*$$

Let's be very clear about these states. The sole electron can exist in either stationary state |+> or |–> indefinitely, absent external interaction. In either of these states, the electron has a precisely defined energy, and is at all times simultaneously around both protons. Yet, if we measure the electron's position, it will never be found around both, it always has a 50% probability of being found entirely near proton #1 and a 50% probability of being found entirely near proton #2.

Alternatively, the electron could be, at time t = 0, entirely in basis state | 2 >. But it will not remain there. Such an electron transitions back and forth between states | 1 > and | 2 >. These states are therefore not stationary, and are not states of definite energy. If we measure the electron's position at time t, the probability of finding it around proton #1 will be $\sin^2(\omega t)$, while the probability of finding it around proton #2 will be $\cos^2(\omega t)$.

3B§15.1 For a **Spin 1/2 Particle with Magnetic Dipole *μ*** in a magnetic field ***B*** that may vary over time, the Hamiltonian is:

$$H_{jk} = \begin{pmatrix} E_0 - \mu B_z & -\mu B_x + i\mu B_y \\ -\mu B_x - i\mu B_y & E_0 + \mu B_z \end{pmatrix}$$

Here, the basis states are:

$| 1 > = \boldsymbol{\mu}$ in the +z-direction

$| 2 > = \boldsymbol{\mu}$ in the –z-direction

3B§16.7 **Hyperfine Splitting**. Due to the four possible orientations of proton and electron spins, hydrogen's ground state splits into a spin 0 singlet state of lower energy and a spin 1 triplet state of higher energy. The spin interaction Hamiltonian H and the energy difference ΔE are:

$$H = A\, \boldsymbol{\sigma}^e \bullet \boldsymbol{\sigma}^p$$

$$\Delta E = 5.87433 \text{ micro-eV}$$

$$f = \Delta E / h = 1420.405{,}751{,}786 \text{ MHz}$$

$$\lambda = c / f = 21.106{,}114{,}051{,}13 \text{ cm}$$

Here, $\boldsymbol{\sigma}^e$ and $\boldsymbol{\sigma}^p$ are the **Pauli spin vector matrices** of the electron and proton respectively. With a lifetime of 10 million years, the famous hydrogen "21-cm" line has a 6×10^{-25} cm **line width** and a resonance Q factor of $3\times10^{+26}$.

3B§16.4 The **Pauli Spin Vector Matrix** for a spin 1/2 particle has three components, each of which is a 2x2 matrix, as given by:

$$\sigma^F = (\sigma^F_x, \sigma^F_y, \sigma^F_z)$$

$$\sigma^F_x = \begin{pmatrix} 0 & 1 \\ 1 & 0 \end{pmatrix} \quad \sigma^F_y = \begin{pmatrix} 0 & -i \\ i & 0 \end{pmatrix} \quad \sigma^F_z = \begin{pmatrix} 1 & 0 \\ 0 & -1 \end{pmatrix}$$

Here, the superscript F denotes the fermion on which the operator acts.

3B§16.**8 Zeeman Splitting** of atomic electron energy levels arises from particle spins interacting with an external magnetic field ***B***. The Hamiltonian is:

$$H = A\, \sigma^e \bullet \sigma^p - \mu_e\, \sigma^e \bullet B - \mu_p\, \sigma^p \bullet B$$

Here, $\boldsymbol{\sigma^e}$ and $\boldsymbol{\sigma^p}$ are the **Pauli spin vector matrices**, and μ_e and μ_p are the magnetic moments of the electron and proton, respectively.

Quantum Operators are sometimes used to determine the values of key quantities. For example, consider Hamiltonian operator H acting on state $|\psi>$:

$$H\, |\, \psi > = E\, |\, \psi >$$

$$E = < \psi\, |\, H\, |\, \psi > \text{ (in a discrete basis)}$$

$$E = \int \psi^*\, H\, \psi\, dV \text{ (in a continuous basis)}$$

Here E is the energy of the particle-wave represented by $|\psi>$.

The operators for the most important physical quantities are:

3C§31.6 **Energy**: $\boldsymbol{H} = -(\hbar^2/2m)\, \partial^2/\partial x^2 + V(x)$

3C§31.6 **Position**: $\boldsymbol{x} = x$

3C§31.6 x-**Momentum**: $\boldsymbol{P_x} = (\hbar/i)\, \partial/\partial x$

3C§31.7 z-**Angular Momentum**: $\boldsymbol{J_z} = \boldsymbol{x}\, \boldsymbol{P_y} - \boldsymbol{y}\, \boldsymbol{P_x}$

3C§31.8 **Operators and Commutators**. In general, matrices and quantum operators **do not commute**. This means two operators acting sequentially on state $|\psi>$ yield two different results depending on which operator acts first. Quantum mechanics precludes the possibility of multiple operators acting on any state simultaneously. An example of two operators that do not commute is:

$$\boldsymbol{x}\, \boldsymbol{P_x}\, \psi(x) - \boldsymbol{P_x}\, \boldsymbol{x}\, \psi(x) = (i\hbar)\, \psi(x)$$

An example of two operators that do commute is:

$$\boldsymbol{y}\, \boldsymbol{P_x}\, \psi(x) - \boldsymbol{P_x}\, \boldsymbol{y}\, \psi(x) = 0$$

The **commutator** of operators A and B is written:

$$[A, B] = A\, B - B\, A$$

If [A, B] = 0, A and B **commute**; if [A, B] is nonzero, they do not commute. This is called the **commutation rule**.

Commutators are a staple of quantum mechanics, and also other advanced fields of physics. Note that angular momentum operators generally do not commute:

$$[J_x, J_y] = i\hbar\, J_z$$

3B§23.2 **Symmetry and Operators**. Feynman says a physical system is symmetric with respect to the operation Q if Q commutes with the system's Hamiltonian H:

$$\text{Symmetry under } Q \leftrightarrow H\, Q = Q\, H$$

3C§31.3 **Operators and Average Values**. The average value of many measurements of a physical quantity Q of a particle-wave described by wave function ψ, in a discrete basis set (D) and in a continuous basis set (C), are given by:

$$D: < Q_\psi > = \Sigma_{jk} < \psi\, |\, j > < j\, |\, \boldsymbol{Q}\, |\, k > < k\, |\, \psi >$$

$$C: < Q_\psi > = \int\int < \psi\, |\, x > < x\, |\, \boldsymbol{Q}\, |\, X > < X\, |\, \psi > dx\, dX$$

Here, $\boldsymbol{Q}$ is the quantum operator representing the physical quantity Q. $< Q_\psi >$ is the ψ-weighted average of Q: the amplitude that $\boldsymbol{Q}$ connects each pair of basis states — j and k in the discrete set, or x and X in the continuous set — weighted by the probability $\psi^*\psi$ of the particle-wave being in those basis states. The last equation reduces to (see **Dirac delta function** in 4B§13.3):

$$C: < Q_\psi > = \int \psi^*(x)\, \boldsymbol{Q}\, \psi(x)\, dx$$

3C§31.9 The **Time-Dependent Average** of a physical quantity represented by operator **A** is given by:

$$d < A_\psi > /dt = < \psi \,|\, (i / \hbar) [H, \mathbf{A}] + d\mathbf{A}/dt \,|\, \psi >$$

The last term is the effect of any time-dependence of operator **A** itself. If **A** is time-independent, this term is zero, leaving only the commutator [H, **A**] term.

4B§13.6 Schrödinger's Equation

3A§9.4 The **Wave Function** ψ in 1-D for a particle-wave with kinetic energy T, momentum p, and in an energy potential V is:

$$\psi = A \exp\{ [(T + V) t - p x] / i\hbar \}$$

3B§22.6 **Schrödinger's Equation** in 3-D for a particle of mass m in a potential V is:

$$i\hbar\, d\psi/dt = -(\hbar^2 / 2m)\ \check{D}^2 \psi + V \psi$$

$$\text{where } \check{D}^2 = \partial^2/\partial x^2 + \partial^2/\partial y^2 + \partial^2/\partial z^2$$

4B§13.7 Atoms and their Electrons Orbits

1A§1.6 **Atoms** are comprised of three types of particles: *electrons*, *protons*, and *neutrons* (see 4B§14.1). Positively charged protons and neutral neutrons — collectively called *nucleons* — have nearly equal masses; each about 1836 times the mass of negatively charged electrons. Nucleons attract one another with the strong force, and are contained within a tiny nucleus. The number of protons in an atom's nucleus is its *element number*, its atomic number Z on the Periodic Table of Elements. Atoms with Z protons may have varying numbers of neutrons; these are called different *isotopes* of the same element. Electrons and protons attract one another with the electric force, thereby holding atoms together. When an atom has equal numbers of protons and electrons, it has zero net charge; when these numbers differ, the atom is charged and is called an *ion*. Electrons orbit nuclei at distances that extend to 100,000 times the radius of the nucleus.

3A§3.6 **Quantization of Electron Waves in Atoms**. Niels Bohr showed that the wavelengths of electrons restrict their allowed orbits around atomic nuclei, as can be stated in two equivalent ways:

1. For the electron wavefunction to be single-valued (as any proper function is) everywhere in its orbit, the orbit's circumference must equal an integral number of wavelengths. (Else ψ would have different values at θ and θ + 2π.) This means:

$$2\pi r = n \lambda, \text{ for some integer } n$$

2. The electron's angular momentum must be an integral multiple of ħ. For a circular orbit, this means:

$$r p = n \hbar$$

The prior two equations are equivalent, as is readily seen by multiplying the first by (p / 2π) and using λ = 2π ħ / p.

3A§3.6 The **Bohr Radius** r_B of the lowest-energy orbit (n=1 in prior equation) of an electron in a hydrogen atom is:

$$r_B = \hbar^2 / m e^2$$

$$e^2 = q_e^2 / 4\pi\varepsilon_0$$

Here, m is the electron mass, and q_e is the electron charge.

3A§3.7 The **Allowed States** of a lone electron around a nucleus of charge Z have radius r_n and energy E_n given by:

$$r_n = n^2 \hbar^2 / \ m Z e^2$$

$$E_n = -Z e^2 / 2r_n$$

$$E_n = -13.61 \text{ eV } Z^2 / n^2$$

Here, n is an integer, the **principle quantum number** that determines the electron's average orbital radius. As usual, m is the electron mass, and q_e is the electron charge. With multiple electrons around a nucleus of charge +Z, electrons in inner orbits partially shield the nuclear charge from electrons in outer orbits, thereby effectively reducing Z.

3A§12.9 **Atoms are Stable**, otherwise you wouldn't be reading my book. Yet classical physics cannot explain why. Classically, orbiting electrons continuously accelerate (changing the direction of their velocity), and must therefore radiate light (see 4B§10.7). Continuously losing energy to radiation, electrons would spiral inward, collapsing the atom.

The Bohr model explains the stability of atoms: once in the allowed orbit of least energy, electrons cannot further reduce their energy; they cannot radiate and cannot spiral inward. The resulting electron shell structure of stable atoms enables all chemical and biological reactions.

3C§29.7 An **Electron in a Hydrogen Atom** is described by a wavefunction $\psi_{n,j,m}(\theta, \phi, \varrho, t)$, where (θ, ϕ, ϱ) are spherical coordinates, and the three integers (n,j,m) are the orbit's **quantum numbers**.

$$\psi_{n,j,m}(\theta, \phi, \varrho, t) = \alpha \; Y \; F \exp\{E_n \, t / i\hbar\}$$

$$Y = Y_{j,m}(\theta, \phi) = \langle j,0| \, R_y(\theta) \, R_z(\phi) \, |j,m\rangle$$

$$F = F_{n,j}(\varrho) = \exp\{-\varrho / n\} \; \Sigma_k \; a_k \; \varrho^{k-1}$$

$$a_{k+1} = 2 \, a_k \; \{k/n - 1\} \, / \, \{k(k+1) - j(j+1)\}$$

$$E_n = -\mu \, e^4 / 2 \, \hbar^2 \, n^2$$

$$\varrho = r / r_B$$

$$r_B = \hbar^2 / \mu \, e^2$$

$$e^2 = q_e^{\,2} / 4\pi\varepsilon_0$$

Here, E_n is the electron energy of orbits with principal quantum number n, α is the normalization factor, and the sum over k is from k = j + 1 to k = n. $Y_{j,m}(\theta, \phi)$ is a **spherical harmonic** that describes the orbit's angular distribution, and $F_{n,j}(\varrho)$ describes the orbit's radial dependence. Also, μ and q_e are the electron mass and charge. $R_y(\theta)$ and $R_z(\phi)$ are respectively the rotation matrices for rotations of angle θ about the y-axis and angle ϕ about the z-axis. Finally, r_B is the Bohr radius, r is the normal radial coordinate, and ϱ is the radial coordinate scaled by the Bohr radius.

Principal quantum number n determines $< 1 / r >$, the orbit's average inverse radius. n can be any integer greater than zero. The maximum n found in known elements is 7.

Orbital angular momentum quantum number j specifies the orbit's total angular momentum: $L = j\,\hbar$. j can be any integer from 0 through n – 1.

Magnetic quantum number m specifies the component of j along any selected axis. m can be any integer from –j through +j.

The coefficients a_k, k = j + 1 to k = n, are defined by the recursive formula given above. When k = n, the numerator equals zero and the ϱ power series terminates; **this is the essential mechanism** that limits electrons to a small set of discrete orbits. If n were not an integer, the ϱ power series would never terminate, and ψ would be infinite at $\varrho = \infty$. This forces n to be an integer, allowing only specific orbits. The first coefficient a_{j+1} is arbitrary, since it is absorbed into the constant α that provides the overall normalization of ψ.

3A§6.4 **Electron Atomic Orbits** with various values of angular momentum quantum number j have letter designations (for historical reasons), and allowed values of the magnetic quantum number m, as follows:

j = 0: "**s**", m = 0

j = 1: "**p**", m = –1, 0, +1

j = 2: "**d**", m = –2, –1, 0 +1, +2

j = 3: "**f**", m = –3, –2, –1, 0, +1, +2 +3

Electron orbits are grouped into **shells** according to their principle quantum number n, with each shell having a definite set of **subshells** of different angular momentum j, according to:

Shell n = 1: 1s

Shell n = 2: 2s, 2p

Shell n = 3: 3s, 3p, 3d

Shell n = 4: 4s, 4p, 4d, 4f

Each allowed combination of n, j, and m quantum numbers can accommodate 2 electrons: one spin up and the other spin down.

Each allowed combination of n and j quantum numbers can accommodate 2×(2j+1) electrons, because there are 2j+1 possible m values.

The maximum capacities of various orbits are:

s-orbits: 2 electrons

p-orbits: 6 electrons

d-orbits: 10 electrons

f-orbits: 14 electrons

Shell n can accommodate $2n^2$ electrons: half spin up, half spin down.

The **ground state** of any atom is the state in which none of its electrons can transition to a lower energy state, because all allowed states with lower energy are fully occupied.

4B§13.8 Light & Matter

3A§11.8 **Light Absorption** by an electric dipole at resonant frequency Ω is governed by:

$$prob = 4\pi^2 \mu^2 \mathbf{I}(\Omega) / (4\pi \varepsilon_0 c \hbar^2)$$

Here, *prob* is the probability per unit time of light being absorbed by dipole μ, $\mathbf{I}(\Omega)$ is the intensity of incident light at frequency Ω, and ħΩ equals the dipole transition energy between the two states: (1) $\boldsymbol{\mu}$ parallel to the light's electric field $\boldsymbol{E}$; and (2) $\boldsymbol{\mu}$ antiparallel to $\boldsymbol{E}$.

3A§6.2 The **Planck Black Body Radiation** spectrum is:

$$\mathbf{I}(\omega) = (\hbar \omega^3 / \pi^2 c^2) / \{\exp(\hbar\omega / kT) - 1\}$$

Here, $\mathbf{I}(\omega)$ is the intensity of light emitted at frequency ω, T is the temperature in Kelvin, and k is Boltzmann's constant. All objects emit **black body radiation** due to their thermal energy. In physics, a "black body" is an object that does not reflect incident light of any frequency.

3C§27.6 **Light Emission by Atoms**. An atom in an excited state can emit a photon and drop to a lower energy state. Energy conservation ensures the photon's energy equals the energy difference between the original and final electron states. Since atomic orbits have a limited number of discrete energy values, emitted photons have a limited number of discrete frequencies. Similarly, an atomic electron can transition to a higher-energy state by absorbing a photon, but only if that photon has one of a limited number of discrete frequencies.

Thus, each type of atom has a **unique spectrum** of light frequencies that it can emit or absorb — a unique fingerprint that allows astronomers to measure the atomic composition of stars and galaxies.

An atom that has **angular momentum quantum number** j = 1 and **magnetic quantum number** m = +1 in the +z-direction can emit a photon in one of two ways.

(1) In **electric dipole radiation**, the atom's parity reverses and the photon state ψ is:

$$| \psi > = \{(1 + \cos\theta) | R, +z > - (1 - \cos\theta) | L, -z >\} / \sqrt{8}$$

(2) In **magnetic dipole radiation**, the atom's parity is unchanged and the photon state ψ is:

$$|| \psi > = \{(1 + \cos\theta) | R, +z > + (1 - \cos\theta) | L, -z >\} / \sqrt{8}$$

Here, θ is the photon's angle relative to the z-axis, | R, +z > is the state of a **RHC** (right hand circularly polarized) photon moving toward +z, and | L, –z > is the state of a **LHC** photon moving toward –z.

1C§35.3 **Atomic Emission Line Widths** are often extremely narrow. Treating atoms as classical harmonic oscillators, Feynman calculates:

$$Q = \lambda / d\lambda$$

$$Q = 3\,\lambda / 4\pi\, r_0$$

$$\text{with } r_0 = e^2 / mc^2 = 2.8179 \times 10^{-15}\text{ m}$$

Here, Q is the ratio of emitted wavelength λ to line width dλ, and r_0 is the **classical electron radius**. We now know the electron radius is actually at least 100,000 times smaller, and might even be zero. The two 589nm sodium lines have Q's of 50 million.

4B§13.9 Electrons in Crystals

Feynman analyzes an electrically neutral crystal with one extra electron. He defines:

$| n >$: basis state of electron at atom n

$| ø >$: electron state vector

A: transition amplitude $| n > \rightarrow | n \pm 1 >$

E_0: electron energy if A = 0

E: electron energy if A not 0

3B§18.3 An **Electron in a 1-D Crystal** with atomic spacing b has a transition amplitude A between neighboring atoms as given by:

$$\text{if } j = m \pm 1: -i\hbar < j | H | m > = A$$

$$\text{otherwise: } -i\hbar < j | H | m > = 0$$

Here, H is the system's Hamiltonian, j and m are any two basis states, and A is the amplitude per unit time for the electron to move from one atom to a neighboring atom. With k being the electron's wave number, the solution of the Hamiltonian equation for any value of k is:

$$\text{let } C_n = < ø | n >$$

$$| ø > = \Sigma_n C_n | n >$$

$$C_n = \exp\{ Et / i\hbar \} \exp\{ ik\, nb \}$$

$$E(k) = E_0 - 2A \cos(kb)$$

The allowed electron energy range, E_0–2A to E_0+2A, is called the **conduction band**.

3B§18.5 The **Electron Effective Mass** in a 1-D crystal with atomic spacing b is:

$$m_{eff} = \hbar^2 / 2A\, b^2$$

Here, A is the transition amplitude defined above, and m_{eff} is the electron's inertial mass that is enhanced by its interactions with the crystal's atom. Inertial mass equals applied force divided by resultant acceleration.

3B§19.1 An **Electron in a 3-D Crystal** is analyzed with the same logic as in a 1-D crystal, with these results:

$$C(x, y, z) = \exp\{Et / i\hbar\} \exp\{ i\boldsymbol{k} \bullet \boldsymbol{r} \}$$

$$E = E_0 - 2A_x \cos(a\, k_x) - 2A_y \cos(b\, k_y) - 2A_z \cos(c\, k_z)$$

Here, along the (x, y, z) axes: (a, b, c) are the atomic spacings, (A_x, A_y, A_z) are the transition amplitudes, and (k_x, k_y, k_z) are the wave numbers.

3B§19.3 **Scattering by Crystal Impurities** involves three waves: the incident wave of amplitude α, a reflected wave of amplitude β, and a transmitted wave of amplitude γ, as given by:

$$\beta / \alpha = -F / \{F - 2i\, A \sin(kb)\}$$

$$\gamma / \alpha = -2i\, A \sin(kb) / \{F - 2i\, A \sin(kb)\}$$

$$|\beta|^2 + |\gamma|^2 = |\alpha|^2$$

Here, E_0 is the electron energy at a normal crystal atom when A=0, and E_0+F is the energy the electron has at an impurity.

3B§20.1 **Semiconductors** are materials that are good insulators at low temperatures, and are modestly conducting at room temperature.

3B§20.2 In **Semiconductors**, electrical current can be carried by both **negative conduction band electrons** and **positive holes**. A hole is the absence of an electron in an atomic orbit where an electron would normally be. The **band gap, E_{gap}**, is the minimum energy required to elevate an atomic electron to the conduction band, thereby creating an electron/hole carrier pair. At room temperature, the band gaps of various pure crystals are:

0.67 eV in germanium

1.11 eV in silicon

1.54 eV in gallium arsenide

5.5 eV in diamond

3B§20.4 The negative and positive **Carrier Densities**, N_n and N_p, in a semiconductor with band gap E_{gap}, are always inversely proportional, as given by:

$$N_n N_p \sim \exp\{ -E_{gap} / kT \}$$

In pure silicon at 20°C:

$$N_n = N_p = 1.08\times10^{+10} \text{ carriers / cm}^3$$

In heavily doped silicon, densities can be:

$$N_n + N_p \sim 10^{+18} \text{ carriers / cm}^3$$

The atomic density of undoped silicon is:

$$5\times10^{+22} \text{ silicon atoms / cm}^3$$

Here, T is temperature in Kelvin, and k is Boltzmann's constant.

3B§20.3 **Semiconductor Doping**. Adding arsenic atoms to a small region of a silicon crystal increases the negative carrier density in that region, making it **n-type**. Adding boron increases the positive carrier density, making that region **p-type**. Abutting one region of each type creates a **p-n junction**, as sketched below.

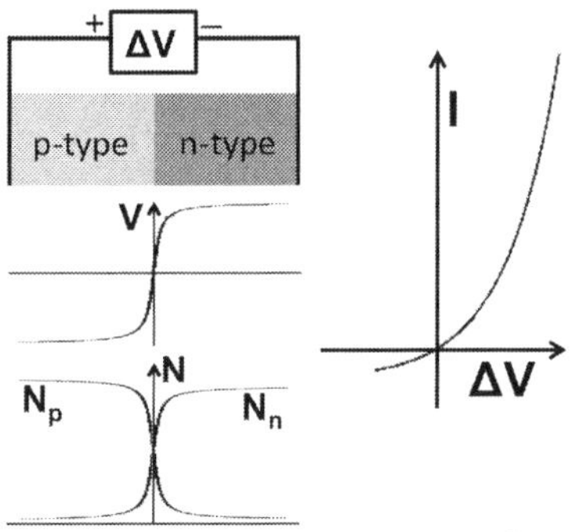

Figure 13-7 p-n Junction as Rectifier

In the two graphs on the left, the voltage V and carrier densities N_p and N_n are plotted vertically for the case of bias voltage $\Delta V = 0$. Plotted horizontally is the lateral position in the silicon crystal. The graph on the right plots net current **I** vertically versus bias voltage ΔV horizontally.

In the p-n junction depicted in Figure 13-7, electrons diffuse through the junction from the higher-electron-density n-side into the lower-electron-density p-side, while holes diffuse into the n-side. Both effects give the n-side a net positive charge and voltage, and give the p-side a net negative charge and voltage.

At equilibrium, and with $\Delta V = 0$, a current $\mathbf{I_0}$ flows in both directions, resulting in zero net current.

3B§20.7 A **Semiconductor Rectifier** conducts current in one direction much more readily than in the opposite direction, as shown in Figure 13-7 in the right-most graph. Net current **I** varies with bias voltage ΔV according to:

$$\mathbf{I} = \mathbf{I_0} \left[\exp\{+q\, \Delta V / kT\} - 1 \right]$$

Here, $\mathbf{I_0}$ is the equilibrium current defined above, q is the hole charge, T is temperature in Kelvin, and k is Boltzmann's constant.

Figure 13-8 shows a sketch of a p-n-p semiconductor transistor.

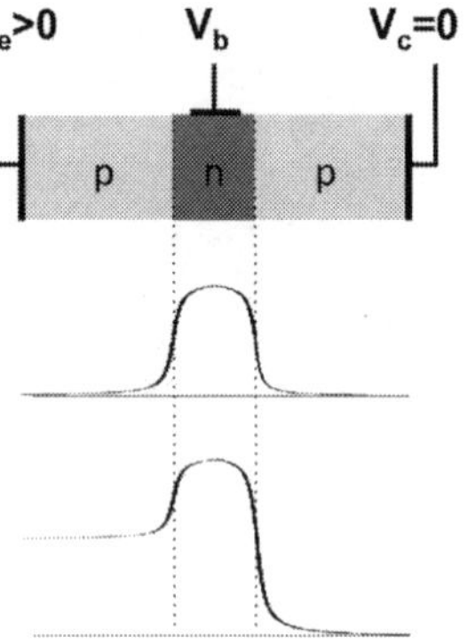

Figure 13-8 p-n-p Transistor

3B§20.7 The **Semiconductor** p-n-p **Transistor** shown above consists of a thin n-type **base** region sandwiched between a p-type **emitter** on the left and a p-type **collector** on the right, with electrodes attached to each region. The upper graph plots voltage vs. position without any external voltages. Here, the n-type region has a higher voltage than the p-type regions, as in the p-n junction above. The lower graph shows voltage vs. position with typical external voltages: $V_c = 0$ and $V_e = +5V$.

Base voltage V_b controls the transistor's operation. When $V_b < V_e$, the left p-n junction is forward biased, as in the rectifier described above, and positive carriers flood through the n-type base and into the collector, with as few as 1% flowing out through the base electrode. Conversely, a high V_b prevents current flow. Since the forward-biased p-n junction is operating on the steep part of the rectifier curve, a 0.1% change in V_b is amplified, resulting in as much as a 20% change in current flow.

In a **p-n-p transistor**, positive holes carry most of the current. Transistors are also fabricated with the opposite doping scheme: a thin p-type base sandwiched between an n-type emitter and collector. In an **n-p-n transistor**, negative electrons carry most of the current.

3B§20.5 The **Hall Effect** produces an electric field $\boldsymbol{E}$ in a heavily doped semiconductor, as given by:

$$\boldsymbol{E} = -\boldsymbol{j} \times \boldsymbol{B} / q\, N$$

Here, $\boldsymbol{j}$ is the current flowing transverse to an external magnetic field $\boldsymbol{B}$, and N and q are the dominant carrier density and its unit charge.

4B§13.10 QM in Magnetic Fields

3C§33.3 **Momentum in a Magnetic Field**. We can simplify many equations by defining two types of momentum that Feynman calls:

"mv-momentum": $\boldsymbol{p} = m\, \boldsymbol{v}$

"p-momentum": $\boldsymbol{p} = m\, \boldsymbol{v} + q\, \boldsymbol{A}$

Here, $\boldsymbol{A}$ is the vector potential of the magnetic field, q and m are the particle charge and mass, and mv-momentum is the normal Newtonian momentum. The quantum mechanical momentum operator becomes:

$$\boldsymbol{P} - q\,\boldsymbol{A}$$

3C§33.2 **Charge in a Magnetic Field.** Absent a magnetic field, the wavefunction of a particle of energy E and momentum $\boldsymbol{p}$ moving an incremental distance $\boldsymbol{ds}$ is multiplied by this factor:

$$\exp\{ (E\,dt - \boldsymbol{p} \bullet \boldsymbol{ds}) / i\hbar\}$$

Adding a magnetic field with vector potential $\boldsymbol{A}$, replaces $\boldsymbol{p}$ with $\boldsymbol{p}+q\boldsymbol{A}$ (see above), changing the multiplicative factor to:

$$\exp\{ (E\,dt - \boldsymbol{p} \bullet \boldsymbol{ds} - q\,\boldsymbol{A} \bullet \boldsymbol{ds}) / i\hbar\}$$

$$= \exp\{ (E\,dt - \boldsymbol{p} \bullet \boldsymbol{ds}) / i\hbar\} \exp\{ iq\,\boldsymbol{A} \bullet \boldsymbol{ds} / \hbar\}$$

The $\boldsymbol{A} \bullet \boldsymbol{ds}$ term adds a phase shift to the wave function. In moving along a path Γ from point S to point F, we integrate the $\boldsymbol{A} \bullet \boldsymbol{ds}$ term in the exponential along Γ. This means the amplitude with no magnetic field and the amplitude with a magnetic field are related by:

$$< F \mid S >_{\text{in}\,A} = < F \mid S >_{A=0} \exp\{ (iq/\hbar) \textstyle\int_\Gamma \boldsymbol{A}\bullet\boldsymbol{ds} \}$$

where:

$\boldsymbol{A}$ is the vector potential ($\boldsymbol{B} = \check{\boldsymbol{D}} \times \boldsymbol{A}$)

$< F \mid S >_{A=0}$ is the $< F \mid S >$ amplitude if $\boldsymbol{A} = 0$

$< F \mid S >_{\text{in}\,A}$ is the amplitude if $\boldsymbol{A}$ not 0

q is the particle's charge

Γ is the path from S to F

4B§13.11 QM at Low Temperatures

3C§33.1 **Superconductivity** is the conduction of electrical current with absolutely zero resistance. Boltzmann's law (see 4B§8.2) says the population of any state whose energy is ΔE higher than the ground state is proportional to:

$$\exp\{ -\Delta E / kT \}$$

At extremely low temperatures, virtually everything enters its lowest energy state. Atoms unable to transition from their ground states cannot interact with and thereby retard flowing conduction electrons.

3C§33.4 **Cooper Pairs** are two electrons flowing in a crystal that act as a single quantum entity. The two spin 1/2 electrons form a composite boson with a total spin of either 0 or 1, as explained by the **BCS theory of superconductivity**. As identical bosons, Cooper pairs have an enhanced amplitude to occupy the same state. Hence, as Cooper pairs start accumulating in their ground state, they overwhelmingly induce all other Cooper pairs to join them.

3C§33.6 The **Meissner Effect** is the exclusion of magnetic fields from the interior of a superconductor.

3C§33.7 The **Quantum of Magnetic Flux** is:

$$\Phi_0 = \pi\hbar / | q_e |$$

Here, Φ_0 is the minimum change in the flux of a magnetic field, and q_e is the electron charge.

3C§33.9 A **Josephson Junction** is formed when two superconductors sandwich a thin insulator. Electrons and Cooper pairs can move through the insulator via the quantum phenomenon of barrier penetration (see 4B§13.2). Applying a voltage of frequency ω and offset V_0 drives a resonant circuit, provided:

$$\omega = - q_C V_0 / \hbar$$

Here, q_C is the charge of a Cooper pair, twice that of one electron.

3C§33.9 A **SQUID** is formed by two parallel **Josephson Junctions**. SQUIDs are the most sensitive devices for measuring magnetic fields, with sensitivities as low as 10^{-15} Tesla.

4B§13.12 The Meaning of Reality

3C§34.2 **Interpretations of Quantum Mechanics**. The formalism of quantum mechanics is solid; its equations are clear, correct, and comprehensive. Its predictions have been validated by tens of thousands of high-precision experiments. Quantum mechanics is one of science's greatest triumphs. While its formalism is clear, its world-view is fuzzy. Physicists have substantially divergent views on its interpretation — what does quantum mechanics really tell us about nature? With its indeterminacy and wave function collapse, quantum mechanics remains incompletely digested more than 100 years after its inception. Many alternative interpretations have challenged the conventional **Copenhagen Interpretation**, but none have become widely accepted, so far.

3C§34.1 **Wave Function Collapse** is the greatest unresolved mystery of quantum mechanics. A particle's wave function can have nonzero amplitudes to simultaneously be in many mutually incompatible states, such as being near each atom in an infinite crystal. Yet, when that particle suffers a substantial external interaction, such as our attempt to measure its position, its wave function collapses — the probability of it being at one specific atom becomes 100%, and the probability of it being at any other atom becomes 0%, with those changes occurring instantaneously and simultaneously throughout all space. Einstein called this the "entirely peculiar mechanism of **action at a distance**". While bizarre, wave function collapse is essential to connecting quantum mechanics with the established reality of classical scientific observation. We have never observed any particle that was simultaneously in two different locations.

3C§27.4 **Positronium** is an "atom" comprised of an electron and an antielectron (a positron). The spin 0 singlet state of positronium decays into two photons of opposite polarization. Quantum mechanics says each photon has a 50% probability of being in each polarization state, and absent an external interaction, we can never know which is which. We say the two particles are **entangled**.

3C34.4 **Entanglement** is the condition of two or more particles sharing a common wave function. In V3p16-11, Feynman says of an entangled state of two particles:

> "describing *only one particle does not define* a [basis] state. Each [basis] state must define the condition of the entire system. You must not think that each particle moves independently ... If there are two particles in nature which are interacting, there is no way of describing what happens to one of the particles by trying to write down a wave function for it alone."

3C§35.1 In 1935, **Einstein, Podolsky, and Rosen** (EPR) employed entangled particles in a thought experiment to demonstrate that quantum mechanics is incompatible with our conventional sense of reality. This is best described in David Bohm's version of **EPR** (3C§35.2): two entangled photons emerge from positronium decay with opposite polarizations; but with each having a 50% probability of being spin up, and a 50% probability of being spin down.

Now imagine physicists Alice and Bob measuring photons' spins along the vertical axis, with Alice located 150m to the left of the positronium source, and Bob 150.3m to the right. Bob detects an entangled photon 1 nsec after Alice. Quantum mechanics says Alice has a 50% probability of measuring spin up and a 50% probability of measuring spin down, but whatever she measures, Bob will measure the opposite.

Einstein would say that defies our concept of reality — photon spins are real, regardless of what humans do. A measurement of the left photon cannot affect the right photon instantaneously, not in less time than it takes light to travel between detectors (300.3m divided by c = 1001 nsec). Yet, experiments prove Alice and Bob **always** detect opposite spins. How does Bob's photon "know" what its spin should be? This is the **EPR paradox**.

3C§35.1 **Locality** and **Realism** are two implicit assumptions of classical physics. Locality denies the possibility of **action at a distance**, saying physical entities interact only when in immediate contact. Realism says physical quantities, such as particles' spins, always have definite precise values independent of their interaction with anything else, including our instruments.

Local Realism says each pair of entangled photons in the Alice / Bob experiment leaves its source with both photons having precise and definite spins that are opposite to one another and that do not change from source to detector.

Carlo Rovelli likens this classical description to a glove maker shipping half a pair of gloves to one customer and half to another — both customers have a 50% probability of receiving a left-handed glove, but the two customers definitely receive opposite gloves. The gloves must be opposite because they were made as a matched pair.

Conversely, quantum mechanics says both gloves were made with semi-thumbs on both sides, and when the first customer opened their package, one glove instantly became entirely right-handed and the other instantly became entirely left-handed.

Classical physics says the photons have definite spins before they are measured, while quantum mechanics says their spins become definite only at the instant they are measured.

How can we possibly know whether or not entangled particles have definite spins **before** we measure them? That question haunted physicists for three decades.

3C§35.5 **Bell's Inequality**, devised in 1964 by John Stewart Bell, resolves this dilemma by measuring spins along different axes. He said Bob should rotate the axis along which he measures spin by angle θ relative to Alice's measurement. Bell showed that any classical (CL) theory based on Local Realism makes a slightly different prediction than quantum mechanics (QM) for a correlation quantity C, the product of Alice's spin measurement and Bob's spin measurement, according to:

$$\text{QM: } C(\theta) = -\cos\theta$$

$$\text{CL: } C(\theta) = -1 + 2\theta / \pi$$

At $\theta = 0$, where Alice and Bob both measure spin along the vertical axis, both theories predict $C = -1$ (opposite spins). But at other angles, the predictions differ. The theories differ most at $\theta = 39.5^\circ$ (0.69 radians) where $\Delta C = 0.21$.

3C§35.6 **Experimental Tests of Bell's Inequality**, from 1972 to 2010, all confirm the prediction of quantum mechanics and strongly reject the prediction of classical physics, by up to 36 standard deviations. Almost all physicists now concede, perhaps painfully, that Local Realism is not a valid description of nature.

3C§35.8 **Relational Quantum Mechanics**, a new world-view proposed in 1994 by Carlo Rovelli, says:

> "Quantum mechanics is a theory about the physical description of physical systems relative to other systems, and this is a **complete description** of the world."

Rovelli says nothing exists in isolation — **the only reality** is how entities relate to one another. He says everything in nature is a quantum system, including physicists and their equipment. He says a "measurement" is an interaction between two quantum systems — the observed and the observer — that results in a new entangled system of both. When Alice measures an entangled photon, a new entangled quantum system results: Alice-measured-photon. This system has equal probability of being Alice-measured-up (A+) and Alice-measured-down (A–). Similarly for Bob: B+ and B–.

Since the measurements occur at widely separated locations, no one can immediately know the results of both measurements. Only at some later time, as constrained by the speed of light, can signals from Alice-measured-photon and Bob-measured-photon combine at a common point. At that instant, the coherent superposition of these entangled states is disrupted, and the total wave function collapses to a state consistent with nature's law of angular momentum conservation: either $|A+, B-\rangle$ or $|A-, B+\rangle$.

Rovelli says there is no paradox.

Chapter 14
Particle Physics

4B§14.1 The Particle Zoo

3C§25.1 **Elementary Particles** are the most fundamental components of nature. We believe they have no internal parts, and are not comprised of anything "smaller". There are 17 different types of elementary particles.

3A§5.1 The **key fact** is: **Every elementary particle of each type is Absolutely Identical** — all electrons are identical; all photons are identical; etc. They are perfectly identical beyond any macroscopic analog. It is not just that physicists cannot measure any differences. Far more importantly, no natural process reveals even the slightest difference between particles of the same type. Quantum mechanics tells us that if this were not precisely true, every atom in the universe would immediately collapse.

The 17 different types of elementary particles are shown here.

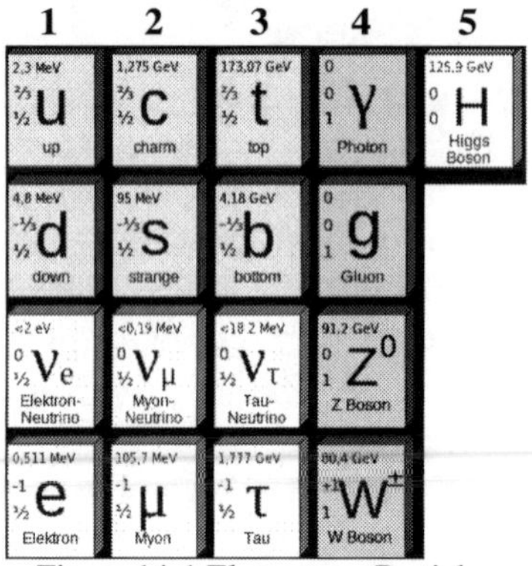

Figure 14-1 Elementary Particles

The 17 types are divided into two very distinct groups: the five bosons that occupy columns 4 and 5 of Figure 14-1, and the twelve fermions that occupy columns 1, 2, and 3.

Each box in this figure describes one particle type, starting with the large one-letter symbol above the particle's name. The three numbers on the left side of each box are: the particle's mass, electric charge, and spin. Electric charges are stated in units of the proton charge. Particle spin is explored in 4B§11.7. The small numbers in the figure may be hard to read, so I will repeat them in the text.

3C§25.1 **Particle Masses** are measured in units of eV/c², with 1 eV (1 electron-volt) being the energy an electron gains traversing a one-volt potential. Particle physicists typically use units in which c = 1, and quote masses in eV, or MeV for million eV, or GeV for billion eV.

From left to right across each row of Figure 14-1, the masses are:

Row 1:	u	c	t	γ	H
	2.4 MeV	1.27 GeV	171.2 GeV	0	125.1 GeV
Row 2:	d	s	b	g	
	4.8 MeV	104 MeV	4.2 GeV	0	
Row 3:	ν_e	ν_μ	ν_τ	Z	
	neutrino masses are discussed below			91.2 GeV	
Row 4:	e	μ	τ	$W^\pm$	
	0.511 MeV	106 MeV	1777 MeV	80.4 GeV	

3C§25.1 **Fermions** are the particles of matter. All elementary fermions have spin 1/2, and are further divided into two groups: six quarks and six leptons.

3C§25.1 The **Six Quarks** occupy rows 1 and 2 of columns 1, 2, and 3 in Figure 14-1. Row 1 contains the up, charm, and top quarks that all have electric charge +2/3. Row 2 contains the down, strange, and bottom quarks that all have charge –1/3. Quarks interact strongly, bonding with one another to form composite particles. The quark names are entirely fanciful. In the micro-world, "up" and "down" are meaningless. Also, no quark is more strange or less charming than the others.

3C§25.1 The **Six Leptons** occupy rows 3 and 4 of columns 1, 2, and 3 in Figure 14-1. Row 4 contains the electron, muon, and tau that all have charge –1. Row 3 contains the electron neutrino, muon neutrino, and tau neutrino that all have zero charge. Leptons interact weakly and do not form composite particles.

3C§25.1 **Fermion Generations**: the three generations of fermions (other than neutrinos) are distinguished by their masses. In each row of fermions, the third generation particles (column 3) are substantially more massive than the second generation (column 2) that are substantially more massive than the first generation (column 1). Third generation particles can **decay** (transform) into second generation particles that can decay into first generation particles. These decays happen rapidly, with lifetimes ranging from millionths of a second to trillionths of a trillionth of a second. First generation particles are stable (they appear to be eternal), because there exist no lighter particles of the right type into which they can decay.

3C§25.1 **Neutrinos** are the most elusive elementary particles. The masses of the three types of neutrinos are poorly determined. We do know that the three masses are all nonzero, all very small, and all different. The values shown in Figure 14-1 are measured in particle reactions. Surprisingly, cosmological measurements provide a much more stringent limit: the sum of the three masses is less than 0.1 eV. In some sense, it is easier to measure the mass of the universe and subtract everything else than to measure the neutrino masses directly.

3C§25.1 The **proton** and **neutron** are not elementary, but are extremely important. A proton is comprised of two up quarks and one down quark; it has charge +1, mass 938.3 MeV, spin 1/2, and is stable (its lifetime is measured to be at least $1.9\times10^{+34}$ years). A neutron is comprised of two down quarks and one up quark; it has zero charge, mass 939.6 MeV, spin 1/2, and its lifetime is 882 seconds.

3C§25.1 The **Four Bosons** in column 4 in Figure 14-1 all have spin 1. These are the exchange particles of forces: the photon for electromagnetism; gluons for the strong force; and Z and $W^\pm$ for the weak force. Estimated lifetimes are 3.16×10^{-25} sec for the $W^\pm$, and 2.64×10^{-25} sec for the Z. In isolation, photons and gluons have infinite lifetimes; both are massless and therefore cannot decay to anything else. Both can be emitted and absorbed by other particles.

3C§25.1 The **Higgs Boson**, alone in column 5 of Figure 14-1, has charge 0, spin 0, and a predicted lifetime of 1.56×10^{-22} sec. Hyped as the "God particle", the Higgs boson is thought to be the **origin of elementary particle masses**. Particles that interact strongly with Higgs are very massive; those interacting weakly with Higgs have small masses. Photons and gluons are massless because they do not interact at all with Higgs. The **composite particle masses** have a quite different source. For example, only about 1% of a proton's mass, the masses of its three quarks, is Higgs related; the other 99% is the kinetic and interaction energy of its quarks. Thus, about 99% of the mass of all normal matter is not due to Higgs, making it seem a bit less Godly.

1B§13.4 **Unstable Particle Masses**. Particles that decay into other particles do not have definite masses. The uncertainty principle stipulates a mass uncertainty — a **mass width** — Δm of:

$$\Delta m = \hbar / \Delta t$$

$$\Delta m = (6.58\times10^{-16} \text{ eV-sec}) / \Delta t$$

Here, Δt is the particle's mean lifetime. For example, the time interval Δt for a mass measurement of a K_1 is limited to its lifetime: 8.954×10^{-11} sec. (See 4B§14.5 for more about kaons.) Hence its mass uncertainty is:

$$\Delta m_{K1} = \hbar / \Delta t = 7.65\text{x}10^{-12} \text{ MeV}$$

This K_1 mass width is about twice the $K_2 - K_1$ average mass difference Δm_{21}, as shown in Figure 14-2.

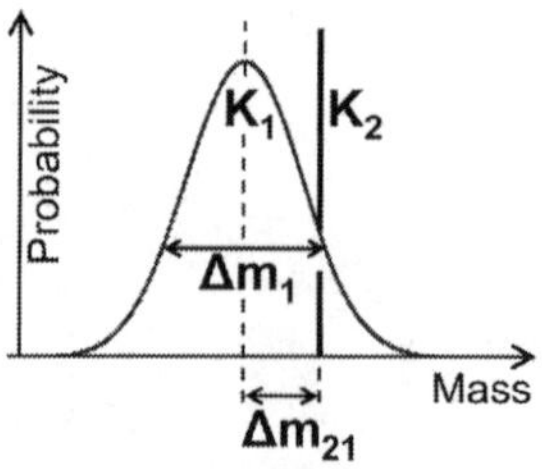

Figure 14-2 K_1, K_2 Mass Distributions

The mass distribution of K_1's mass is the resonance curve above. The K_2 mass distribution is indicated by the bold vertical line above. The K_2 lifetime is $5.116{\times}10^{-8}$ sec, making its mass width $1.3{\times}10^{-14}$ MeV, about 600 times smaller than the K_1's.

3C§25.1 **Antiparticles:** each type of fermion has a corresponding antiparticle, not shown in Figure 14-1. Particles and their antiparticles have the same mass, but all their other quantum properties have opposite polarities. For example, the electric charge of the antielectron is +1.

Since all of their quantum properties cancel one another, a particle / antiparticle pair can be produced from pure energy, energy without mass. Conversely, when a particle and its antiparticle collide, they may **annihilate** — both completely vanish, their masses convert into pure energy, and no trace remains that either entity ever existed. The resulting pure energy may become a pair of photons, or combinations of new particles and their antiparticles.

Bosons are their own antiparticles; or if you prefer, the particle / antiparticle distinction does not apply to bosons.

3C§27.4 **Parity** is a fundamental characteristic of each particle. Mirror reflection, or the inversion of all coordinate axes, multiplies a particle's wavefunction by the particle's parity quantum number P. P is either +1 ("even") or –1 ("odd"). Each elementary fermion has even parity (+1), while its antiparticle has odd parity (–1). The parity of a combination of particles equals the product of the parities of all its constituents multiplied by $(-1)^J$, where J is the combination's total orbital angular momentum.

4B§14.2 Particle Conservation Laws

3C§26.1 The net number of **Each Type of Fermion is Conserved** in every strong and electromagnetic interaction. The total number of fermions minus antifermions of each of the twelve types is always conserved in all processes driven by either of these forces. This rule almost certainly applies to gravitational interactions as well, but this is not experimentally confirmed.

3C§26.1 **Bosons are Not Conserved**. The number of bosons, either within one type or summed over all types, is not conserved.

3C§26.1 The **Weak Force Can Change Fermion Types**. Changing one type of fermion into another type enables nuclear fusion, the process that powers the stars and creates the elements heavier than helium. Since changing fermion types requires the weak force, these reactions proceed at much lower rates (less probability per second) than reactions driven by the strong or electromagnetic forces. The ability of the weak force to change one fermion into another is limited, as sketched in Figure 14-3.

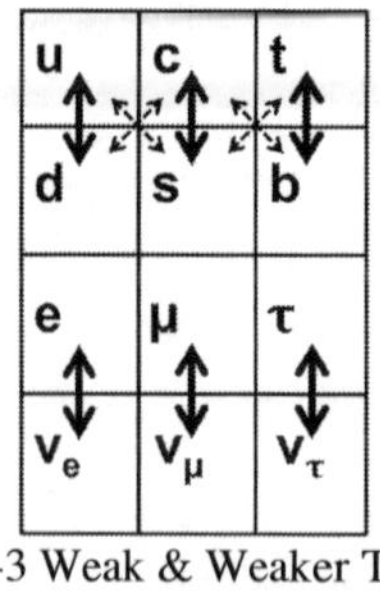

Figure 14-3 Weak & Weaker Transitions

The bold arrows above indicate transitions between fermion types that proceed at normal weak force rates. These include transitions between up and down quarks, and between electrons and electron neutrinos. The dashed arrows indicate transitions between fermion types that the weak force enables, but at much lower rates. These include transitions between up and strange quarks, and between down and charm quarks. Note that there is no arrow between u and c: the weak force cannot directly transform a charm quark into an up quark. It can very weakly transform c into d, and then weakly transform d into u — but that's super weak.

We also know that transitions occur between different types of neutrinos. Those reactions are not well understood at this time, but they most probably involve the weak force.

None of these transitions changes the total number of quarks minus antiquarks, nor the total number of leptons minus antileptons. In appropriate circumstances, all these transitions can proceed in either direction. But, none of these transitions can occur in isolation.

When a down quark transitions to an up quark, electric charge increases by +1. That transition can only occur in tandem with another reaction that changes electric charge by –1, such as an electron neutrino transitioning to an electron. Additionally, energy must be conserved in all reactions. Since a charm quark has more mass than a down quark, a down-to-charm transition must occur in tandem with another reaction that supplies the required energy.

4B§14.3 Force as Particle Exchange

3B§13.5 The **Modern Theory of Forces** is the particle-exchange model of **Quantum Field Theory**. This model repudiates the prior notion of action-at-a-distance that postulates objects interact directly and instantaneously even when separated by great distances.

Newton's theory of gravity does not explain how the Sun pulls on Earth 93 million miles away. Nor does Maxwell's theory of electromagnetism explain the field-mechanism through which the Sun's electrons repel Earth's electrons.

The particle-exchange model says the force between any two objects A and B is the sum of the forces between each elementary particle in A and each elementary particle in B. Furthermore, it says the force between two elementary particles arises from those particles exchanging a third particle: a **virtual particle**.

For example, the force between two electrons arises from the electrons exchanging a virtual photon. This is schematically represented in the **Feynman diagram** shown below, where time is the vertical axis, and electron-electron separation is the horizontal axis. These diagrams are schematic not quantitative — they only purport to identify what occurs, and the sequence of occurrences.

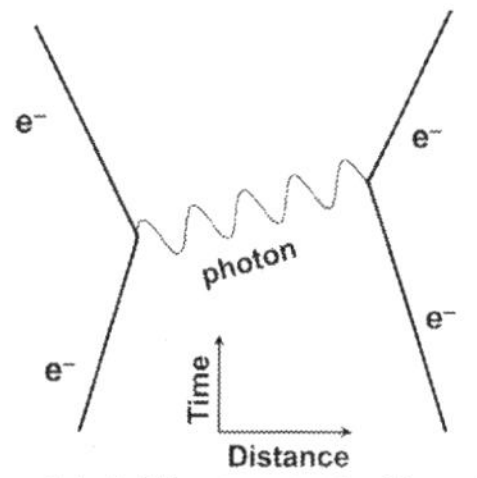

Figure 14-4 Electrostatic Repulsion

In the Feynman diagram above, as two electrons approach one another, the left electron emits a virtual photon that is later absorbed by the right electron. This exchange transfers energy and momentum, resulting in a "force".

In V3p10-7, Feynman says: "We do not 'see' the photons inside the electrons before they are emitted or after they are absorbed, and their emission does not change the 'nature' of the electron."

Exchange particles, a photon in this case, are called **virtual particles** to distinguish them from real particles. Virtual particles are not directly observable and they may have exotic properties, including negative energy, negative mass, and imaginary momentum.

To understand why exchange particles must be exotic, consider what this reaction looks like in the initial rest frame of the left electron in the above figure. There, the initial state is a stationary electron with zero kinetic energy; its total energy equals mc^2, where m is the electron mass. If the electron emits a photon of energy E_γ, the electron must recoil with some kinetic energy T_e. The total energy would then be mc^2 plus T_e plus E_γ. That sum must be greater than the initial energy mc^2, in violation of energy conservation.

Physicists typically resolve this apparent conflict by saying that the exchanged virtual photon can never be directly observed, and is promptly absorbed by a real particle. Because they are unobservable, virtual particles can have any energy, any mass, and any momentum.

An alternative view invokes the uncertainty principle, saying that energy and momentum are not conserved at each vertex of a Feynman diagram, but are eventually conserved in the final states. The final states must be comprised of real, observable particles that can be measured with very large values of Δt and Δx. Such measurements mandate very small values of ΔE and Δp, thereby ensuring energy and momentum conservation eventually if not instantaneously.

The particle-exchange model very successfully describes the weak force. Shown below is the Feynman diagram for neutron decay.

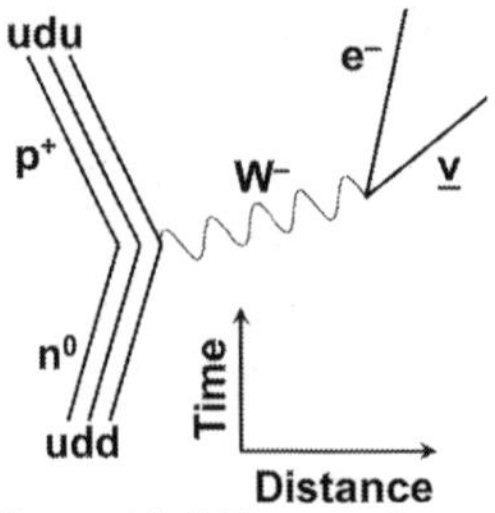

Figure 14-5 Neutron Decay

Neutrons are comprised of one up quark and two down quarks (udd). In the above diagram, one of the neutron's d quarks transitions to a u quark plus a virtual W^- boson, a carrier of the weak force. The new u quark combines with the original neutron's remaining u and d quarks, thus forming a proton. The W decays to an electron and an anti-electron-neutrino. Recall from above these key masses:

u quark mass: 2.4 MeV

d quark mass: 4.8 MeV

W boson mass: 80.4 GeV

When the d quark transitions to a u quark, the energy it releases is less than 0.003% of the mass of the W. This means the virtual W is "far off the mass shell", an extreme deviation from a real W. This is fundamentally why this decay is "weak", and proceeds at such a low rate. The lifetime of a particle that can decay via the strong force is typically trillionths of a trillionth of a second. The neutron lifetime is 15 minutes, 10^{27} times longer.

3B§13.6 The **Range λ of a Force** due to the exchange of a particle of mass m is limited by the uncertainty principle, as given by:

$$\lambda = \hbar c / mc^2 = 197 \text{ MeV fermi} / mc^2$$

Shown below are different configurations of two interacting nucleons, with various separations, spin orientations, and orbital angular momenta (indicated by dashed lines).

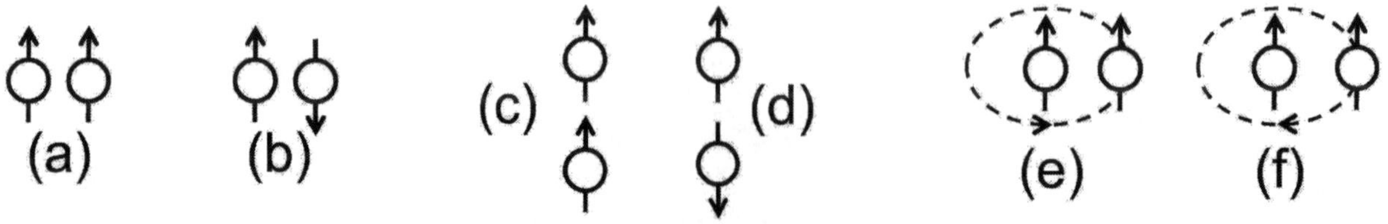

Figure 14-6 Nucleon-Nucleon Interactions

2A§8.5 The **Strong Force is Complicated**. The strong force between nucleons (protons and neutrons) varies in every possible way except one. In Figure 14-6, each possible orientation of spin and physical separation — (a), (b), (c), and (d) — result in different force strengths. Different spin and angular momentum orientations — (e) and (f) — also result in different force strengths.

The only way in which the strong force is less complicated than it could possibly be is that the force between two nucleons is the same for two protons, two neutrons, or one of each.

3B§13.7 **Quarks and the Strong Force**. The **Standard Model** of particle physics says the strong force arises from quarks exchanging **gluons** with one another. Gluons are massless, chargeless, spin 1 bosons. The force between nucleons (protons and neutrons), what we used to call the "strong force" fifty years ago, is merely a residual effect of the much stronger interaction between the quarks within those nucleons. Today, physicists often identify the quark-quark interaction as the "strong force", and identify the residual effect between nucleons as the "nuclear force".

The residual "nuclear force" has a very limited range. Its strength drops precipitously when nucleons are separated by more than 1 to 2 times the proton radius.

The "strong force" is much more complex than electromagnetism, and is not yet fully understood. Unlike all other forces, the strong force **does not diminish with increasing distance**.

Indeed, the strength of the strong force as a function of quark-quark separation D is quite bizarre. Its strength seems to be very weak, perhaps even zero, at D = 0; it seems to increase linearly with D, and plateau at an enormous strength for all D greater than the size of a proton.

At that plateau, the strong force between two quarks is estimated to be 10,000 newtons, about the weight of a one-ton mass on Earth's surface.

Think about that: the strong force between two quarks equals the gravitational force between all 10^{52} quarks in Earth and all 10^{30} quarks in a one-ton mass — a ratio of 10^{82}.

3B§13.7 **Quarks are Inseparable**. Any attempt to pull two quarks apart fails, because the energy required to overcome the strong force is sufficient to spontaneously create a new quark-antiquark pair. A failed effort to pull two quarks apart is shown schematically below.

Figure 14-7 Failure to Separate Quarks

Our effort to separate quarks begins with three quarks at the top of Figure 14-7, and ends at the bottom. The field energy builds as the u and ud quarks separate, as represented by the increasing number of dashed field lines. Eventually that energy becomes large enough (2×4.8 MeV) to convert into a down-antidown quark pair, increasing the number of particles to five. Three quarks combine to make a neutron (udd), and the remaining two combine to make a pion ($u\underline{d}$). This is why individual quarks are never observed: they are too strongly attracted to one another to ever go it alone.

4B§14.4 Particles in Fields

2C§30.1 and 1C§37.3 A **Particle in a Magnetic Field *B*** orbits in the plane perpendicular to ***B***, according to:

$$R = p / q B$$

Here, R is the orbital radius, q is the particle's charge, and p is the particle's momentum within the orbital plane. If q is the proton charge, and B is measured in tesla, R in meters, and p in GeV/c:

$$p = 0.3 B R$$

2C§30.6 **Weak Focusing** of charged particles by dipole magnets depends on the magnet's **field index** n. In a particle accelerator, with vertical magnetic fields and particles orbiting in a horizontal plane, n is defined by:

$$n = (dB/dR) / (B/R)$$

$$n = (dB/B) / (dR/R)$$

If $n > -1$, particles focus radially

If $n < 0$, particles focus vertically

If $-1 < n < 0$, particles focus both radially and vertically. Here, B is the magnitude of the magnetic field, and R is the radius of particle orbits.

2C§30.9 A **Constant Force** F applied to an object of mass m eventually accelerates that object to speeds near c, the speed of light. Let's assume position x and velocity v are both zero at time t = 0. Recall that **proper time** τ is the time measured by a clock moving with the object, while t is the time measured in the stationary reference frame. We assume F is constant in the object's rest frame. These equations apply:

$$\text{let: } a = F / m$$

$$\text{let: } u = a\,t / c$$

$$\text{let: } \beta = v / c$$

$$\text{let: } \gamma = 1 / \sqrt{(1 - \beta^2)}$$

$$\gamma^2 = 1 + \gamma^2 \beta^2$$

$$\tau = t / \gamma$$

$$v = a\,\tau = a\,t / \gamma$$

$$\gamma = \sqrt{(u^2 + 1)}$$

$$\beta = u / \gamma = u / \sqrt{(u^2 + 1)}$$

$$x = (c^2 / a)\,\{ -1 + \gamma \}$$

$$x = (c^2 / a)\,\{ -1 + \sqrt{(u^2 + 1)} \}$$

$$u = \sinh(a\,\tau / c)$$

$$\gamma = \cosh(a\,\tau / c)$$

$$\beta = \tanh(a\,\tau / c)$$

$$x = (c^2 / a)\,\{-1 + \cosh(a\,\tau / c) \}$$

2D§38.2 The **Landé g-factor** relates an object's angular momentum $\boldsymbol{J}$ to its magnetic dipole moment $\boldsymbol{\mu}$, as given by:

$$\boldsymbol{\mu} = g\,(q / 2m)\,\boldsymbol{J}$$

Here, q is the charge and m is the mass of the object. For orbital angular momentum, g=1. For elementary particles, g is called the **gyromagnetic ratio**. For the most important particles, the g-values are listed below, along with the measurement uncertainties in their right-most two digits in parentheses.

proton: g_p = 5.585,694,702 (±17)

electron: g_e = 2.002,319,304,361,46 (±56)

muon: g_μ = 2.002,331,841,8 (±13)

neutron: g_n = –3.826,085,45 (±90)

For neutrons, we use the proton charge for q in the equation for μ.

4B§14.5 Neutral Kaons

3C§26.3 The **Neutral Kaon Two-State System**: the K^0 and its antiparticle the $\underline{K}^0$ are not two independent particles. Instead, their decay processes unite them as the basis states of one combined two-state system. Both particles can decay to two pions; this enables the following very weak transition from particle to antiparticle:

$$K^0 \leftrightarrow \pi + \pi \leftrightarrow \underline{K}^0$$

This nonzero transition amplitude between basis states results in stationary states that are mixtures of K^0 and $\underline{K}^0$. Neutral kaons were the first discovered example of a general phenomenon: **neutral particle oscillations**, also observed with B^0s, D^0s, and neutrinos.

In many two-state systems (see 4B§13.5), all components of the Hamiltonian are real numbers. This occurs when the particles in those systems do not decay — when the total number of particles is constant. By contrast, neutral kaons do decay. Their Hamiltonian therefore includes imaginary amplitudes that cause its solutions to decrease exponentially over time.

Let's select basis states $|+>$ for a K^0 and $|->$ for a $\underline{K}^0$. The Hamiltonian, with complex amplitudes, is:

$$H_{++} = E_0 - i\beta$$

$$H_{+-} = a - i\alpha$$

$$H_{-+} = a - i\alpha$$

$$H_{--} = E_0 - i\beta$$

where:

$$a - i\alpha = < K^0 | \underline{K}^0 > = < \underline{K}^0 | K^0 >$$

Since kaons decay, and we have not accounted for their decay products, the total probability of this system decreases exponentially over time, and quantum probability **Principle #7** ($< A | B > = < B | A >^*$) does not apply.

The "stationary" solutions of the Hamiltonian equation are the sum and the difference of the basis states that we call $| K_1>$ and $| K_2>$.

$$| K_1> = (| K^0> - | \underline{K}^0>) / \sqrt{2}$$

$$| K_2> = (| K^0> + | \underline{K}^0>) / \sqrt{2}$$

The time-dependent amplitudes are:

$$K_1\text{: } A \exp\{E_1 t / i\hbar\} \exp\{ -(\beta - \alpha) t / \hbar\}$$

$$K_2\text{: } B \exp\{E_2 t / i\hbar\} \exp\{ -(\beta + \alpha) t / \hbar\}$$

The energy levels of the "stationary" states are:

$$E_1 = E_0 - a$$

$$E_2 = E_0 + a$$

With decaying particles, no state is completely "stationary." What we are referring to are those states whose basis state amplitudes maintain constant ratios, the states that maintain their essential character as their overall magnitudes decrease exponentially.

Mass measurements of K_1 and K_2 yield E_0 and a.

$$E_0 = 497.65 \text{ MeV}$$

$$2a = 3.5\text{x}10^{-12} \text{ MeV}$$

Measured lifetimes yield β and α.

$$t_1 = 8.954\times10^{-11} \text{ sec}$$

$$t_2 = 5.116\times10^{-8} \text{ sec}$$

$$\beta / \hbar = 5.594\times10^{+9} / \text{sec}$$

$$\alpha / \hbar = 5.574\times10^{+9} / \text{sec}$$

The tiny mass difference 2a determines the K_1/K_2 oscillation period $t_{\Delta m}$.

$$2a / \hbar = 5.32\times10^{+9} / \text{sec}$$

$$t_{\Delta m} = \hbar / 2a = 1.88\text{x}10^{-10} \text{ sec}$$

The three time constants have these relationships:

$$t_{\Delta m} = 2.1 \, t_1$$

$$t_2 = 571 \, t_1$$

Since the decay time t_1 is less than the oscillation time $t_{\Delta m}$, this system is an over-damped harmonic oscillator. K_1 rapidly vanishes, leaving only K_2.

What all this means is that a K^0 produced in a strong reaction has a 50% probability of decaying into a pair of pions, and a 50% probability of becoming a symmetric superposition of K^0 and $\underline{K}^0$ — its original self and its antiparticle in equal measures: a quantum version of Dr. Jekyll and Mr. Hyde. A $\underline{K}^0$ produced in a strong reaction has exactly the same fate.

3C§26.4 **Kaon Regeneration**, a beautiful demonstration of quantum state mixing, is called the **Pais-Piccioni effect** for its discoverers Abraham Pais and Oreste Piccioni. A neutral kaon beam produced by strong interactions may initially be an equal mixture of K_1 and K_2, but after many K_1 lifetimes, all that remains are K_2's that are equal mixtures of K^0 and $\underline{K}^0$. When that beam enters matter, its $\underline{K}^0$ component is preferentially absorbed. The depleted beam is once more a K_1 / K_2 mixture. (If the $\underline{K}^0$ component were completely eliminated, the beam would be equal parts K_2 and K_1.) The result is: the long-vanished K_1 is reborn.

K_2 has a much longer lifetime than K_1, because the latter decays to two pions, while the former must decay to three pions at a much-reduced rate. This difference is due to **CP-conservation**. CP symmetry is the exchange of matter and antimatter. Like parity P, most particles are eigenstates of definite CP, with eigenvalues of either +1 or –1. A K_1 has CP +1, a K_2 has CP –1, and a pion has CP –1. Two pions have CP $(-1)^2 = +1$, matching the K_1. Three pions have CP $(-1)^3 = -1$, matching the K_2.

Physicists long believed CP was a universal symmetry that was conserved in all reactions.

3C§26.5 **CP-Violation**: in 1964, Cronin and Fitch observed long-lived neutral kaons decaying to two pions, in violation of CP conservation. Later, my own thesis experiment discovered a CP-violating asymmetry in long-lived neutral kaon decays to:

$$\pi^+ \mu^- \underline{v}_\mu \text{ versus } \pi^- \mu^+ v_\mu$$

If nature were perfectly matter-antimatter symmetric, the rates of the above decays would be exactly equal. But in fact, the rates differ by 0.3%.

Our current understanding attributes CP-violation in kaon decays to an asymmetry of unknown origin in the eigenstates, rather than an asymmetry in the decay process itself. We now define the decay eigenstates to be the K-Short and K-Long, as given by (ignoring normalization):

$$| K_S> = | K_1> + \varepsilon | K_2>$$

$$| K_L> = | K_2> + \varepsilon | K_1>$$

$$\text{with } \varepsilon = (2.228 \pm 0.011)\times 10^{-3} \times \exp\{ i\,(43.5\pm 0.5)^o \}$$

1D§49.11 In 1956, **Madame Wu** shocked physicists with her remarkable discovery of **parity violation** in weak interactions. Yet, there was some consolation in the fact that parity violation is a 100% effect. Nature was surprising, but decisive. By comparison, CP-violation is even more distressing. If ε's phase angle were 45°, its real and imaginary components would be equal — why are they almost but not exactly equal? And why is nature 99.7% CP-symmetric?

Bizarre as it seems, CP-violation plays a critical role in cosmology. Our universe began with pure energy that created equal amounts of matter and antimatter. One second after the Big Bang, CP-violation enabled matter to gain a slight edge: for every one billion particles of antimatter, there were one billion **and one** particles of matter. The billions annihilated one another, producing photons, and leaving the one per billion of excess matter to make all the stars, planets, and people that exist today.

Fifty years later, we still have no comprehensive explanation for the origin of this tiny asymmetry.

Chapter 15
Problem-Solving Tricks & Caveats

1C§30.2 In V1p26-3, Feynman says:

"The real *glory* of science is that we can *find a way of thinking* such that the law is *evident*."

The essence of all these problem solving "tricks" is finding that way of thinking — the way of viewing a problem that makes its solution simplest. Unfortunately, there is no "magic bullet", no single technique that solves all problems. Physicists fill our mathematical toolboxes with wrenches, drills, and duct tape; when a challenge arises, we rummage to see what works.

4B§15.1 Reductionism & Holism

Reductionism is a cornerstone of Western science. We break systems down into ever-smaller pieces, confident that it is easier to understand the pieces individually and then reassemble them. Eastern philosophies stress holism: focusing on the whole first and its parts later. As with most things in life, the best approach is a judicious balance combining elements of both.

4B§15.2 Where To Start

Perhaps the best way to begin solving a problem is to focus on its dominant features, and ignore all the details. Ask yourself: What symmetries are present? Which fundamental principles, such as conservation of energy or angular momentum, might play major roles? Which coordinate axes or reference frames might work best?

Symmetries are often powerful constraints, quickly eliminating broad classes of wrong answers, and highlighting a system's conserved quantities. A left-right symmetry can literally eliminate half the problem.

Conservation laws often make complex problems much simpler. The exact motion of a pendulum is hard to calculate (and maybe of little interest), but we can easily know where and when it stops.

Coordinate systems are yours to choose. Any system will suffice eventually, but coordinate systems that match a problem's symmetries can greatly facilitate its solution. To analyze the motion of a spinning cylinder, aligning your coordinate system to its central axis eliminates two of three velocity components.

Reference frames are another free choice. Analyzing particle interactions, or the motion of 3-D solids, is greatly simplified in their center of mass frame.

4B§15.3 Simplify, Simplify, Simplify

Feynman recommends attacking a complex problem in stages. Start with the simplest possible version of the problem, make every simplifying approximation, focus on the most important effect, and neglect all else.

2A§11.6 In V2p11-10 Feynman says: "This is one of the tricks of theoretical physics. One does a different problem because it is easier to figure out the first time—then when one understands how the thing works, it is time to put in all the complications."

When you face a tough three-dimensional problem, trying to solve a 1-D version.

When a problem has many independent variables, look for solutions in which all but one of the variables is held constant.

When a problem is nonlinear, try to solve a linear version.

Look for ways to manipulate a challenging equation so that at least some of it matches another equation that you know how to solve.

If you cannot solve an equation for an unknown function f(x), replace f with a **Taylor series** or a **Fourier series** (see 4B§3.7), as shown here:

$$\text{Taylor: } f(x) = a_0 + a_1 x + a_2 x^2 + \dots$$

$$\text{Fourier: } f(x) = a_0 + a_1 \sin x + a_2 \sin(2x) + \dots$$

You may find that the above coefficients are related by simple equations.

3C§29.2 **Factoring a Function** is another promising technique. In deriving the electron wavefunction $\psi(\varrho)$ for spherically symmetric orbits in hydrogen, Feynman obtains this daunting equation:

$$\partial^2(\varrho\,\psi)/\partial\varrho^2 = -(\varepsilon\,\varrho + 2)\,\psi$$

Feynman then says ψ must decrease exponentially at large radius ϱ, because a bound electron has negative energy far from the nucleus. He factors out this exponential dependence with this substitution:

$$\varrho\,\psi = \exp\{-\beta\,\varrho\}\,\{\Sigma_n\, a_n\, \varrho^n\}$$

He then solves for β and for the Taylor series coefficients a_n (see 4A§8.1).

2A§6.2 **Vector Equations** can always be written out in terms of components, and solved as three separate 1-D problems. In V2p6-4, Feynman says:

> "The first time we encounter a particular kind of problem, it usually helps to write out the components to be sure we understand what is going on. There is nothing inelegant about putting numbers into equations, and nothing inelegant about substituting the derivatives for the fancy symbols [like $\check{\boldsymbol{D}}$]. In fact, there is often a certain cleverness in doing just that."

4B§15.4 Separation of Variables

A challenging function of multiple variables becomes tractable if its variable dependencies can be separated.

3C§29.5 In deriving the electron wavefunction $\psi(t, r, \theta, \phi)$ for atomic orbits with angular momentum, Feynman obtains this equation:

$$-i\hbar\, d\psi/dt = (e^2/r)\,\psi + (\hbar^2/2\mu)\,\{\partial^2(r\,\psi)\partial r^2\}/r$$
$$+ (\hbar^2/2\mu)\,\{\partial(\sin\theta\,\partial\psi/\partial\theta)\,\partial\theta\}/r^2\sin\theta + (\hbar^2/2\mu)\,\{\partial^2\psi/\partial\phi^2\}/r^2\sin^2\theta$$

Feynman simplifies this by separating ψ's functional dependencies with this substitution:

$$\psi = Y_{j,m}(\theta, \phi)\, F_j(r)\, \exp\{Et/i\hbar\}$$

The resulting equations for Y and F are still challenging, but are now solvable.

4B§15.5 Algebraic Tricks

2D§40.4 **Graphical Solutions** are sometimes possible even when analytical solutions are not. In analyzing paramagnetism, Feynman obtains this unsolvable equation:

$$M = N\,\mu\,\tanh(x)$$

$$\text{with: } x = (\mu\,H/kT) + (\mu\,\lambda\,M/\varepsilon_0 c^2\,kT)$$

He defines a variable β, and rewrites x as:

$$\text{let: } \beta = \tanh(x) = M/N\,\mu$$

$$x = C + \beta/D$$

$$\text{with: } C = \mu\,H/kT$$

$$\text{with: } D = \varepsilon_0 c^2\,kT/\mu^2\,N\,\lambda$$

Hence, we have two equations relating β and x:

$$\beta = (x - C)\,D$$

$$\beta = \tanh(x)$$

The solution lies where both equations yield the same result — where their graphs intersect in the figure below.

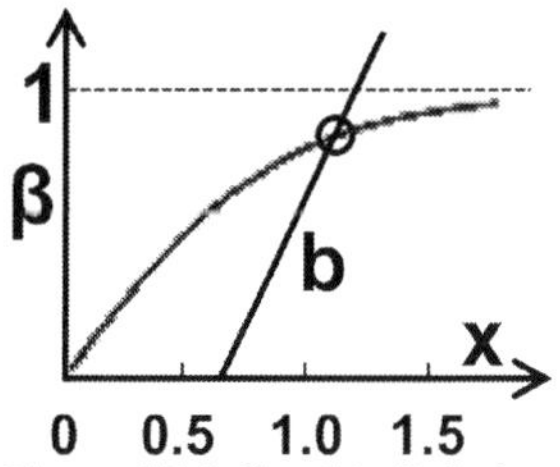

Figure 15-1 Graphic Solution

Here, the line labeled b represents $\beta(x) = (x - C)\,D$, and the curve represents tanh(x). For any set of H, T, μ, λ, and N, the resulting M is determined by where the line and the curve intersect, as circled in the figure.

2B§18.9 The electrical energy stored in two inductive coils is given by:

$$U = L_1\, j_1^2/2 + L_2\, j_2^2/2 + M\, j_1\, j_2$$

Feynman relates the mutual inductance M to the two coil inductances L_1 and L_2 by cleverly rewriting this equation as:

$$U = L_1\,(j_1 + M\, j_2 / L_1)^2 / 2 + (L_2 - M^2 / L_1)\, j_2^2 / 2$$

Stored energy can never be negative; hence, U must be >=0 for any combination of currents and inductances. In the special case of:

$$j_1 = -\,M\, j_2 / L_1$$

the first term in the prior equation is zero, which means the second term must be >=0. Thus, for any nonzero j_2, we have:

$$(L_2 - M^2 / L_1) >= 0$$

$$L_1\, L_2 >= M^2$$

For any L_1 and M, we can always choose currents j_1 and j_2 that satisfy the "special case" condition. Therefore the last equation must always hold.

4B§15.6 Linear Is Simpler

1D§44.6 In V1p49-7, Feynman joyfully describes the theory of linear systems as: "the most general and wonderful principle of mathematical physics."

1B§14.3 Any physical system governed by a linear differential equation with constant coefficients can be solved with exponentials with complex exponents. These solutions typically oscillate, and may decline exponentially, as functions of space and time. Any behavior of a linear system, however complex, can be represented as a linear sum of exponential solutions.

4B§15.7 Infinity is Limitless

Sometimes infinity is simpler than a finite number. A wave propagating along an infinite string never hits an end — no boundary conditions complicate its motion.

3B§18.2 For an electron added to a row of identical atoms, the simplest solution is for the case of an infinitely long row without beginning or end.

3B§21.7 We can extend the simple infinite solution to a ring of N atoms by adding the requirement that the solution repeats every N atoms.

3B§21.8 We can also extend the simple infinite solution to a finite row of atoms numbered 1 through N by adding the requirement that the solution has zero amplitude at atom #0 and atom #N+1.

3B§18.1 Feynman says the most effective approach is not to ask what happens to an electron placed on atom #j, but rather to search for patterns that propagate through the row of atoms as waves, like the displacement of violin strings.

4B§15.8 Trig Tricks

1C§32.8 Feynman frequently employs the following trig identities, particularly when analyzing interfering waves:

$$\cos A + \cos B = 2 \cos\theta \cos\phi$$

with $\theta = (A + B) / 2$

with $\phi = (A - B) / 2$

$$\cos(2\theta) = 2 \cos^2\theta - 1$$

$$1 - \cos\theta = 2 \sin^2(\theta/2)$$

$$\sin\theta = 2 \sin(\theta/2) \cos(\theta/2)$$

$$\sin\theta / (1 - \cos\theta) = \cos(\theta/2) / \sin(\theta/2)$$

4B§15.9 Calculus Tricks

There is no analytical procedure for evaluating an arbitrary integral — only numerical integration can solve any integral with a well-defined integrand. One must learn to integrate from experience and from successful techniques discovered by others, like Feynman.

4B§3.4 **Integration by Parts** is a useful trick in evaluating integrals with two functions u and v:

$$\int u \, dv = u\, v - \int v \, du$$

1B§16.5 **Squaring to Simplify**. The following integral for Q is not analytically calculable, but Q^2 is.

$$Q = \int_{-\infty}^{+\infty} \exp(-w^2) \, dw$$

$$Q^2 = \{ \int \exp(-x^2) \, dx\} \{ \int \exp(-y^2) \, dy\}$$

$$Q^2 = \int_{-\infty}^{+\infty} \int_{-\infty}^{+\infty} \exp(-x^2 - y^2) \, dx \, dy$$

Switching to polar coordinates, yields:

$$Q^2 = \int_0^{2\pi} \int_0^{\infty} \exp(-r^2) \, r \, dr \, d\theta$$

$$Q^2 = 2\pi (-1/2) \{ \exp(-r^2)|_0^{\infty} \}$$

$$Q^2 = -\pi \{0 - 1\}$$

$$Q = \sqrt{\pi}$$

1D§45.6 **Completing The Square**. The Fourier transform of a Gaussian involves this frightening integral:

$$\int_{-\infty}^{+\infty} \exp\{ -x^2/2\sigma^2 - ikx \} \, dx$$

The trick is adding the right constant to the exponent to make it a perfect square. We seek a constant A that satisfies:

$$- x^2/2\sigma^2 - ikx - A^2 = - (x/\sigma\sqrt{2} + A)^2$$

$$- ikx = - 2x \, A / \sigma\sqrt{2}$$

$$A = ik \, \sigma / \sqrt{2}$$

With $u = x/\sigma + ik\sigma$, the integral becomes:

$$\int_{-\infty}^{+\infty} \exp\{ -x^2/2\sigma^2 - ikx - A^2\} \exp\{+A^2\} \, dx$$

$$= \int_{-\infty}^{+\infty} \exp\{-u^2/2\} \exp\{-k^2\sigma^2/2\} \, \sigma \, du$$

$$= \exp\{-k^2\sigma^2/2\} \, \sigma \sqrt{(2\pi)}$$

2D§36.1 Let's **find the v that minimizes Q** in:

$$Q = \int_a^b v^2\, dt$$

We rewrite the velocity function v(t) in terms of its time average < v > as:

$$\text{let: } \Delta t = t_b - t_a = \int_a^b dt$$

$$\text{let: } \langle v \rangle = \int_a^b v\, dt / \Delta t$$

$$v^2 = (v - \langle v \rangle)^2 + 2v \langle v \rangle - \langle v \rangle^2$$

$$Q = \int_a^b \{(v - \langle v \rangle)^2 + 2v \langle v \rangle - \langle v \rangle^2\}\, dt$$

$$Q = \langle v \rangle^2 \Delta t + \int_a^b \{(v - \langle v \rangle)^2\, dt$$

The first term is constant; hence minimizing Q means minimizing the integral of a quantity that cannot be negative. Evidently, its minimum occurs when v(t) = < v > for all t.

1B§13.5 Feynman shows that some products of derivatives can be simplified:

$$2\, x\, dx/dt = d\{x^2\}/dt$$

$$2\, (dx/dt)\, (d^2x/dt^2) = d\{\, (dx/dt)^2\, \}/dt$$

2C§25.4 To calculate group velocity $d\omega/dk_z$ from:

$$\omega = \sqrt{\{\, c^2 k_z^2 + \Omega^2\}}$$

Feynman uses the trick of calculating its simpler reciprocal, as show by:

$$k_z = (1/c)\, \sqrt{[\omega^2 - \Omega^2]}$$

$$dk_z/d\omega = (1/c) / \sqrt{[1 - \Omega^2/\omega^2]}$$

$$v_{gp} = d\omega/dk_z = 1 / (dk_z/d\omega)$$

$$v_{gp} = c\, \sqrt{[1 - (\Omega / \omega)^2]}$$

4B§15.10 Summing Series

1A§6.9 **Zeno's paradox** remained unsolved by the ancient Greeks. Yet with a small trick, we can easily sum the infinite series S for |x| < 1.

$$S = 1 + x + x^2 + x^3 + \dots$$

$$x\, S = x + x^2 + x^3 + \dots$$

$$S = 1 + x\, S$$

$$S = 1 / (1 - x)$$

1B§20.3 Now consider this tricky infinite sum for |z| < 1:

$$\Sigma_J \{J\, z^J\}$$

We rewrite the sum as:

$$\Sigma_J \{J\, z^J\} = z + 2z^2 + 3z^3 + 4z^4 + \dots$$

$$= z\, (1 + z + z^2 + z^3 + \dots)$$

$$+ z^2\, (1 + z + z^2 + z^3 + \dots)$$

+ infinitely more rows

In each row above, the infinite series equals 1 / (1–z). That leaves the infinite sum of rows:

$$\Sigma_J \{J\, z^J\} = z\, (1 + z + z^2 + z^3 + \dots) / (1 - z)$$

$$\Sigma_J \{J\, z^J\} = z / (1 - z)^2$$

4B§15.11 Caveats

2C27.1 In V2p26.2, **Feynman says**: "Whenever you see a sweeping statement that a tremendous amount can come from a very small number of assumptions, you always find that it is false. There are usually a large number of implied assumptions that are far from obvious if you think about them sufficiently carefully." This is also a sweeping statement that is sometimes false.

Feynman advises care in manipulating operators. Operators are not numbers or simple variables; **operators are governed by special algebraic rules**. Here are some examples:

1A§6.3	Derivative dx/dt is not equal to x/t
3C§31.4	$\boldsymbol{a} \bullet \boldsymbol{b} = \boldsymbol{a} \bullet \boldsymbol{c}$ does not imply $\boldsymbol{b} = \boldsymbol{c}$
3C§31.4	$\boldsymbol{x} < x \mid \psi > = < x \mid \alpha >$ does not imply $\boldsymbol{x} \mid \psi > = \mid \alpha >$
2A§2.9	$\check{\boldsymbol{D}}\, \psi \times \check{\boldsymbol{D}}\, \phi$ is not equal to $\check{\boldsymbol{D}} \times \check{\boldsymbol{D}}\, (\psi\, \phi) = 0$
2A§2.4	$T\, \check{\boldsymbol{D}}$ is not equal to $\check{\boldsymbol{D}}\, T$

2A§2.9 **Vector Operators in non-rectilinear coordinates** have complex forms, listed in Appendix 5, that require great care.

2B§22.7 In V2p22-12, Feynman highlights a pitfall in employing **complex quantities in nonlinear applications**. Complex functions are very effective in analyzing periodic systems. In an oscillating electrical circuit, we can represent current J as the real part of $\boldsymbol{J}=J_0\exp(i\omega t)$, and voltage V as the real part of $\boldsymbol{V}=V_0\exp(i\omega t)$. But, power JV **is not** equal to the real part of $\boldsymbol{JV}$, because JV is quadratic rather than linear. Compare these results:

$$\text{real}\,\{\boldsymbol{J}\} \bullet \text{real}\,\{\boldsymbol{V}\} = J_0\, V_0 \cos^2(\omega t)$$

$$\text{real}\,\{\boldsymbol{J}\,\boldsymbol{V}\} = J_0\, V_0\, [\cos^2(\omega t) - \sin^2(\omega t)]$$

We need the former, not the latter; power equals (real current) multiplied by (real voltage).

1B§22.2 Another pitfall involves **partial derivatives**. Compare these thermodynamic equations for the change in the kinetic energy Ŧ of a gas:

$$dŦ = dT\, (\partial Ŧ/\partial T)_V + dV\, (\partial Ŧ/\partial V)_T$$

$$dŦ = dQ + dW = dQ - P\, dV$$

Feynman cautions us not to identify pressure P with $-(\partial Ŧ/\partial V)_T$. The final term in the lower equation is the change in Ŧ for a change in volume, at constant pressure. The final term in the upper equation is the change in Ŧ for a change in volume, at constant temperature. Those two quantities are different. With partial derivatives, one must be careful to note what is allowed to change and what is not.

Index

Appendices are located after 4A Chapter 20

Other Great Books by Robert Piccioni

The Feynman Simplified Series — ebooks

- ***4A***: ***Math For Physicists,*** all the math needed to master the Feynman Lectures, and more.
- ***4B***: ***The Best of Feynman,*** Feynman's most important principles and equations
- ***1A, 1B, 1C and 1D*** cover Feynman's Volume 1
- ***2A, 2B, 2C and 2D*** cover Feynman's Volume 2
- ***3A, 3B, and 3C*** cover Feynman's Volume 3

The Feynman Simplified Series — paperbacks

- ***Part 1*** covers Feynman Volume 1
- ***Part 2*** covers Feynman Volume 2
- ***Part 3*** covers Feynman Volume 3
- ***Part 4***: ***Math For Physicists,*** and ***The Best of Feynman***

Printed Books, each top-rated by Amazon readers:

- *Everyone's Guide to Atoms, Einstein, and the Universe*
- *Can Life Be Merely An Accident?*
- *A World Without Einstein*

The Everyone's Guide Series of short ebooks

- *Einstein: His Struggles, and Ultimate Success*, plus
    - *Special Relativity*; 3 Volumes
    - *General Relativity*: 4 Volumes, introduction to Differential Topology
- *Quantum Mechanics*: 5 Volumes, introduction to Entanglement
- *Higgs, Bosons, & Fermions…*, introduction to Particle Physics
- Cosmology
    - *Our Universe*: 5 Volumes, everything beyond the Sun
    - *Our Place in the Universe*, a gentle overview
    - *Black Holes, Supernovae & More*
    - *We are Stardust*
- *Timeless Atoms*
- *Science & Faith*

Meet the Author

Congratulations for embarking on a most exciting adventure—the quest to understand Nature.

I'd like to tell you something about myself and share some stories.

First, the obligatory bio: I have a B.S. in physics from Caltech, a Ph.D. in high-energy physics from Stanford University, and was on the faculty of Harvard University. Now "retired", I teach at the Osher Institutes at UCLA and CSUCI, where students honored me as "Teacher of the Year." In between, I ran eight high-tech companies, and hold several device patents in medical, semiconductor, and smart-energy fields.

My goal is to help more people appreciate and enjoy science. One doesn't have to be a world-class musician to appreciate great music, and I believe the same is true for science — everyone can enjoy the exciting discoveries and intriguing mysteries of our universe. I've given over 400 presentations on science to general audiences of all ages and interests, and have written 4 printed books and 40 ebooks. My books have won national and international competitions, are among the highest rated books on Amazon.com, and three have been the 1st, 2nd, and 3rd best selling books in their categories.

I knew Richard Feynman well before entering Caltech. He was a friend and colleague of my father, Oreste Piccioni. On several occasions, Feynman drove from Pasadena to San Diego to sail on our boat and have dinner at our house. I like to say that I taught Feynman how to sail and he taught me how to play pool.

At Caltech, I was privileged to learn physics from the greatest scientist of our age. I absorbed all I could. His style and enthusiasm were as important as the facts and equations. The highest-ranking professors rarely teach undergraduates. But Feynman realized traditional introductory physics did not well prepare students for modern physics. Therefore, he created a whole new curriculum and personally taught Caltech's freshman and sophomore physics classes in 1961-62 and 1962-63.

The best students thrived on a cornucopia of exciting frontier science, but many others fared poorly. Although Caltech may be the world's top science school, about half its elite and eager students drowned in Feynman's class. Even a classmate, who received the Nobel Prize in Physics decades later, struggled in this class. Feynman once told me that as he walked through campus, students sometimes gave him the "stink eye" — he added: "Me thinks he didn't understand angular momentum,"

Some mundane factors contributed to the high failure rate: Feynman's book wasn't written yet; class notes came out weeks after the lectures; and students' traditional helpers (teaching assistants and upperclassmen) didn't understand physics the way Feynman taught it. But the biggest problem was that so much challenging material flew by so quickly. Like many elite scientists, Feynman's teaching mission was to inspire those who might become the next generation's scientific elite. At the end of the two-year course, Feynman estimated that in a 180-student class, "there were one or two dozen students who—very surprisingly—understood almost everything in all of the lectures."

My goal is to reach the other 90%. It's a great shame that so many had so much difficulty with the original course—there is so much great science to enjoy. I hope to help change that and bring Feynman's genius to a wider audience.

Please let me know how I can make ***Feynman Simplified*** even better—contact me at my website:

www.guidetothecosmos.com

While you're there, check out my other books and sign-up for my newsletters.

21606085R00161

Made in the USA
San Bernardino, CA
04 January 2019